INTERNATIONAL SERIES IN
NATURAL PHILOSOPHY

General Editor: D. ter HAAR

Volume 101

Flare Stars

OTHER PERGAMON TITLES OF INTEREST

Books

CLARK & STEPHENSON:
The Historical Supernovae

ELGAROY:
Solar Noise Storms

GINZBURG:
Theoretical Physics and Astrophysics

HEY:
The Radio Universe, 2nd Edition

MEADOWS:
Stellar Evolution, 2nd Edition

PACHOLCZYK:
Radio Galaxies

REDDISH:
Stellar Formation

SAHADE & WOOD:
Interacting Binary Stars

SOLOMON & EDMUNDS:
Giant Molecular Clouds in the Galaxy

Journals*

Chinese Astronomy

Planetary and Space Science

Vistas in Astronomy

*Free specimen copy available on request.

A full list of titles in the International Series in Natural Philosophy follows the index.

Flare Stars

REVISED AND ESSENTIALLY UPDATED BY THE AUTHOR

G. A. GURZADYAN

Garny Space Astronomy Laboratory
Armenian SSR, USSR

EDITED BY

YERVANT TERZIAN

Cornell University, NAIC
Ithaca, N.Y.

TRANSLATED BY

E. A. van HERK-KLUYVER

The Netherlands

PERGAMON PRESS

OXFORD · NEW YORK · TORONTO · SYDNEY · PARIS · FRANKFURT

U.K.	Pergamon Press Ltd., Headington Hill Hall, Oxford OX3 0BW, England
U.S.A.	Pergamon Press Inc., Maxwell House, Fairview Park, Elmsford, New York 10523, U.S.A.
CANADA	Pergamon of Canada, Suite 104, 150 Consumers Road, Willowdale, Ontario M2J 1P9, Canada
AUSTRALIA	Pergamon Press (Aust.) Pty. Ltd., P.O. Box 544, Potts Point, N.S.W. 2011, Australia
FRANCE	Pergamon Press SARL, 24 rue des Ecoles, 75240 Paris, Cedex 05, France
FEDERAL REPUBLIC OF GERMANY	Pergamon Press GmbH, 6242 Kronberg-Taunus, Pferdstrasse 1, Federal Republic of Germany

First Edition 1980

British Library Cataloguing in Publication Data

Gurzadyan, G A
Flare stars. - Revised ed. - (International
series in natural philosophy; vol. 101).
1. Flare stars
I. Title II. Terzian, Yervant III. Series
523.8'446   QB843.F55   79-41746
ISBN 0-08-023035-0

In order to make this volume available as economically and as rapidly as possible the authors' typescripts have been reproduced in their original forms. This method has its typographical limitations but it is hoped that they in no way distract the reader.

*Printed and bound in Great Britain by
William Clowes (Beccles) Limited, Beccles and London*

Contents

Preface to the English Edition

This English edition of my book differs in many ways from the Russian version.
First, the entire observational material has been renewed and supplemented, the
empirical relations found between the various flare parameters have been made more
precise and systematic. After the appearance of the Russian edition of this book
in 1973, the X-ray radiation of flare stars, predicted by the fast-electron hypothesis, was discovered. Therefore a separate chapter is devoted to X-ray radiation
in the present book. Another chapter deals with the theory of the transition
radiation; its importance, at least for T Tauri type stars, and peculiar objects,
has become more evident. Chapter 6 and 7, in which the photometric and colorimetric flare data are analyzed on the basis of the fast-electron hypothesis, have
been considerably extended; and the most recent photoelectric observations by
Moffett, Kunkel, Cristaldi and Rodono, Osawa et al. have been included. The theory
of the chromosphere of flare stars has been redeveloped in more detail (Ch. 9).
New results have appeared concerning the general laws, for the behavior of flare
stars in stellar associations (Ch. 11), and for T Tauri type stars (Ch. 10). The
latest observational data have shed new light on the problems of radio emission and
flare dynamics (Chaps. 13 and 15). The great cosmogonic importance of stellar
flares, in particular for understanding the nature and the true character of
energy sources within the stars, have become more evident (Ch. 16).

Important progress has been made in the domain of photoelectric, photographic and
spectral observations of stellar flares. However, further progress is still
expected from observations in the ultraviolet and X-rays from space laboratories.
Infrared observations will also be very valuable; one of the important predictions of the fast-electron hypothesis, the possibility of a "negative" infrared
flare, remains to be verified.

Finally I would like to note the initiative and enthusiasm displayed by my astrophysics colleagues during the preparation of the English edition of this book;
in particular D. ter Haar, Arthur Beer, Yervant Terzian, the last being responsible for editing the book; and also Sir B. Lovell, G. Haro, G. H. Herbig,
T. J. Moffett, B. W. Boop, S. R. Spangler, S. Cristaldi, M. Rodono, K. Kodaira
and others who sent me illustrations and other material. The translation was
done faultlessly by Mrs H. A. van Herk-Kluyver. To all these the author
expresses his deep gratitude.

G. A. Gurzadyan
May 1979
Yerevan, Armenian SSR

Editor's Preface

In the spring of 1976 while visiting Armenia I became aware of the important work that G. A. Gurzadyan had written on flare stars. Soon I was convinced that his Russian book on <u>Flare Stars</u> would be very valuable if translated into English. In 1977 arrangements were made with Pergamon Press for publishing the translation of Gurzadyan's book. The general translation was performed by Mrs. H. A. van Herk-Kluyver in Holland and the entire editing and the preparation of the manuscript were performed by the editor at Cornell University, Ithaca, New York.

The author has carefully revised and supplemented the 1973 Russian edition of his book. His work now presents up-to-date observational data on flare stars, as well as pioneering new theoretical explanations. Although his "fast electron hypothesis" has not been widely accepted, he describes numerous convincing examples to prove its validity. Many individuals assisted in the preparation of the English manuscript, including Judy Marcus, Andrea Schmidt, Velma Ray and Mary Roth.

Yervant Terzian
June 1979
Ithaca, New York

From the Preface by the Author to the Russian Edition

Flare stars are thought to be one of the enigmas of present-day astrophysics and theoretical physics. It can only be stated that the flare phenomenon itself--a sudden, strong, but short-lived increase of the star's brightness and an almost equally fast return to its normal state--is closely related with the internal structure of the star and the nature of the sources of energy within the star. There is also no doubt about the enormous evolutionary significance of this phenomenon: the flare activity occurs in young and newborn stars and decreases as the development of the star proceeds. The flare seems to be the main way of getting rid of superfluous internal energy during the formation and evolution of the star, in an analogous way as the outflow or ejection of gaseous matter is the principal means of getting rid of superfluous stellar mass. Hence it follows that flare stars do not constitute a special physical category of stars. The flare phenomenon, or the state of flare activity, is a period through which every, or nearly every, star of the main sequence passes at an early stage of its life.

Of decisive importance, for understanding the nature of the flare, is the radio-emission by flare stars. It proves that at the time of the flare relativistic electrons appear in large quantities.

The formation of a dense short-lived cloud consisting of relativistic particles, near the star, above its photosphere, must lead to the generation of radiation at other frequencies as well, in particular at optical wavelengths. Here the main role is played by the inverse Compton effect--inelastic collisions of photons with relativistic electrons, and also by the bremsstrahlung of these particles. Such a cloud must also be a source of powerful X-ray emission. The present monograph is devoted to an explanation of the main results of the theory of the stellar flare based on this conception, which will be called the "fast electron hypothesis."

1973, Moscow

From the Preface to the Russian Edition

The monograph by G. A. Gurzadyan on "Flare Stars" is devoted to one of the most essential and modern problems of astrophysics. As observational data are acccumulated and empirical laws appear, it becomes more and more evident that all dwarf stars, in the initial period of their lives, pass through a rather long lasting phase during which the occurrence of flares may be the most important peculiarity.

This monograph has two merits: 1. It gives equal attention to the flare stars in the neighborhood of the sun, and to those in associations and relatively young clusters. This is quite correct, since in the case of the nearby flare stars we can study well the character of the physical processes which take place in them, and studying the flare stars in stellar aggregates we obtain more extensive information about the evolutionary changes and many statistical data. 2. The entire material is exposed from one point of view, namely that of the "fast electron hypothesis". For the explanation of the spectral distribution of the energy in the radiation of most of the flares the inverse Compton effect is assumed.

One of the main conclusions of this hypothesis, namely that high energy particles play an important role in flares and that the total energy, produced at the time of the outburst which causes the flare, exceeds by many times the energy of the optical radiation of the flare, is, in my opinion, undoubtedly true.

If this is the case, the views of the author of this book may serve as a good starting point for further developments of the theory. The book is justified if it draws the attention of young investigators to the wide range of interesting problems connected with the physical processes which occur in the early stages of stellar development.

V. A. Ambartsumian
1973

CHAPTER 1
General Information about Flare Stars

1. DISCOVERY OF THE FLARE PHENOMENON IN STARS

The phenomenon of star flares was discovered accidentally. In 1924 Hertzsprung [1], when looking at plates taken by himself, in the direction of the constellation Carina, noticed that the brightness of one faint star--DH Car--had increased by nearly two magnitudes. After careful comparison with other plates of this star obtained earlier, Hertzsprung concluded that in this case the star's brightness had increased rapidly in an unusually short interval of time. It is interesting to note that even then, Hertzsprung was certain that this phenomenon was not the same as the appearance of a nova; he even tried to explain it as due to a minor planet falling on the star. Later more flares of the star DH Car were recorded and today it is catalogued among flare stars.

For a long time Hertzsprung's discovery was forgotten. After more than twenty years, on December 7, 1947, an event occurred which started the study of a new type of objects now known as flare stars. The American astronomer Carpenter photographed a red dwarf star L 726-8, now called UV Cet, by the chain method, i.e. on one plate he obtained five images of the star, each with four minutes exposure time. He intended to use the plates for a determination of the trigonometric parallax of this star, which was known for its large proper motion. But already a quick look at the plate showed that something unusual had happened to this star: the second image proved to be very bright, but in the following images the brightness decreased gradually, reaching nearly its normal value on the last, fifth, image. Luyten [2] who has published this story, estimates that the increase in brightness was 2^m.7 This means an increase of the star's brightness by a factor of 12 during three minutes. Wishing to stress that this is something quite unprecedented, Luyten remarks that even in a supernova outburst the increase in brightness in three minutes is scarcely 5%. Finally, long before this, astronomers knew cases of irregular fluctuations in the brightness of some stars. For instance, such a phenomenon was discovered by Van Maanen [3] in the stars BD +44°2051 (B) and Ross 882. The latter star is today one of the well known flare stars, but with another name--YZ CMi. But in these and other cases it was only known that irregular brightness fluctuations really existed. The plate taken by Carpenter was the first and entirely convincing observational evidence that a new phenomenon exists in the world of stars--an unpredictable, exceptionally rapid and strong increase in brightness and nearly as rapid a return to the initial state. In this connection it should be pointed out that Luyten even then fully understood the extraordinary importance of this phenomenon, when he expressed the opinion that its explanation must be directly related to a larger problem--that of the stellar energy sources.

Soon a chain reaction took place in the discovery of flare stars. In 1949 Gordon and Kron found from photoelectric observations [4] an outburst of AD Leo; in 1950 Thackeray [5] discovered an outburst of the star nearest to the Sun, V645 Cen, from a spectrogram taken at the time of the flare; in 1953 Winterhalter [6] noticed a flare of EV Lac. Further, in 1957 Joy [7] discovered the first certain flare of

YZ CMi; Kron et al.[8] communicated an outburst of V 1216 Sgr, etc. Moreover some
flare stars are found from inspection of existing plates. This happened, e.g. with
WX UMa, the first certain flare for which an amplitude of about two stellar magnitudes seems to have been found by Van Maanen [9] in 1945, i.e. two years before the
discovery of the flare of UV Cet.

It soon turned out that all known flare stars mainly belong to spectral class M and
that they have exceptionally large proper motions, and hence must be near to the
Sun. In fact all these stars of magnitude 10 to 13 proved to be red dwarfs at dis-
tances of only a few parsecs from the Sun. However strange it may seem, the process
of discovering new flare stars in the neighborhood of the Sun continues even today.
Thus, e.g. since 1972 flare activity of the stars G 447 = Ross 128 [10], EQ Her =
G 517 [11], two anonymous stars in Aquila [12] and Lupus [13] has been discovered.
In 1975 Haro et al. [14] communicated the discovery of two flare stars (V = 12.2
and V = 13.4) known as G 009-008 and G 040-026 which are undoubtedly close to the
Sun. Quite incidentally, when taking objective prism spectra of stars Bond [15]
discovered in 1976 a flare star with an exceptionally large amplitude, $\sim 6^m$, in
Pisces.

2. FLARE STARS IN THE NEIGHBORHOOD OF THE SUN

The first list of flare stars was made in 1960 by Joy [15]; it contained 20 objects.
In the beginning of 1970 they numbered 50 [16] and in the middle of 1976 more than
70. A list of these stars is given in Table 1.1. In most of the cases, the outburst was observed on direct photographs, sometimes from photoelectric observations,
and rarely from visual estimates. In some cases the flare was found from characteristic changes of the continuous spectrum or from strengthening of the emission lines
in the star's spectrum.

From the data in Table 1.1 and also from additional information about these stars
the following conclusions may be drawn:

1. Nearly all flare stars belong to spectral class M, or rather to the late subclasses of M. Only five stars ($\sim$ 7%) of this list proved to belong to class K.
There are no flare stars at all belonging to a spectral class earlier than K.

2. All flare stars given in Table 1.1 are near the Sun--within a distance of 20
parsecs; 30 of them, i.e. nearly half, are closer than 10 parsecs and 14 of them
are closer than 5 parsecs. The nearest flare star is V645 Cen, at a distance of
1.3 parsecs only. Then follow UV Cet (2.7 parsecs), V1216 Sgr (2.9 parsecs), G 447
= Ross 128 (3.3 parsecs), etc. As yet the furthermost star is Butler's star. However, its estimated distance of 80 parsecs is not based on a parallax determination,
but simply on the adopted absolute magnitude (M = +6), usual for stars of class K0.

The comparison of the distances of the flare stars with their spectral classes leads
to the conclusion that they are all red dwarfs.

3. At least half of the stars given in Table 1.1 are known to be double--
spectroscopic, photographic and even visual double systems; in the last case the
distances between the components are so large that interaction between them is
scarcely probable. In some cases (YY Gem) the flare star is at the same time a
spectroscopic and an eclipsing system.

4. The flare stars are more or less evenly distributed in the neighborhood of the
Sun. Anyway, their relatively small number and their closeness to the Sun do not
yet allow us to distinguish with certainty any regularities in their apparent or
space distributions.

TABLE 1.1   UV Cet Type Flare Stars

	Gliese's Number	Name of Star	α(1950)	δ(1950)	m_V	Sp	M_V	Distan. pc	Ref.
1	2	3	4	5	6	7	8	9	10
1	15 A	BD +43°44	00^h 15^m.5	+43° 44.4	8.07	dM1e	10.32	3.6	1,23
2	15 B	CQ And	00 15.5	+43 44.4	11.04	dM6e	13.29	3.6	51
3	22 A	BD +66°34	00 29.3	+66 57.8	10.51	dM2.5e	10.42	10.4	46
4		Butler's star	00 58.1	-73 13.4	10.6	K0	6.0?	80?	4,5
5	51	Wolf 47	01 00.1	+62 05.8	13.66	dM7e	13.81	9.3	3
6	54.1	LPM 63	01 09.9	-17 16	11.6	dM5e	11.5	7.0	6,50
7	65 A	L 726-8	01 36.6	-18 12.7	12.45	dM5.5e	15.27	2.7	7,8
8	65 B	UV Cet	01 36.6	-18 12.7	12.95	dM5.5e	15.8	2.7	7,8
9	83.1	G 3-33	01 57.5	+12 50.1	12.27	dM8e	13.91	4.7	9
10	103	CC Eri	02 32.5	-44 00.6	8.7	dK7e	8.4	11.4	45
11	164	Ross 28	03 09.2	+52 29.7	13.2	M5	12.3	15.1	2
12	166 C	40 Eri C	03 13.1	-07 44.1	11.17	dM4.5e	12.73	4.8	6,49
13	206	Ross 42	05 29.5	+09 47.3	11.50	dM4e	10.73	14.2	43
14	207.1	V 371 Ori	05 31.2	+01 54.8	11.68	dM3e	10.8	15.2	11
15	229	BD -21°1377	06 08.5	-21 50.6	8.13	dM1e	9.33	5.7	51
16	234 A	Ross 614	06 26.8	-02 46.2	11.07	dM7e	13.08	4.0	12,49
17	234 B	--	06 26.8	-02 46.2	14.4		16.4	4.0	49
18	--	PZ Mon	06 45.8	+01 16.6	10.8	dK2e	7.1	16	13,14
19	268	AC +38°23616	07 06.7	+38 37.5	11.48	dM5e	12.62	5.9	46
20	278 C	YY Gem	07 31.6	+31 58.8	9.07	dMo.5e	8.26	14.5	40,41
21	285	YZ CMi	07 42.1	+03 40.8	11.2	dM4.5e	12.29	6.0	15
22	--	BD +33°1646B	08 05.7	+32 56	11.0	dMe	8.6	33	2
23	--	G 009-008	08 28.8	+19 34.0	12.2	M5e	--	--	37
24	--	G 040-026	08 28.8	+19 34.0	13.4	M5e ?	--	--	37
25	388	AD Leo	10 16.9	+20 07.3	9.43	dM4.5e	10.98	4.9	16
26	398	L 1113-55	10 33.5	+05 22.7	12.61	M4e	11.7	15.2	43
27	406	CN Leo = Wolf 359	10 54.1	+07 19.2	13.53	dM8e	16.68	2.35	17,40
28	412 B	WX UMa	11 03.0	+43 46.7	14.53	dM53	15.88	5.4	15
29	--	DH Car	11 12.8	-61 29.3	14.9	K2	--	--	18.19
30	424	SZ UMa	11 17.5	+66 07.0	9.32	dM1	9.70	8.5	2
31	447	Ross 128	11 45.2	+01 01.0	11.10	dM5	13.50	3.3	44
32	473 A	Wolf 424 A	12 30.8	+09 17.6	13.16	dM5.5e	14.98	4.3	12,50
33	473 B	Wolf 424 B	12 30.8	+09 17.6	13.4	M7	15.2	4.3	49,23
34	493.1	Wolf 461	12 58.1	+05 57.1	13.34	M5e	--	10.1	43
35	494	DT Vir	12 58.3	+12 38.7	9.79	dM2e	9.4	12.1	20,49
36	516 A	VW Com	13 30.3	+17 04.2	12.00	dM4e	11.0	16.0	2
37	516 B	--	13 30.3	+17 04.2	12.3	dM4e	11.3	16.0	49

TABLE 1.1 (continued)

	Gliese's Number	Name of Star	α(1950)		δ(1950)		m_V	Sp	M_V	Distan. pc	Ref.
1	2	3	4		5		6	7	8	9	10
38	577	EQ Her = CD-72°1700	13	32.1	-08	05.1	9.34	dK5e	8.0	18.8	47,48
39	540.2	Ross 845	14	10.4	-11	47.2	13.5	dM5.5e	12.8	13.8	43
40	551	V 645 Cen	14	26.3	-62	28.1	11.05	dM5e	15.45	1.31	2
41	569	DM+16°2708	14	52.1	+16	18.3	10.20	dM0e	10.1	10.4	35,49
42	--	Anon (Lupus)	15	05.7	-48	46	--	--	--	--	38
43	616.2	DM +55°1823	16	16.0	+55	23.8	9.96	dM1e	8.9	16.3	4,40
44	644 A	Wolf 630 = V 1054 Oph	16	52.8	-08	14.7	9.76	dM4.5e	10.79	6.2	12,49
45	644 B	--	16	52.8	-08	14.7	9.8	dM4.53	10.8	6.2	49
46	699 A	Ross 868	17	17.9	+26	32.8	11.36	dM4e	11.25	10.5	12,40
47	699 B	Ross 867	17	17.9	+26	32.8	12.92	dM5e	12.81	10.5	33,42
48	719	BY Dra	18	32.7	+51	41.0	8.6	dK6e	7.6	14.1	24,52
49	729	V 1216 Sgr	18	46.8	-23	53.5	10.6	dM4.5e	13.3	2.9	25
50	735	V 1285 Aql	18	53.0	+08	20.3	10.07	dM2e	9.9	10.9	36
51	752 B	BD +4°4048 B	19	14.5	+05	04.7	17.38	dM5e	18.57	5.8	26
52	781	Wolf 1130	19	20.1	+54	18.2	11.9	dM3e	10.8	17.0	17,28
53	--	G 208 - 44	19	52.3	+44	17.3	15.4	--	15.07	4.7	39
54	--	G 208 - 45	19	52.3	+44	17.3	16.6	--	15.65	4.7	39
55	791.2	G 24-16	20	27.4	+09	31.2	13.06	dM6e	13.2	9.4	9
56	--	S 5114	20	33.3	-70	04.2	14.8	M5	--	--	29
57	799 A	AT Mic	20	38.7	-32	36.6	10.83	dM4.5e	11.09	8.8	25,50
58	799 B	--	20	38.7	-32	36.6	10.9	dM4.5e	11.2	8.8	49
59	803	AU Mic	20	42.1	-31	31.1	8.61	dM0e	8.87	8.8	6,50
60	815 A	AC +39°57322	20	58.1	+39	52.7	10.26	dM3e	9.8	12.5	40,1
61	815 B	--	20	58.1	+39	52.7	12.7	--	11.8	12.5	40,49
62	852	Wolf 1561	22	14.7	-09	03.0	13.5	dM4.5e	13.6	9.7	43
63	860 A	BD +56°2783	22	26.2	+57	26.8	9.85	dM4	11.87	4.0	49
64	860 B	DO Cep	22	26.2	+57	26.8	11.3	dM4.5e	13.3	4.0	25,50
65	867 B	L 717-22	22	36.0	-20	52.8	11.45	dM4e	11.8	8.3	35
66	873	EV Lac	22	44.7	+44	04.6	10.2	dM4.5e	11.65	5.1	32
67	--	S 10113 And	23	20.9	+52	35	15.5				34
68	--	Anon (Pisces)	23	29.1	-03	01.7					53
69	896 A	DM +19°5116	23	29.5	+19	39.7	10.38	dM4e	11.33	6.4	2
70	896	EQ Peg	23	29.5	+19	39.7	12.4	dM6e	13.4	6.4	25,33
71	908	BD +1°4774	23	46.6	+02	08.2	8.98	dM2e	10.19	5.7	

1. A. H. Joy, Stellar Atmospheres, Ed. J. L. Greenstein, University of Chicago Press, 1960.
2. M. Petit, DOB 9, Sept. 1955.
3. H. L. Johnson, W. W. Morgan, Ap. J. 117, 323, 1953.
4. M. Petit, Ciel et Terre 70, 407, 1954.

TABLE 1.1 (continued)

5. A. D. Andrews, PASP 79, 368, 1967.
6. W. E. Kunkel, Private commun.
7. A. H. Joy, M. L. Humason, PASP 61, 133, 1949.
8. W. J. Luyten, Ap. J. 109, 532, 1949.
9. W. E. Kunkel, IBVS 294, 1968.
10. C. Hoffmeister, IBVS 126, 1966.
11. A. A. Wachmann, Beob. Zirk., A.N. 21, 25, 1939.
12. W. E. Kunkel, "Flare Stars," Doctoral Thesis, Texas, 1967.
13. L. Munch, G. Munch, Bol. Obs. Tonant. y Tacub. 13, 36, 1955.
14. S. Gaposhkin, Bol. Obs. Tonant. y Tacub. 13, 39, 1955.
15. A. Van Maanen, Ap. J. 91, 505, 1949; PASP 57, 216, 1945.
16. K. Gordon, G. Kron, PASP 61, 210, 1949.
17. H. U. Sandig, A.N. 280, 39, 1951.
18. E. Hertzsprung, BAN 1, 87, 1924.
19. S. Tapia, IBVS 286, 1968.
20. W. P. Bidelman, Ap. J. Suppl. 7, 175, 1954.
21. N. I. Shakhovskaya, IBVS 362, 1969.
22. A. D. Thackeray, M.N. 110, 45, 1950.
23. R. E. Gershberg, Bamberg Variable Star Coll. No. 15, Bamberg, 1972.
24. D. M. Popper, PASP 65, 278, 1953.
25. H. Joy, Non Stable Stars, Ed. G. H. Herbig, Cambridge, 1957, p. 31.
26. G. H. Herbig, PASP 68, 531, 1956.
27. Hansen J. Vinter, Circ. Union Astr. In., No. 1692, 1959.
28. A. H. Joy, Ap. J. 105, 101, 1947.
29. G. S. Mumford, PASP 81, 890, 1969.
30. W. J. Luyten, H.B. No. 830, 1925.
31. C. Hoffmeister, Mitt. Veränd. Sterne, Sonnenberg, No. 490, 1960.
32. N. E. Wagman, Harvard Ann. Card. No. 1226, 1953.
33. P. Roques, PASP 66, 256, 1954.
34. C. Hoffmeister, IBVS 203, 1967.
35. G. Asteriadis, L. N. Mavridis, IBVS 712, 1972.
36. T. J. Moffett, M.N. 164, 11, 1973.
37. G. Haro, E. Chavira, G. Gonzales, IBVS 1031, 1975.
38. S. Suryadi, IBVS 975, 1976.
39. S. Cristaldi, M. Rodono, Astron. Astrophys. 48, 165, 1976.
40. T. J. Moffett, Ap. J. Suppl. 29, 1, 1974.
41. M. Plavec, Z. Pekny, M. Smetanova, BAC 11, 180, 1950.
42. N. I. Shakhovskaya, W. Sofina, IBVS 730, 1972.
43. W. E. Kunkel, IBVS 748, 1972.
44. T. A. Lee, D. T. Hoxie, IBVS 707, 1972.
45. R. E. Nather, J. Harwood, IAU Cir. No. 2434, 1972.
46. B. R. Petterson, Astron. Astrophys. 41, 87, 1975.
47. S. Mello Parazz, IAU Cir. No. 2482, 1972.
48. I. C. Busko, C. Torres, IBVS 939, 1974.
49. N. I. Shakhovskaya, IAU Colloq. No. 15, 1971.
50. W. E. Kunkel, IAU Sympos. No. 67, 1974.
51. N. I. Shakhovskaya, Izv. Crimean Astr. Obs. 45, 124, 1972.
52. S. Cristaldi, M. Rodono, Astron. Astrophys. 12, 152, 1971.
53. H. E. Bond, IBVS 1160, 1976; Sky and Telesc. 52, 180, 1976.

If the spatial concentration of flare stars in the Galaxy is taken to be everywhere constant, their total number must be of the order of 10^8.

5. Most of the flare stars near the Sun are fainter than 10^m; only about one fifth of the stars in Table 1.1 are brighter than 10^m. The brightest ones are G 229, G 15 A (V $\approx$ 8), the faintest one is G 752 B (V = 17.4). We note that among the flare stars found in stellar associations (see below) objects fainter than 21^m occur.

6. The absolute brightness of flare stars covers a very wide range--from M_V = 7.1 for the star PZ Mon to M_V = 18.57 for G 752 B = BD +4°4048 B. In other words, flare activity can be found in stars whose absolute luminosities differ by a factor of 40,000.

Next to the flare stars whose flare activity has been established with certainty (Table 1.1) also exist stars which are suspected of flare activity and stars of which the facts about the flares are not sufficiently certain and require further confirmation; these stars--about twenty--are listed in Table 1.2. Sometimes communications appear about flares of stars belonging to earlier spectral classes (F - B). These objects--only three-four--have not been studied well, and the reliability of the recorded flares is doubtful; therefore such stars have not been included in Tables 1.1 and 1.2. However, in principle, an outburst of stars of earlier types is not impossible. The nature of the flare itself in such stars differs from what we have in cool dwarfs. In this sense "hot" flare stars form a special group of objects requiring separate consideration; this will be done later (Ch. 8, § 5).

3. FLARE STARS IN STELLAR ASSOCIATIONS

A new and particularly important phase in the history of the study of flare stars started with the fundamental work of Haro [17], the first--in 1953--to discover flare stars in stellar associations and young stellar clusters. The importance of this discovery for stellar evolution and stellar physics can hardly be overestimated. Haro pointed out that the flare process is inherent to the youngest stars, thus to the process of formation and development of stars, and is directly related to the physical processes occurring in the nuclei of stars which are not yet quite formed. At the same time this phenomenon has been discovered in cases where the star is no longer young and lies far beyond the boundaries of the associations and clusters where it was born.

During the period 1953-1976, by the efforts of Haro and his collaborators in the observatory at Tonanzintla (Mexico), and astrophysicists from other observatories-- Asiago (Italy), Byurakan (USSR) and others--a great number of flare stars in stellar associations and open clusters were discovered; we shall consider this in detail in Chapter 11. The youngest association--that in Orion--proved to be the most interesting one; more than 300 flare stars were discovered in it. Flare stars have also been discovered in the Pleiades cluster, a system of "middle" age, the clusters NGC 2264, Praesepe and others. The total number of flare stars in stellar associations and clusters up until 1975 is of the order of 1000. Simple statistical considerations show that their real number in associations and clusters must be considerable, even of the same order as the total number of stars in associations.

Hence, it appears that each star, in an early stage of its life, must pass through a state of flare activity.

The last statement may, if proved correct, become particularly valuable for the early stages of stellar evolution.

TABLE 1.2 Suspected UV Cet Type Flare Stars

	Name of Star	α (1950)	δ (1950)	m_{pg}	Sp	Reference
1	2	3	4	5	6	7
1	V Psc	00^h 12^m8	$+06°$ $23\!.8$	12	M1 V	1
2	SVS 1989	00 32.6	+39 53	18.5	--	2
3	SVS 1729	00 40.0	+40 14.2	16.8	--	3
4	QZ Per	03 14.8	+37 23.2	17.5	--	4
5	VSV 1849	04 27.1	+23 47	14.5	--	5
6	G 451 B	11 50.1	+38 04.6	--	--	6
7	G 177.4	12 53.7	+51 12.1	16.1	--	7
8	Anon	12 55.2	-65 33	14.7	--	8
9	Anon	13 02.5	-61 56	12.7	--	8
10	Anon	13 30.1	-61 58	15.5	--	8
11	G 526	13 43.2	+15 09.7	8.50	dM4e	8
12	V 475 Her	17 18.1	+25 15	15.8	--	4
13	G 258-7	17 27.4	+67 01.1	16.3	--	7
14	Anon (Serpens)	18 14.9	-10 14	12.6	dK5	10, 11
15	V 2354 Sgr	18 27.1	-24 53.5	13.5	--	4
16	Anon (Aquila)	18 55.9	+06 06.5	12	M3	12
17	KZP 4661	19 20.1	+28 14	--	--	13
18	V 868 Cyg	19 25.1	+28 41.1	--	--	13
19	SVS 1532	19 25.8	+28 20.4	--	--	13
20	FV Vul	19 36.6	+27 29.0	14.1	M:	4
21	KO Vul	19 55.3	+28 57	17.5	--	4
22	AK Mic	21 03.6	-40 14.0	16.5	--	14
23	FZ And	23 03.3	+52 51.4	15.5	--	4

1. M. Petit, DOB No. 9, Sept. 1955.
2. A. K. Alksnis, A. S. Sharov, Astron. Zirc. USSR No. 800, 1973.
3. A. S. Sharov, A. K. Alksnis, Astron. Zirc. USSR No. 600, 1970.
4. P. V. Kukarkin et al., Gen. Catalogue of Var. Stars, Moscow, 1969.
5. V. P. Tsesevich, Astron. Zirc. USSR No. 733, 1972.
6. P. Van de Kamp, Low Luminosity Stars, Ed. S. Kumar, 1968, p. 199.
7. H. L. Giclas, R. Burnham, N. C. Thomas, Bull. Lowell Obs. 203, 1965.
8. N. Sanduleak, IBVS 275, 1968; Ap. J. 155, 1121, 1969.
9. W. E. Kunkel, CR 255, 1275, 1962.
10. F. M. Steinon, IBVS 545, 1971.
11. D. J. MacConnell, IBVS 570, 1971.
12. N. B. Kurochkina, Astron. Zirc. USSR No. 451, 1967.
13. B. Hidejat, M. U. Akyol, IBVS 623, 1972.
14. C. Hoffmeister, Mitt. Veränd. Sterne, Sonnenberg, No. 490, 1960.

4. THE MORPHOLOGICAL COMPOSITION OF FLARE STARS

According to all the data the flare stars of UV Cet type in the neighborhood of
the Sun do not form a homogeneous group with regard to age or their membership in
some type of stellar population. Thus, e.g. the flare stars YZ CMi and AT Mic
(= G 799 A = CD -32°16135) are in all probability young objects, since in their
quiet state, i.e. outside flares, they lie almost above the M_V vs (R - I) diagram
[22]. UV Cet itself and also Wolf 424 are members of the moving cluster, the Hyades
[23]. On the other hand, the stars Wolf 630, Wolf 359, BD +43°44 A and B, 40 Eri C,
CQ And, SZ UMa, and also Ross 128 = G 447, belong to stellar population I [24].

The examples given, and also the fact that flare stars occur in large numbers in
stellar associations and young clusters, show that an equal flare activity may
sometimes be observed in stars of widely differing ages, at least by one order of
magnitude. At the same time it is quite evident that in old dwarf stars the flare
phenomenon has the character of a remnant from a former state.

The cosmogonic significance of the flare phenomenon in general and of flare stars
in particular will be considered in the following chapters as the general flare
theory based on the hypothesis of fast electrons will be developed.

5. DEFINITION OF A FLARE

How should a flare be defined? How can it be distinguished from ordinary irregular
fluctuations of a star's brightness? Evidently we need quantitative criteria here.
The definition of a flare generally used, as a "rapid and strong increase of brightness" contains subjective elements. This problem has been discussed repeatedly.
A number of observers [17-20] drew attention to so-called "secondary" or "slow"
brightness fluctuations with small amplitudes. It is often hard to distinguish a
flare with small amplitude from "secondary" luminosity fluctuations. A flare cannot
be characterized only by the value of its amplitude. Cases are known of insignificant brightness fluctuations in various stars which cannot be considered as flares
because they last too long. The length of time during which the star shows increased luminosity cannot be used to characterize the flare either. As a criterion
for the definition of a flare the ratio of these two quantities--the value of the
increase of brightness and the duration of the flare, i.e. the rate of increase of
luminosity, should be used. Haro [21], who has considered this problem in detail,
reaches the conclusion that a rate of increase of about 0^m005 s^{-1} should be considered as the minimum value defining a flare. This means that an eruptive variable which increases its luminosity at a rate of 0^m3 min^{-1} cannot be classified as
a flare star.

As an additional criterion for the definition of a flare the lightcurve after
maximum may be used. In fact, the descending branch of the lightcurve has a quite
definite form, characteristic for flare stars only, and this form does not depend
on the amplitude of the flare nor on its duration; it depends only on the nature
of the flare itself (cf. Ch. 15).

6. THE LIGHTCURVE. TWO TYPES OF FLARES

The course of variations of the star's brightness during the outburst is called
the lightcurve. When the flare is recorded photographically the lightcurve is obtained by the chain method--a series of images of the star, with short but strictly
equal exposure times, is taken on one plate. If the star is faint this leads to an
underestimate of the real amplitude of the increase in brightness; in these cases
the measures give only a lower limit of the amplitude.

Matters are different in cases where the flare is recorded photoelectrically. With a time constant of the order of a few seconds and in some cases fractions of a second, the photoelectric method makes it possible to follow many details of an outburst which cannot be done photographically. Besides, the photoelectric method is the only means of recording very shortlived (of the order of one minute and less) and rapidly developing flares.

Usually the lightcurve is given as a graph showing the dependence on time of the quantity

$$i = \frac{I_{f+o} - I_o}{I_o} = \frac{I_f}{I_o} \tag{1.1}$$

which represents the "flow of the outburst," i.e. the flow of excess radiation I_f, radiated at a given moment in the development of the flare, expressed in units of the flow I_o from the star in normal non-excited state.

Often the fraction of the "flow of the outburst" is given in stellar magnitudes, i.e.

$$\Delta m = 2.5 \lg(i + 1) = 2.5 \log \frac{I_f + I_o}{I_o} \tag{1.2}$$

Examples of lightcurves of flares will be given in the following chapters. Here we consider only one property, common to all lightcurves.

Notwithstanding the large diversity in the shapes of the lightcurves, two principal types--I and II--predominate. They differ from each other by the absolute values of the main parameters of the lightcurve--t_b and t_a, where t_b is the interval of time from the start of the flare until maximum, and t_a is the time from maximum until complete re-establishment of the original brightness of the star. These parameters are shown in the scheme of the lightcurve given in Fig. 1.1.

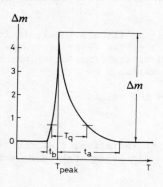

Fig. 1.1. Lightcurve of an outburst of a star. T_{peak} corresponds to the moment of the maximum of the flare.

The numerical values of t_b and t_a for the two types of lightcurve mentioned are:

Type I:

t_b = from a few seconds to a few minutes t_a = from a few minutes to about one hour

Type II:

t_b = 30 minutes or more t_a = 3 to 10 hours

The lightcurve of type I is in general characteristic for flare stars of UV Cet type, and that of type II, for flare stars connected with stellar associations and young clusters.

Haro [21] mentions cases where one and the same star in an association or a cluster may one time show an outburst of type I and another time of type II. This is observed most often in very young clusters (Orion, NGC 2264). This does not hold, however, for stars of UV Cet type; they never show an outburst of type II. Therefore it may be assumed that stars in which the flare phenomenon has the character of a remnant cannot show outbursts of type II.

One other property draws attention: although the absolute values of the parameters t_a and t_b in types I and II differ by more than an order of magnitude, the ratio t_b/t_a for the two types is nearly the same. These two types differ from each other mainly by the absolute amount of energy emitted in the outburst, but the nature of the outburst is the same in the two cases. The fact that flares of type II occur most often in star associations and young clusters may be taken as proof that particularly strong and long-lasting outbursts are proper to very young, not yet quite formed, stars.

7.   THE PARAMETERS OF THE LIGHTCURVE

A general description of the lightcurve has been given above. Here we shall enumerate those quantities or parameters which, taken from the lightcurve, may characterize various aspects of the flare itself (cf. also [31]).

1.   The amplitude of the flare Δm. This is the excess radiation at maximum brightness expressed in stellar magnitudes.

The value of the flare amplitude can be found from equation (1.2). Depending on the spectral range in which the flare is recorded--in photographic, in U, B or V light, etc.--the amplitudes of the flares are considered in photographic light, Δm_{pg}, in U, B and V radiation--ΔU, ΔB, ΔV, respectively.

2.   The time t_b--from the start of the outburst until maximum. This quantity is sometimes of the order of a few seconds and is therefore the hardest to determine.

3.   The time t_a--from the maximum of the outburst until its end. Together with the value of the amplitude of the flare Δm, the parameters t_b and t_a give the possibility of finding the rate of increase of brightness before maximum and the rate of the decrease of brightness after maximum, respectively, in stellar magnitudes per second.

4.   The duration of the outburst T_q--here q indicates the (lower) level, in fractions of the maximum brightness, from which the length of the flare is counted (Fig 1.1). Often q is ~ 0.2; then $T_{0.2}$ is the length of the time during which the brightness is more than 20% of the brightness at maximum.

5.   The "flare integral" P. This value is found from the relation

$$PI_o = \int (I_{f + o} - I_o)dt , \qquad (1.3)$$

where the right-hand part is the integral of the excess radiation, i.e. the integral of the lightcurve, the radiation I_o of the non-disturbed star being subtracted.

The parameter P has the dimension of time (minutes) and yet it characterizes the energetic strength of the outburst, for $P I_o$ is the total energy radiated by a quiet star with constant radiant power I_o. P has the same physical significance as the equivalent width of spectral lines when their intensities are found. Therefore we may call P the "equivalent time of the flare," given by the relation:

$$P = \int \left(\frac{I_{f + o}}{I_o} - 1 \right) dt . \qquad (1.4)$$

6. The criterion for the reality of the outburst $3\sigma(mag)$. Here $\sigma(mag)$ is the standard error caused by accidental fluctuations--or "noise":

$$\sigma(mag) = 2.5 \log \frac{I_0 + |\sigma|}{I_0} \qquad (1.5)$$

The quantity σ may vary from one night to the next. In fact the minimal value of the amplitude and therefore the reliability of the flare is determined by the quantity $\sigma(mag)$. In particular, the outburst may be thought to be real when its amplitude Δm is larger than $3\sigma(mag)$.

8. THE RATE OF DEVELOPMENT OF THE FLARE

To understand the true nature of a flare it is highly important to know the extreme values of the various parameters. Among those, the velocity of the increase of the brightness from the moment the flare appears until the maximum is reached, has a particular significance.

Generally, for most of the flare stars of UV Cet type, the rate of development of the flare is of the order of 0^m05-0^m1 sec^{-1}. However, cases are known in which the brightness increased with considerably larger velocity. For example, in a number of consecutive flares of UV Cet on October 10, 1975 [25], in several cases the brightness increased by a factor of 19 in 10 seconds, i.e. the rate of increase was about 0^m6 sec^{-1}. There were cases of short flares in which the brightness increased four times in one second, i.e. the rate was 1^m5 sec^{-1}.

In this respect, the record among the observed outbursts appears to be held by the exceedingly strong flare found by Jarrett and Gibson [26] on September 22, 1974 in a survey of UV Cet where the brightness increased 420 times in 31 seconds (!), i.e. 13.5 times per second, or with a rate of 2^m8 sec^{-1}. The duration of the flare itself was, in this case, also among the longest ones--2 hours 44 minutes.

9. THE FLARE AS AN ACCIDENTAL PHENOMENON

The flare phenomenon in a given flare star has an accidental character and can be represented with sufficient accuracy by a Gaussian distribution:

$$p(n,m) = \frac{R^n(m)}{n!} \ e^{R(m)} , \qquad (1.6)$$

where $p(n,m)$ is the probability that n flares brighter than magnitude m occur per unit time, and $R(m)$ is the average frequency of flares brighter than m per unit time.

The parameter $R(m)$, or $R(U)$ if the flares are considered in U-radiation, can be represented with sufficient accuracy in the following form [27, 28, 24]:

$$R(U) = e^{\alpha(U - U_0)}, \quad (flares/hour) \qquad (1.7)$$

where U is the brightness of the star at flare maximum in U-radiation, U_0 is the level of brightness at which the flare frequency is one per hour; this quantity is different for different stars and is found from statistical considerations of the observational results. But U_O can also be different for one and the same star depending on the fluctuations of its total activity.

Let us give some examples [28]. For UV Cet, U_O = 13.0, and for YZ Cet U_O = 17.5. For EV Lac, AD Leo, and YZ CMi the quantities U_O are about equal; they are 14.9, 14.5 and 14.0, respectively.

As regards the coefficient α in (1.7), this does not vary much from one star to another; on the average α = 1.00 ± 0.05 [24].

10. ENERGY DISTRIBUTION IN THE SPECTRA OF FLARE STARS

Flare stars belong to the late spectral classes, and their continuous spectra are distorted by numerous absorption lines and bands of atoms and molecules. Moreover they are very faint stars, which makes it very difficult to obtain and measure their spectrograms. Therefore we know very little about the energy distribution in the spectra of flare stars in their undisturbed state, outside flares. The few data about this problem which we have at our disposal have mainly been obtained by color-imetric measurements. From these data the following conclusions may be drawn:

1. In the color-color diagram U – B vs B – V flare stars of UV Cet type, i.e. those in the neighborhood of the Sun, seem normal; they are either on the main sequence or slightly above it; in the latter case the star has an ultraviolet excess.

2. There are no indications for the existence of a significant infrared excess in flare stars of UV Cet type, such as is the case for T Tauri stars. Judging from the results of infrared measurements by Mendoza [29], the energy distribution in the long wavelength region of the spectrum ($\sim$ 1μ and longer) of flare stars nearly coincides with the Planck curve for the energy distribution for effective temperatures of the star of 2800–3000 K. As an example confirming this statement, in Fig. 1.2

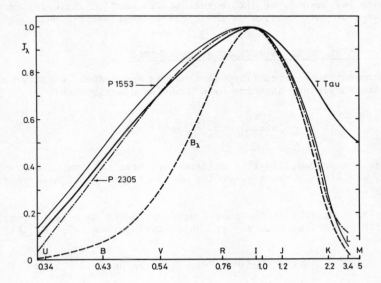

Fig. 1.2. Energy distribution of the flare stars Haro 146 (P 1553) and Haro 77 (P 2305), the star T Tau and a black body for T = 2900 K (B_λ), in the wavelength interval 3400 Å (U) to 50,000 Å (M). The intensity at 10,000 Å (1μ) has been taken as unity.

energy distribution curves in the continuous spectra of two flare stars--Haro 146
(= P 1553) and Haro 77 (= P 2305)--are given; also for comparison, that of the star
T Tau is shown. The Planck curve corresponding to a temperature 2900 K is drawn as
well. In this case, these flare stars have an excess of energy in the short wave-
length part of the spectrum, which indicates some activity, of the type of that of
T Tau stars, even in the quiet non-flare state.

Usually the occurrence of an infrared excess in some star is interpreted as evidence that it is surrounded by a dense gas and dust cloud. Thus, the absence of
infrared excess in flare stars of UV Cet type may be taken as an indication that
they are not actually surrounded by such clouds. However, the conclusion that
there is no infrared excess cannot be considered as definitive. Anyway we have no
sufficient evidence for excluding the possibility of a small infrared excess in
some of the UV Cet stars. (See Chapter VIII).

With regard to the stars in stellar associations and clusters, objects with infrared excess do occur (T Tau type).

We may conclude that the flare phenomenon occurs in a wide range of objects--from
new-born stars, like objects of T Tau type, to stars of "middle" age, like those
of UV Cet type; whereas the infrared excess occurs in a quite narrow range of objects, stars of T Tau type, it disappears when the star outgrows its "infancy."

For further discussion it is essential that in the derivation of theoretical relations the normal undisturbed spectrum of the star in the infrared region be represented by Planck's formula, for a given effective temperature.

11. FLARE FREQUENCY

By flare frequency we understand the average number of flares taking place in a
certain interval of time; for example, during one hour or one day. Sometimes the
flare frequency is given by means of the average time interval between two consecutive outbursts.

The flare frequency for a given star depends strongly on the adopted amplitude of
the flares: the smaller the amplitude, the higher the frequency. The minimum value
of the amplitude for which a flare can still be reliably observed is different for
different observers and depends on many factors--the diameter of the mirror in the
telescope, the method and special properties of the recording apparatus, the brightness and spectral class of the star, the quality of the images, the state of the
night sky, etc. Therefore comparative analyses should be based on observational
material which are as homogeneous as possible. Notwithstanding these difficulties,
at present we have a sufficiently correct picture of flare frequency, at least for
stars of UV Cet type. It is based on analysis [30] of more than 1,000 outbursts
of about 20 flare stars in the neighborhood of the Sun. Here we present results
based on an analysis of the most homogeneous observations by Kunkel [28] in U-
radiation, and by Moffett [32] in U, B and V-radiation, for a large number of flare
stars. The total number of flares recorded for individual stars (UV Cet, CN Leo,
Wolf 630) is more than one hundred.

Table 1.3 gives the flare frequency f--the number of flares with amplitude $\geqslant 0^m_.1$
per hour for each star in the spectral range indicated (U, B or V). At the same
time, the values of f_U, f_B and f_V from Moffett's data [32] or only f_U from Kunkel's
data [28] are shown for each star. The same table gives, in the second column, the
absolute photovisual magnitudes of the stars considered, taken mainly from the
catalog of Gliese [33].

TABLE 1.3   Flare Frequencies f (Flares/Hour) in U, B and V Radiation as a Function of Absolute Luminosity of the Flare Star

Star	M_V	f_U (1)	f_U (2)	f_B (2)	f_V (2)
CN Leo	16.68	3.4	3.5	2.7	0.9
UV Cet	15.27	4.9	1.7	1.4	1.1
Wolf 424	14.31	--	4.1	3.0	0.75
40 Eri C	13.73	2.2	--	--	--
Ross 614	13.08	4.0	--	--	--
YZ CMi	12.29	0.7	1.2	0.75	0.43
EV Lac	11.50	--	0.35	0.33	0.13
EQ Peg	11.38	--	0.70	0.56	0.14
AT Mic	11.09	1.8	--	--	--
AD Leo	10.98	--	0.42	0.08	0.0
Wolf 630	10.79	1.3	--	--	--
AU Mic	8.87	0.9	--	--	--
YY Gem	8.36	--	0.16	0.16	0.13
EQ Her	8.00	0.16*	--	--	--

(1)   From data of Kunkel, _Ap. J. Suppl._, 25, 1, 1973.
(2)   From data of Moffett, _Ap. J. Suppl._, 29, 1, 1974.
*   From data of Busko, Torres, _IBVS_, No. 939, 1974.

From data actually available about flare frequency, the following laws can be established (cf. also Table 1.3):   a) The flare frequency f_U, f_B or f_V increases with decreasing absolute magnitude of the star M_V.   For the faintest stars (CN Leo, UV Cet), the flare frequency is highest--on the average 3-4 flares in U radiation per hour.   At the same time, the brightest stars (YY Gem, EQ Her) show one outburst in about four hours, i.e. nearly 15 times less.

To illustrate, Fig. 1.3 shows the dependence of the flare frequency in U radiation on M_V, constructed from the data in Table 1.3 (second and third columns).

Notwithstanding the large scatter of the points in Fig. 1.3, a certain empirical relation between f_U and M_V can be noticed. It can be represented in the following form:

$$\log f_U = -1.78 + 0.148 \, M_V , \qquad (1.8)$$

or, if we introduce the interval P_U between the outbursts (in hours), we have:

$$\log P_U = 1.78 - 0.148 \, M_V . \qquad (1.9)$$

These relations can be used in particular to find the probable frequency or the probable time interval between consecutive

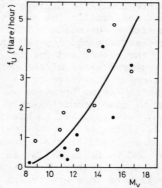

Fig. 1.3.   Dependence of flare frequency f_U (flares/hour) on absolute luminosity of the star, M_V, from observational data of Kunkel (circles) [28] and Moffett (dots) [32] (for all flares with $\Delta U \geq 0.1$).

outbursts of an unknown star which is suspected of flare activity.

b) For all stars, independent of spectral class, absolute luminosity, value of the flare amplitude, etc., the following relation holds:

$$f_U > f_B > f_V \ .$$

(1.10)

There has not been a single case where this rule has been violated. However, it becomes less pronounced when we pass from stars with small absolute luminosity to those with large luminosity.

As a rule, the flare frequency in U radiation is several times higher than in V radiation, sometimes by one order of magnitude. Therefore, when seeking new flare stars or wanting to confirm the flare activity of a star the observations should certainly be made in U light. On the other hand, U observations have their own difficulties: in this spectral range flare stars are considerably, sometimes by 3^m, fainter than in V radiation, and hence a more sensitive recording apparatus is needed.

c) The flare frequency of a given star is not constant; it may change depending on fluctuations in its general flare activity. For example, from the data of Kunkel [28], in 1966 the flare frequency of UV Cet was $f_U = 3.6$ fl/hr (all flares with amplitude $\Delta U \geqslant 0.05$); in 1968 it increased to $f_U = 7.0$ fl/hr and this value was maintained also in the following year, 1969. From the same data the flare frequency of another star--Wolf 630--was 1.6, 1.9 and 1.8 fl/hr for 1967, 1968, and 1969 respectively, but in 1970 it increased by a factor of nearly two and a half: $f_U = 4.8$ fl/hr.

In the case of another flare star--V 1216 Sgr ($M_V = 13.3$), Cristaldi and Rodono [34] in 1971 did not find a single flare during a period of 28.5 hours, and therefore $f_U < 0.035$ fl/hr. In 1973 they found $f_U = 0.035$ fl/hr. In June 1974 eight outbursts of this star were observed with frequency $f_U = 0.15$ fl/hr [36], and in July $f_U = 0.35$ fl/hr [37].

d) A somewhat different picture is observed in the case of flare stars which are members of stellar associations and clusters; the data concerning these stars are given in Chapter 11. Comparing these data with those in Table 1.3, we see that the flare frequency in associations is more than one order of magnitude smaller than for flare stars near the Sun. However, this may not be true since observational conditions for flare stars in aggregates are such (photographs with exposure times of 5-10 minutes) that flares with amplitudes smaller than $0.6-0^m\!.7$ practically can not be observed. Whereas the data in Table 1.3 for flare stars of UV Cet type refer to outbursts with amplitudes $\geqslant 0^m\!.1$. As a consequence of this selection, no appropriate comparison between the flare frequencies of the two categories of stars is possible.

e) Matters are more difficult with respect to the dependence of flare frequency and the average duration of an outburst, on spectral class. Deriving this dependence from the observations of stars of UV Cet type is practically impossible, if only because most stars belong to spectral type M5-M6. It is different in associations where the spectral range of flare stars is wider, from K0 to M5. From incomplete data [38] a substantial increase of the duration of the outburst is observed towards the earlier spectral classes; whereas in stars of class M4-M6 the outburst lasts about 20 minutes, for stars of class K6-K8 the average duration of the flare is 90 minutes.

For flare stars of UV Cet type, Kunkel [39] established an analogous dependence in a somewhat different form. This is a relation between the absolute magnitude of

the star, M_V, and gradient of the decrease of brightness after maximum, G, expressed in "magnitude/minute" as shown in Table 1.4 from data of about five flare
stars of UV Cet type.

TABLE 1.4   Dependence of Gradient G on Absolute Magnitude

Star	M_V	Log G	n
Wolf 359	16.87	−0.24 ± 0.10	6
UV Cet	15.27	−0.28 ± 0.09	15
40 Eri C	13.73	−0.55 ± 0.14	6
YZ CMi	12.29	−0.68 ± 0.09	5
AD Leo	10.98	−0.87 ± 0.17	4

f)   The data in Table 1.3 about flare frequencies refer to flares with all possible
amplitudes larger than 0^m1. Frequencies $\phi(\Delta U)$ of flares with given amplitude ΔU
may be determined numerically from the equation:

$$\phi(\Delta U) = f_U F(\Delta U), \tag{1.11}$$

where $F(\Delta U)$ is the distribution function of the amplitudes of the flares in U radiation (cf. §13). Relations analogous to (1.11) may be written down for determining
the frequency of flares with a given amplitude in B and V radiations.

The function $F(\Delta U)$ decreases with increasing ΔU. Therefore, the function $\phi(\Delta U)$
also decreases. For example, with the data in Tables 1.3 and 1.8 we find from
(1.11) that the average interval of time between two consecutive outbursts in U
radiation, with amplitude larger than 1^m, is about 40 minutes for UV Cet and about
24 hours for AD Leo.

12.   THE AMPLITUDE OF THE FLARES

There exist certain empirical regularities in the sizes of the amplitudes of star
flares [30]. Let us consider some of them:

a)   The dependence of the amplitude on spectral range. For many years synchronous,
three-color observations of star flares in U, B and V radiation were unique. Until
the end of the 1960's only a few cases of synchronous recordings of flares in UBV
light were known. The first of these refers to a flare star in the Pleiades--
H II 1306, for which Johnson and Mitchell [40] obtained, in 1957, lightcurves of
one outburst. Later appeared the first three-color recordings of outbursts of
AD Leo [41], EV Lac [42, 43], DH Dra [44], and of a very strong flare of S 5114 [45].

The situation changed radically after 1970 when at several observatories--Catania
(Italy), Kyoto (Japan), MacDonald (USA)--systematic synchronous observations in UBV
were started of flare stars. Nowadays, we have at our disposal vast material referring in particular to the amplitudes ΔU, ΔB, and ΔV for a number of flare stars
acquired by Cristaldi and Rodono [46, 47], Osawa et al. [48], and Moffett [32].
The analysis of flare amplitudes given below is based mainly on this observational
material--of about 1000 flares--sufficiently homogeneous, at least within the limits
of each of these series of observations. Table 1.5 shows examples of amplitudes
of flares, one strong and one weak for each star, found from synchronous UBV observations, for a number of flare stars. The quantities ΔU, ΔB and ΔV have been

TABLE 1.5 Examples of Amplitudes of Flares for Synchronous Observations in UBV Radiation for a Number of Flare Stars

Star	ΔU	ΔB	ΔV
UV Cet	3^m47	1^m48	0^m50
"	0.72	0.12	0.04
AD Leo	5.01	2.93	1.65
"	0.60	0.18	0.03
EV Lac	1.54	0.43	0.16
"	0.32	0.08	0.03
YZ CMi	3.49	1.74	0.78
"	0.29	0.05	0.02
CN Leo	4.06	1.85	0.70
"	0.42	0.11	0.07
Wolf 424	2.52	0.82	0.24
"	0.47	0.07	0.05
EQ Peg	2.40	0.79	0.28
"	0.45	0.10	0.06
YY Gem	1.97	0.60	0.06
"	0.69	0.15	0.06

TABLE 1.6 Average Flare Amplitude in U, B and V Radiation as a Function of Absolute Magnitude M_V

Star	M_V	$\overline{\Delta U}$ (1)	$\overline{\Delta B}$ (2)	$\overline{\Delta V}$ (2)
CN Leo	16.68	1^m32	0^m29	0^m11
UV Cet	15.27	1.32	0.42	0.24
Wolf 424	14.31	0.42	0.07	0.02
40 Eri C	13.73	0.42	--	--
Ross 614	13.08	0.37	--	--
YZ CMi	12.29	0.97	0.28	0.12
EV Lac	11.50	1.00	0.27	0.10
EQ Peg	11.38	0.81	0.21	0.07
AT Mic	11.09	0.62	--	--
AD Leo	10.98	0.37	--	--
Wolf 630	10.79	0.27	--	--
AU Mic	8.87	0.06	--	--

(1) Mean of data in [28, 32, 48].
(2) From data in [32].

taken, for AD Leo from [48], and for the other stars from [32]. From these data
a relation between the values of the amplitudes follows, namely:

$$\Delta U > \Delta B > \Delta V \quad . \tag{1.12}$$

It should be stressed that this inequality always holds for all flare stars, strong
and weak flares alike. Moreover, this rule is valid not only for individual flares
but also, statistically, for any number of independent flares, which have been recorded in one case, say in U radiation, in others in B radiation, and in still
others in V radiation [30]. Such data are given in Table 1.6, where $\overline{\Delta U}$, $\overline{\Delta B}$, $\overline{\Delta V}$
are statistical averages of the amplitudes in U, B, V respectively. It is seen
that the mean values $\overline{\Delta U}$, $\overline{\Delta B}$ and $\overline{\Delta V}$ found from these series for a given star also
obey (1.12), independent of the absolute magnitude of the star.

b) Dependence of flare amplitude on absolute luminosity of the star. Available
data show that the average amplitude of the outbursts decreases with increasing
absolute luminosity of the star in the normal state. This is illustrated by the
graph in Fig. 1.4, made up from the data in Table 1.6 (third column) and also from
[30]. The dependence of $\overline{\Delta U}$ on M_V can be represented approximately by the following
empirical formula:

$$\overline{\Delta U} = -2.05 + 0.225 \; M_V \quad . \tag{1.13}$$

Larger average flare amplitudes in stars of low absolute luminosity imply larger
relative amounts of energy liberated during the outburst. However, we cannot deduce from this how the absolute amount of energy emitted during the flare depends
on the absolute luminosity of the star (cf. Ch. 11).

c) Dependence of the flare amplitude on the spectral class of the star. As
mentioned above, flare stars belong to spectral classes K – M; class G occurs quite
seldom (apparently only in stellar associations and young clusters). The average
value of the flare amplitude is found to be different for stars of different spectral classes: it is smallest in the early subclasses of K and largest in the subclass M5. As may be seen from the flare stars in the Pleiades, where the spectral
classes are known for most of the members, this dependence can be represented with
sufficient reliability in quantitative form (Table 1.7).

TABLE 1.7   Dependence of Maximum Amplitude of Flares in U
Radiation on Spectral Class (Pleiades)

	K3–K4	K5–K7	M
ΔU_{max}	$1\overset{m}{.}6$	$3\overset{m}{.}3$	$6\overset{m}{.}5$
n	4	27	52

In Fig. 1.5, another form of this relation is shown, where the maximum flare amplitudes observed, ΔU_{max}, in flare stars of various spectral classes in the Pleiades
[49], are shown.

From the last line in Table 1.7, yet another dependence can be found: the increase
of the number of flare stars n from the early subclasses of K to M. Evidently it
must be kept in mind that the determination of the spectral class of stars in the
interval K – M is in no way influenced by observational selection.

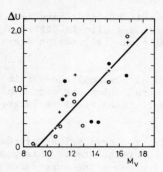

Fig. 1.4. Dependence of average flare amplitude in U radiation, $\overline{\Delta U}$, on absolute magnitude of the star, M_V, from observational data of Kunkel (circles) [28], Moffett (dots) [32] and other sources (crosses) [16].

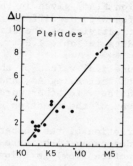

Fig. 1.5. Dependence of maximum amplitude of the flare, ΔU_{max}, on spectral class for flare stars in the Pleiades.

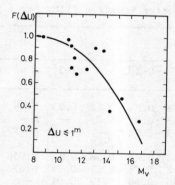

Fig. 1.6. Dependence of the distribution function of flare amplitudes, $F(\Delta U)$, on absolute luminosity of the star, M_V, for flares with $\Delta U \leqslant 1$ (Table 1.8).

13. THE DISTRIBUTION FUNCTION OF THE AMPLITUDES OF FLARES

Let $n(\Delta m)$ be the number of recorded outbursts of a given star with amplitudes between Δm and $\Delta m + 1$. The distribution function of flare amplitudes is denoted by $F(\Delta m)$, and is given by the equation:

$$F(\Delta m) = \frac{n(\Delta m)}{\Sigma n(\Delta m)} \tag{1.14}$$

where $\Sigma n(\Delta m)$ is the total number of flares recorded in a given spectral range of observation.

The value of the function $F(\Delta m)$ is determined numerically by observations and for each star separately. Table 1.8 gives the values of the function $F(\Delta m)$ found in this way for a number of flare stars in U, B and V radiation, taken mainly from the results given in [30]. It should be pointed out that the numerical values of the function F(Δm) given in this table, are not very reliable; for many reasons they are found to be different for different observers, and also in different periods of activity of the star. The following conclusions [30] may be drawn from the data in Table 1.8:

a) The numerical values of the distribution function of the amplitudes F(Δm) decrease with increasing Δm, and do so more slowly in U radiation and more rapidly in B radiation. In other words, the gradient of the function F(ΔB) is larger than the gradient of the function F(ΔU).

b) The fraction of the number of flares corresponding to large values of the amplitudes decreases with increasing absolute luminosity of the star. This property of the function F(Δm) is illustrated in Fig. 1.6, which shows the dependence of F(ΔU) on M_V, for values of the flare amplitudes in U radiation smaller than 1^m. By far, most of the flares--up to 80-90% and more--in such bright stars as AD Leo, Wolf 630 or AU Mic, correspond to amplitudes smaller than 1^m, whereas in the case of faint stars--CN Leo (Wolf 359) or UV Cet--flares with ΔU > 1^m are less than half of the total number.

TABLE 1.8 Distribution Function F(ΔU) for a Number of Flare Stars

Star	M_V	Spectral Range	N	Δm						References
				0-1	1-2	2-3	3-4	4-5	5-6	
CN Leo	16.68	U	138	0.28	0.42	0.23	0.06	0.01	--	30,28,32
		B	63	0.95	0.05	--	--	--	--	32
		V	63	1.00	--	--	--	--	--	32
UV Cet	15.27	U	613	0.48	0.34	0.10	0.056	0.016	0.008	30,28,32
		B	142	0.76	0.17	0.04	0.03	--	--	30, 32
		V	95	0.80	0.19	0.01	--	--	--	30, 32
Wolf 424	14.31	U	11	0.36	0.55	0.09	--	--	--	32
40 Eri C	13.73	U	38	0.89	0.11	--	--	--	--	28
Ross 614	13.08	U	35	0.91	0.09	--	--	--	--	28
YZ CMi	12.29	U	87	0.71	0.20	0.07	0.01	0.01	--	30,28,32
		B	122	0.85	0.12	0.02	0.01	--	--	30, 32
EV Lac	11.50	U	50	0.68	0.24	0.08	--	--	--	30,28
		B	64	0.81	0.14	0.05	--	--	--	30, 32
EQ Peg	11.33	U	27	0.81	0.18	0.01	--	--	--	32
AD Leo	10.98	U	69	0.73	0.16	0.10	--	0.01	--	32,30,48
		B	56	0.91	0.05	0.04	--	--	--	32,30
AT Mic	10.09	U	24	0.92	0.08	--	--	--	--	28
Wolf 630	10.79	U	119	0.98	0.02	--	--	--	--	28
AU Mic	8.87	U	31	1.00	--	--	--	--	--	28

REFERENCES

1. E. Hertzsprung, BAN 2, 87, 1924.
2. W. J. Luyten, Ap. J. 109, 532, 1949.
3. A. Van Maanen, Ap. J. 91, 505, 1949.
4. K. Gordon, G. Kron, PASP 61, 210, 1949.
5. A. D. Thackeray, M.N. 110, 45, 1950.
6. N. E. Wagman, HAC No. 1225, 1953.
7. A. J. Joy, Non-Stable Stars, Ed. G. Herbig, Cambridge, 1957, p. 31.
8. G. Kron, S. Gascoigne, H. White, A.J. 62, 214, 1957.
9. A. Van Maanen, PASP 57, 216, 1945.
10. T. A. Lee, D. T. Hoxie, IBVS No. 707, 1972.
11. I. C. Busko, C. A. Torres, G. R. Quast, IAU Circ. No. 2482, 1972; IBVS
    No. 939, 1974.
12. B. Hidajat, M. U. Akyol, IBVS No. 623, 1972.
13. S. Suryadi, IBVS No. 975, 1975.
14. G. Haro, E. Chavira, G. Gonzalez, IBVS No. 1031, 1975.
15. A. H. Joy, Stellar Atmospheres, Ed. J. L. Greenstein, The University of
    Chicago Press, 1960.
16. G. A. Gurzadyan, Flare Stars, Nauka Press, Moscow, 1973.
17. G. Haro, W. W. Morgan, Ap. J. 118, 16, 1953.
18. G. Haro, E. Chavira, Vistas in Astronomy, VIII, 1965, p. 89.
19. P. Roques, PASP 70, 310, 1958.
20. P. F. Chugainov, Izv. Crimean Obs. 33, 215, 1965.
21. G. Haro, Stars and Stellar Systems, Vol. VII, "Nebulae and Interstellar
    Matter," Ed. B. M. Middlehurst and L. H. Aller, 1968, p. 141.
22. O. J. Eggen, Ap. J. Suppl. 16, 49, 1968.
23. O. J. Eggen, PASP 81, 553, 1969.
24. W. E. Kunkel, PASP 82, 1341, 1970.
25. A. H. Jarrett, J. B. Gibson, IBVS No. 1105, 1976.
26. A. H. Jarrett, J. B. Gibson, IBVS No. 979, 1975.
27. W. E. Kunkel, IBVS No. 315, 1968.
28. W. E. Kunkel, Ap. J. Suppl. 25, 1, 1973.
29. E. E. Mendoza, Ap. J. 143, 1010, 1966.
30. G. A. Gurzadyan, Bol. Obs. Tonatn. y Tacub. 31, 41, 1969.
31. A. D. Andrews, P. F. Chugainov, R. E. Gershberg, V. S. Oskanian, IBVS No. 326,
    1969.
32. T. J. Moffett, Ap. J. Suppl. 29, 1, 1974.
33. W. Gliese, Veröff des Astron. Rechen-Instituts Heidelberg, No. 22, 1969.
34. S. Cristaldi, M. Rodono, IBVS No. 602, 1971.
35. S. Cristaldi, M. Rodono, IBVS No. 835, 1973.
36. G. Felix, IBVS No. 943, 1974.
37. A. H. Jarrett, G. Grabner, IBVS No. 968, 1974.
38. G. Haro, E. Chavira, Bol. Obs. Tonant. y Tacub. 12, 3, 1955.
39. W. E. Kunkel, Low Luminosity Stars, Ed. S. S. Kumar, 1968, p. 195.
40. H. L. Johnson, R. I. Mitchell, Ap. J. 127, 510, 1958.
41. G. Abell, PASP 71, 517, 1959.
42. P. F. Chugainov, Izv. Crimean Obs. 40, 33, 1969.
43. S. Cristaldi, M. Rodono,"Non-Periodic Phenomenon in Variable Stars,"
    Ed. L. Detre, 1969, p. 51.
44. S. Tapia, IBVS No. 286, 1968.
45. G. S. Mumford, PASP 81, 890, 1969.
46. S. Cristaldi, M. Rodono, IBVS No. 525, 1971.
47. S. Cristaldi, M. Rodono, Astron. Astrophys. Suppl. 10, 47, 1973.
48. K. Osawa, K. Ichimura, K. Okida, H. Koyano, Y. Shimuzu, IBVS No. 524, 608,
    1971; 615, 635, 1972; 790, 1973; 876, 1974.
49. G. Haro, Bol. Inst. Tonantzintla 2, 3, 1976.

CHAPTER 2
The Hypothesis of Transformation of Infrared Photons

1. THE NONTHERMAL NATURE OF THE CONTINUOUS EMISSION

A new stage in the understanding of the nature of flares and the production of continuous emission in unstable stars and anomalous nebulae started in 1954, when V. A. Ambartsumian came to the conclusion, from detailed analysis of the available observational material, that in all known cases the additional energy emitted by the star can not be of thermal origin [2]. Two facts should be specially stressed: the exceptionally rapid increase in brightness of the star and the very large amount of energy produced by the star in a very short interval of time.

When the increase in brightness is connected with thermal radiation it must be due either to an increase of the dimensions of the photospheric layers of the star or to an increase of the stellar temperature. In the first case, to explain an increase in brightness of several times in a time interval of the order of ten seconds the star must have increased its dimensions by a factor of two at least, during this time. In order to obtain this result the photospheric layers should expand with a velocity of the order of some tens of thousands of kilometers per second, which is completely excluded; even without considering the fact that it is impossible for a gaseous medium to expand with such a velocity, the emission lines observed at the time of the flare do not show any significant Doppler shift. V. A. Ambartsumian believes that these facts cannot be explained by any known processes of thermal character. Only one conclusion is left: the production of such improbable quantities of energy must be connected with nuclear processes, occurring in the outer regions of the star's atmosphere and on a large scale. Furthermore, V. A. Ambartsumian excludes the possibility that these are nuclear processes known to us and thinks that "by their character they are quite different from the known processes of production of nuclear energy, and, in particular, from the thermonuclear reactions. The fact that the production has the character of a flare indicates that masses of matter in nuclear-unstable states are transported from the inner to the outer layers. On the other hand, since this phenomenon is observed in young stars, it is natural to assume that the transported mass consists of pre-stellar matter of high density, i.e. it is matter in a quite particular, hitherto unknown state" [2].

V. A. Ambartsumian does not suggest a concrete mechanism responsible for the flare and for the continuous emission. Perhaps this was not necessary at the time. Quite properly not yielding to the temptation to "prove" some theoretical model for the interpretation of these phenomena, V. A. Ambartsumian stressed the essential points and brought the whole problem to more general considerations. These considerations are of fundamental value for an understanding of phenomena connected with stellar evolution, and the formation and development of stars, of which the flare and the continuous emission are temporary events.

2. MAIN PROPERTIES OF FLARES IN UNSTABLE STARS

The properties and regularities of the flare phenomenon are more or less common to all flare stars independent of whether they lie in the neighborhood of the Sun or in stellar associations. Some of these properties were noticed already at an early stage of the study of stellar flares [1, 2, 3].

The most important properties and regularities which can be derived from the available data about the flare and the phenomenon of continuous emission are the following [4]:

1. The unforeseen and sudden increase in brightness in photographic, and in particular, in ultraviolet radiation may be considered as the fundamental property of the flare. The increase of brightness, i.e. the flare amplitude, is usually 1 to 2 stellar magnitudes. However, in some cases it reaches 6 to 7^m in ultraviolet light. Sometimes, in flare stars in stellar associations, an amplitude in U-radiation of more than 8^m has been found, corresponding to an increase in luminosity of two thousand times.

2. At the time of the outburst, additional radiation is produced in the form of continuous emission, superposed on the normal spectrum of the star. The increase of intensity of the spectrum occurs chiefly in the photographic and ultraviolet radiation; in the visual region it is smaller. Sometimes, in not very powerful flares, an increase in brightness is observed in U and B radiation with hardly any increase in V radiation. This property can be written in the following form: $\Delta U > \Delta B > \Delta V$. It is important to stress that not a single case has been observed where this rule has been violated. At the same time it means that during the flare the star is bluer. The changes of the spectrum last only a very short time. The star's temperature at the time of the flare does not appear to change much.

3. The increase in brightness of the star during the flare is very rapid: during a time of the order of one minute the star's luminosity can increase by a factor of several tens. Sometimes this increase takes place during ten seconds or even less. The fall in brightness after maximum is slower, but still relatively quick-- in some few minutes the star falls back to its original luminosity. The original spectrum of the star is also re-established very rapidly and completely.

Apart from these there exists a class of objects--stars of T Tauri type--in which the production of the continuous emission occurs practically at a constant rate and lasts long periods of time. In other words, the "flare" in stars of T Tauri type has a quite stable character. In all other respects--amplitude of the brightness fluctuations, character of the continuous spectrum, behavior of emission lines, etc.--they do not differ from normal flare stars.

4. The relative strengthening of the spectrum in photographic or ultraviolet radiation can be characterized by the quantity $\Delta B/\Delta V$ or $\Delta U/\Delta B$. These ratios are for purely thermal processes somewhat larger than unity. For instance, for fluctuations in photospheric temperature between 2500 K and 6000 K this ratio is equal to 1.25, whereas for practically all flares this ratio is more than 2, and in some cases it reaches 4 or more.

5. The flare phenomenon is observed as a rule in dwarf stars of late spectral classes, chiefly M0 - M6, sometimes K5 - M0 and very rarely G5 - K0. It has also been found that the flare frequency is larger for stars of later spectral classes.

6. As a rule, during the outburst the emission lines occurring in the normal spectrum are strengthened, and new emission lines appear, with high ionization potential, which are not normally observed in the star's spectrum. Strengthening occurs mainly

in the hydrogen lines. The new emission lines which appear quite often during
flares are the neutral helium lines 4471 HeI, 4026 HeI, sometimes the line of
ionized helium 4686 HeII, and also the fluorescence lines of FeI, FeII and CaII.
Forbidden lines, among which the most typical ones—those of [OII], [OIII], etc.—
are never observed in stellar flares. Sometimes the line 4069 [SII] is seen and
rarely the lines of [FeII].

As regards the absorption lines, during the flare they are covered by continuous
emission, hence become much weaker, and more vague, and in some cases they disappear completely.

7. It is very characteristic that no significant variations in the star's brightness are observed in the infrared, even in strong flares.

8. The flare in optical light is accompanied by a flare at radio frequencies.

9. There is a definite regularity in flare frequency:
    a) the flare frequency increases with decreasing absolute luminosity.
    b) for a given star the flare frequency is higher in U radiation, lower in B
        and still lower in V radiation.

These properties and regularities are most typical for all flare stars. In the
following chapters they will be considered in more detail.

The problem is to find a mechanism for the excitation of stellar flares which gives
a natural explanation for these regularities and properties.

3.   THE HYPOTHESIS OF TRANSFORMATION OF INFRARED PHOTONS

The fact that the flare and the appearance of continuous emission occur mainly in stars
of late spectral classes leads to the idea that the flare may be related to a
peculiarity in the spectral distribution of the radiation of low-temperature stars.
This peculiarity may be in particular that the maximum of the radiation of these
stars lies in the infrared wavelength region ($\sim$ 10,000 Å), and hence the number of
photons in the photographic region is a very small fraction of the total number of
infrared photons.

Fig. 2.1 gives computed curves showing the distribution of the number of photons
N_λ as a function of λ (in arbitrary units) for the various spectral classes from
G5 to M5, on the assumption that the distribution follows Planck's law with effective temperature T_{eff}. The hatched regions show the relative number of photons
N_B in the photographic region (3500 – 5000 Å). Then the ratio N_B/N was found
for each spectral class, where N is the total number of photons in the region
from 5000 Å to infinity.

The relative number of photons in photographic light decreases quite rapidly with
decreasing effective temperature of the star (Table 2.1). For T_{eff} = 5500 K, e.g.
N_B is about 10% of the total amount of infrared photons, for T_{eff} = 3600 K it is
1%, and for T_{eff} = 2800 K less than 0.2%.

The relative number of ultraviolet photons N_U (3000-3900 Å) decreases even more
rapidly with decreasing T_{eff}: the ratio N_U/N is about 3% for G5 stars, about 0.2%
for M0 stars and 0.02% for class M5. The numerical values given for N_B/N and N_U/N
should, for several reasons, be considered to be lower limits of the true values.

Thus in the infrared spectral region for stars of late classes there are so many
photons that the ones in the photographic and ultraviolet regions are a negligible

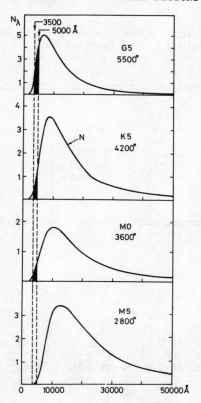

Fig. 2.1. Distribution of the number of photons, N_λ, in the visual (3500–5000 Å) and infrared (up to 50,000 Å) regions of the spectra of stars of classes G5 – M5. The scale for N_λ is in arbitrary units.

fraction of the total. This fact, which is quite evident and well known, is stressed here in particular because in certain circumstances this stock of infrared photons may be used for the excitation of the optical outburst. For this it suffices to have some mechanism giving additional energy to the infrared photons, so that they become photons of higher frequencies. This transformation need take place for only a few percent of the infrared photons in order to strengthen the radiation in the short wavelength region of the spectrum by many tens or hundreds of times. Such a mechanism may be found in the inverse Compton effect—inelastic collisions of electrons with energies much exceeding their rest energy ($E > mc^2$), with long-wavelength photons. These electrons are not thermal, but we must not think them to be extremely relativistic either; therefore we shall provisionally call them fast electrons. We assume that due to certain processes such electrons may appear above the photosphere spontaneously and practically all of a sudden, which is the essential character of the flare (cf. Chapter 16). Here it is not required, for the explanation of the observed increased brightness in the short-wavelength region of the spectrum, that new luminous photons are generated. The whole excess of energy in the short-wavelength region of the spectrum is obtained by the transition of part of the infrared photons of the photosphere to the short-wavelength region, and the additional energy of the photon is taken from the energy of the fast electron in the form of Compton losses.

Hence, if we think there exists around the star a shell or layer consisting of fast electrons, then the radiation incident from the side of the photosphere with a Planckian distribution leaves the outer boundary of this layer with a completely different spectral distribution. In particular the maximum of the radiation emitted by this medium will be strongly displaced to the side of short wavelengths. The exact shape of this distribution may be found by solving the equations of energy transfer through the medium of the fast electrons, which will be shown in Chapter 4.

Now we shall briefly consider some properties of the inverse Compton effect from the physical point of view. The mechanism proposed for the excitation of flares in stars will be called for short the "fast electrons hypothesis."

TABLE 2.1 The Relative Number of Photons in the Photographic (N_B/N)
and Ultraviolet (N_U/N) Regions of the Spectrum for
Various Spectral Classes.

Spectral Classes	T_{eff}(K)	N_B/N	N_U/N
G5	5500	0.083	0.028
K5	4200	0.028	0.0062
M0	3600	0.010	0.0018
M5	2800	0.0018	0.0002
M6	2500	0.00055	0.00003

REFERENCES

1. G. Haro, W.W. Morgan, Ap. J. 118, 16, (1953).
2. V.A. Ambartsumian, Non-Stable Stars, Ed. G.H. Herbig, Cambridge, (1957),
        p. 177.
3. A.H. Joy, PASP 66, 5, (1954).
4. G.A. Gurzadyan, Astrofizika 2, 217, (1966).

CHAPTER 3
The Inverse Compton Effect

1. COLLISION OF A PHOTON WITH A THERMAL ELECTRON

Generally a collision of a photon with a <u>thermal</u> electron results in a) a change
in the direction of propagation of the photon; b) a decrease of its frequency.
The efficiency of scattering of the photon in the direction ϕ after the encounter
with a thermal electron (Fig. 3.1) is equal for all wavelengths and is given by
the expression:

$$d\sigma_T(\phi) = \tfrac{1}{2}\left(\frac{e^2}{mc^2}\right)^2 (1 + \cos^2\phi)d\Omega \ , \tag{3.1}$$

where m is the mass of the electron, $d\Omega = 2\pi\sin\phi d\phi$. From (3.1) we may write for
the spatial coefficient of scattering for one electron

$$\sigma_T = \int d\sigma_T(\phi) = \frac{8\pi}{3}\left(\frac{e^2}{mc^2}\right)^2 = 6.65 \times 10^{-25} \ cm^2. \tag{3.2}$$

The relation for σ_T is called Thomson's formula; it characterizes the scattering
of photons by electrons. Sometimes this process is called Thomson scattering,
and σ_T--the effective scattering cross-section, which is constant for all wavelengths, from the radio region up to γ-radiation. In the optical and radio regions
Thomson scattering can take place on free electrons only. In the case of γ-rays
it can happen on free electrons as well as on electrons which are bound to atoms
or ions.

After each act of scattering of a photon by a thermal electron its wavelength
increases by the quantity

$$\Delta\lambda = \frac{h}{mc} (1 - \cos\phi) \ . \tag{3.3}$$

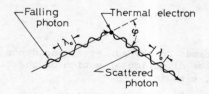

Fig. 3.1. Collision of a photon with a thermal electron

27

The quantity h/mc is called the Compton wavelength and is indicated by Δ; it is equal to

$$\Delta = \frac{h}{mc} = 0.0242 \text{ Å} \quad . \qquad (3.4)$$

The Compton length does not depend on the wavelength of the radiation falling on the electron. But the relative Compton length increases with decreasing wavelength. Thus, for instance, for γ-rays ($\lambda < 0.01$ Å) it is comparable with the wavelength itself, for X-ray radiation ($\lambda \sim 1$ Å) it is about 1%, and for the optical region ($\lambda \sim 5000$ Å) it disappears completely. Therefore in the following, when talking about scattering of optical photons by thermal electrons, we shall assume that the frequency of the photon before and after scattering remains the same.

2. COLLISION OF A PHOTON WITH A RELATIVISTIC ELECTRON. THE INVERSE COMPTON EFFECT.

The situation is completely different when the photon collides with a <u>relativistic</u> electron. In this case the frequency of the primary photon is changed after the collision.

For the first time the problem of the interaction of photons with relativistic electrons was posed in the 1930's in connection with the problem of energy losses of cosmic rays on their passage through the interstellar medium. The first results were obtained by Breit and Wheeler [1], later Follin [2], Donahue [3], Felten and Morrison [4]. However, with sufficient completeness this problem was considered in the well-known work of Feenberg and Primakoff [5], which appeared in 1948. A collision of a relativistic electron with a photon with very small energy as compared with the energy of the electron itself (hν << E) results in the electron transferring part of its energy to the photon. By this the electron suffers so-called "Compton" losses [6] and the energy of the photon increases, i.e. the frequency becomes higher (Fig. 3.2). This is the inverse of what happens in the classical Compton effect, where energy of the photon is transferred to the electron. Therefore this effect--scattering of electrons by photons, i.e. transfer of energy of the electrons to photons--is sometimes called the "inverse Compton effect." In the most general case, the frequencies of the photon before (ν_0) and after (ν) the collision with the relativistic electron are connected by the following relation

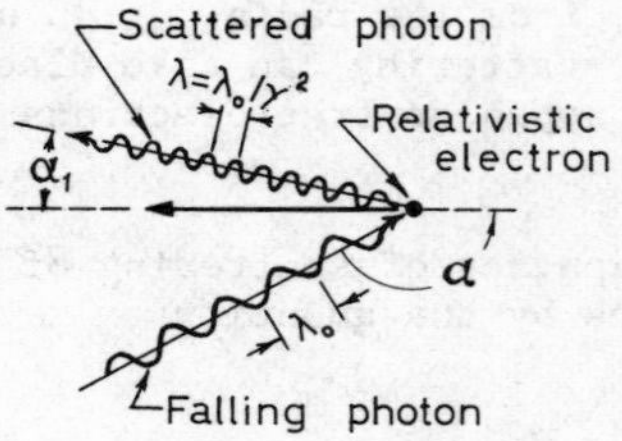

$$\nu = \nu_0 \gamma^2 f(\alpha, \alpha_1) \quad , \qquad (3.5)$$

where

$$\gamma = \frac{E}{mc^2} \quad , \qquad (3.6)$$

Fig. 3.2. Collision of a photon with a relativistic electron.

and the function $f(\alpha, \alpha_1)$ depends on the angles of incidence α and reflection α_1 of the photon with respect to the direction of motion of the electron (Fig. 3.2) and has the following form (see, e.g. [6])

$$f(\alpha, \alpha_1) = \frac{(1 + \beta\cos\alpha)(1 - \beta\cos\alpha_1')}{1 + \gamma h\nu/mc^2 \ (1 + \beta\cos\alpha)[1 - \cos(\alpha' + \alpha_1')]} \quad , \qquad (3.7)$$

where $\beta = v/c$, v--the velocity of the electron, and α' and α_1' are related to α and α_1 by the usual relativistic formulae of aberration of light (dashes indicate the quantities in the "fixed" reference system, i.e. the system connected with the observer on the Earth)

$$\tan \alpha' = \frac{\sin \alpha}{\gamma(\cos\alpha + \beta)} \; ; \tag{3.8}$$

$$\tan \alpha_1' = \frac{\sin \alpha_1}{\gamma(\cos\alpha_1 + \beta)} \; . \tag{3.9}$$

From (3.9) it follows in particular that for values of the energies of fast electrons $\gamma \sim 3$, the angle α_1' will be quite small. This means that after the collision with the electron the photon will be propagated with changed frequency practically in the direction of motion of the electron.

For thermal photons and for energies of fast electrons of the order $\gamma \sim 3$ we have $\gamma \frac{h\nu}{mc^2} \ll 1$. Then we have instead of (3.7)

$$f(\alpha, \alpha_1) = (1 + \beta\cos\alpha)(1 - \beta\cos\alpha_1') \; . \tag{3.10}$$

If the stream of photons speeds in the opposite direction or nearly so of the motion of the relativistic electrons (frontal encounter) we find from (3.10), with $\beta = 1$,

$$f(\alpha, \alpha_1) \approx 1 \; . \tag{3.11}$$

It is easily seen that equation (3.11) holds on the average also for the case of isotropic radiation, or isotropic distribution of the electrons over all directions.

Finally as soon as we talk of applying these considerations to the conditions of stellar atmospheres, then there are always a certain number of photons for which the condition $f(\alpha, \alpha_1) \approx 1$ will be fulfilled. In these cases we shall consider the "effective number of fast electrons" or the "effective stream of electrons," taking part in inelastic collisions. Thus the relation between the frequencies of the photon before the collision with the relativistic electron (ν_o) and after it (ν) has the following form, which is sufficiently accurate for practical purposes

$$\nu \approx \nu_o \gamma^2 \; . \tag{3.12}$$

From this it follows that always when the condition $h\nu \ll mc^2$ is fulfilled, the scattered photon has after the collision with the fast electron a wavelength γ^2 times shorter than before the collision.

The relation (3.12) will be used in our further calculations. This holds as long as the energy of the photon is not comparable with that of the electron. In practice, for electron energies $\gamma \sim 2$ to 3, the validity of formula (3.12) can easily be extended up to wavelength 0.01 Å, i.e. the region of hard X-ray radiation.

3. THE EFFECTIVE CROSS-SECTION FOR A COLLISION OF A PHOTON WITH A RELATIVISTIC ELECTRON

When the energy of the photon $h\nu$ is of the same order or larger than mc^2, i.e. when $\nu \gtrsim 10^{20}$ s^{-1}, or $\lambda \lesssim 0.03$ Å, corresponding to hard x-ray radiation or gamma photons, the value of the effective cross-section for the collision of a photon with a relativistic electron depends already on the energy of the electron and is given by the well-known formula of Klein-Nishina.

However, in application to the phenomena connected with flare stars, we will be interested in scattering processes of <u>optical</u> and <u>infrared</u> photons, with energies (1 – 3 MeV) considerably smaller than mc^2, with relativistic electrons. In that case the effective cross-section σ_s for scattering of a photon by the collision with an electron with energy $\gamma(> mc^2)$ is given by the following relation [3, 5]

$$\sigma_s \approx \gamma^2 \, \sigma_T \, , \tag{3.13}$$

i.e. it increases rapidly with increasing γ. Formally this relation holds for $\lambda h\nu << mc^2$ In our case where the energy of the fast electrons ($\gamma^2 \sim 10$) this corresponds with photon energy smaller than 10^4 eV.

4. THE DOUBLE INVERSE COMPTON EFFECT

In the inelastic collision of one photon with a relativistic electron it is possible that two or more photons of approximately equal energy are emitted. In that case we speak of double, triple, etc., inverse Compton effect.

The double Compton effect was predicted theoretically by Heitler and Nordheim [8]. They estimate that the ratio of the effective scattering cross-section for double Compton effect σ_{II} to that for normal Compton effect σ_T must be of the order 1/137. Later, detailed calculations gave for this ratio the value 0.4×10^{-4} [9] for the case when two photons are scattered in directions perpendicular to each other, and the energy of the electron is of the order 10^6 eV.

The possibility of double Compton effect was first shown experimentally by Cavanagh [10]. He found $\sigma_{II}/\sigma_T \sim 10^{-4}$, in good agreement with the value predicted theoretically. Thus, the effective cross-section for the double Compton effect, and therefore also for the double inverse Compton effect, is at least three to four orders of magnitude smaller than the effective cross-section for the ordinary Compton effect. Thus in conditions as in the stellar atmospheres, and in particular, in processes of flare excitation, the double inverse Compton effect cannot play any significant role.

5. POLARIZATION OF THE RADIATION

The radiation scattered after the collision with a relativistic electron--the "Compton radiation"--will have a certain preferred direction. It will be concentrated within an angle α'_1 depending on the electron energy γ, around the direction of motion of the electron. For sufficiently large values of γ the quantity α'_1 is of the order:

$$\alpha'_1 \approx \frac{1}{\gamma} = \frac{mc^2}{E} \, . \tag{3.13}$$

From what has been said it follows that the Compton radiation must be polarized.
In fact, as shown by Milburn [11], the degree of polarization for the elementary
scattering can be quite high--more than 60%.

As we shall see later, all additional radiation emitted by the star during the
flare is entirely of Compton origin. Thus the polarization of the star light during the flare may be expected to differ from zero. However, on account of the
isotropy of the propagation of the electrons, and the isotropy of the radiation,
the observed degree of polarization may be small.

The problem of polarization of radiation in processes due to the inverse Compton
effect was considered by a number of theoreticians [12-15]. The most interesting
case is that of isotropic distribution of the electrons, considered by Bonometto
et al. [15]; it is shown that in this case the radiation leaving the medium will
be depolarized--up to 50% in the case of circularly polarized radiation and up to
75% for linear polarization, if the original radiation is polarized completely.
In the case that the original radiation is not polarized (for instance, the photospheric radiation of a star) the radiation leaving the medium of fast electrons
will not be polarized either.

6. FAST ELECTRONS

The observed peculiarities of outbursts in flare stars can be explained on the
assumption that all additional radiation emitted by the star during the flare is
of Compton origin. More precisely this means:

a) the radiation energy of the flare is taken wholly from the energy of relativistic electrons;

b) the flare phenomenon itself--a rapid and strong increase of the star's bright-
ness--leads to rapid and intensive production or generation of relativistic
electrons above the star's photosphere;

c) the elementary process causing the flare is the inverse Compton effect--inelas-
tic collisions of infrared photons of the ordinary photospheric radiation of
the star with relativistic electrons, causing transformation or drift of infrared photons to the region of higher energy photons;

d) in the visual region of the spectrum practically no new photons are formed during the flare; the number of short-wavelength photons appearing during the
flare is exactly compensated by the number of infrared photons of the photosphere resulting from inelastic collisions with electrons. Further, for the
explanation of the observed properties of stellar flares it suffices to have
electrons with energies somewhat higher than the proper electron energy, i.e.
$\gamma \sim 2$ to 3. Such electrons are also relativistic. However, in the following
we shall call them "fast electrons," having in mind relativistic electrons with
a given value of energy, namely $\gamma \sim 3$ or $\gamma^2 \sim 10$.

A cloud, a layer or shell consisting of fast electrons may appear above the photosphere at a certain distance of the star's surface. The electrons themselves will
move practically with the speed of light. At the same time their motion will be
determined by the magnetic field of the star. Since the energy of the fast electrons is relatively small ($E \sim 10^6$ eV), even a weak magnetic field at a considerable
distance from the star's surface can influence their motion. In these circumstances
there will always be found electrons undergoing a frontal or practically frontal
encounter with photons leaving the photosphere; in that case the maximum efficiency
of the phenomenon of the inverse Compton effect will be reached. Besides, an

envelope of fast electrons will expand considerably faster than would be the case in an expansion of an arbitrary gaseous envelope even if the star has a magnetic field, which plays a "conserving role."

A cloud or shell of fast electrons will be electrically neutral or nearly neutral, since it may contain protons and other nuclei having no influence whatever on the processes connected with the exchange of energy between fast electrons and photons.

REFERENCES

1. G. Breit, J. A. Wheeler, Phys. Rev. 46, 1087, 1934.
2. J. W. Follin, Phys. Rev. 72, 743A, 1947.
3. T. M. Donahue, Phys. Rev. 84, 472, 1951.
4. J. E. Felten, P. Morrison, Phys. Rev. Lett. 10, 453, 1963.
5. E. Feenberg, H. Primakoff, Phys. Rev. 73, 449, 1948.
6. V. L. Ginzburg, S. I. Syrovatsky, Origin of Cosmic Rays, Moscow, 1963.
7. J. E. Felten, P. Morrison, Ap. J. 146, 686, 1966.
8. W. Heitler, L. Nordheim, Physica 1, 1059, 1934.
9. F. Mandl, Phys. Rev. 87, 1131, 1952.
10. P. Cavanagh, Phys. Rev. 87, 1131L, 1952.
11. R. H. Milburn, Phys. Rev. Lett. 10, 74, 1963.
12. B. Neumke, H. J. Meister, Zs. f. Phys. 192, 162, 1966.
13. J. F. Dolan, Space Sci. Rev. 6, 579, 1967.
14. F. R. Arytunjan, I. I. Goldman, V. A. Tymanyan, JEPT 18, 218, 1964.
15. S. Bonometto, P. Cazzola, A. Saggion, Astr. Astrophys. 7, 292, 1970.
16. Y. S. Efimov, Izv. Crimean Obs. 41, 373, 1970.
17. K. A. Grigorian, M. A. Eritzyan, Comm. Byurakan Obs. 42, 41, 1970.

CHAPTER 4

Transfer of Radiation through a Medium of Fast Electrons

1. THE STATEMENT OF THE PROBLEM

The sudden and rapid increase of a star's brightness--the flare--is caused by a sudden and rapid appearance of electrons with energies of the order 10^6 eV above the star's photosphere. The photosphere itself does not suffer significant excitation; its temperature remains practically the same. The fast electrons form a short-lived shell at some distance from the photosphere. The total energy of this layer is determined by the total number of fast electrons in it, and the optical efficiency, by its optical depth due to Thomson scattering processes.

Radiation with a given spectral distribution falls from the direction of the photosphere on the inner boundary of this layer. This may be a Planck distribution, corresponding to a given effective temperature of the star. The elementary process of interaction of a photon with a fast electron during the collision leads to the electron giving part of its energy to the photon; as a result the frequency of the photon becomes higher. The fact that the layer contains other types of high energy elementary particles, e.g. protons, makes no difference; the coefficient for Thomson scattering by relativistic protons is six orders of magnitude smaller than that for scattering by fast electrons.

In this chapter the problem of energy transfer through the medium of fast electrons will be considered. First we treat the simplest, one-dimensional problem; then the results are found in explicit form. Later we shall apply this model to a real photosphere and various energy distributions of the fast electrons.

2. THE EQUATION OF RADIATION TRANSFER

Consider a plane-parallel layer, consisting only of fast electrons, all with the same energy γ (Fig. 4.1). Let N_e be the effective number of fast electrons in a column with cross-section 1 cm^2, and τ_0 the total effective optical depth of this layer for the process of inelastic Thomson scattering; then

$$\tau = \sigma_s \int_0^z n\,dz\ ;\quad \tau_0 = \sigma_s \int_0^{z_0} n\,dz\ = \sigma_s N_e \tag{4.1}$$

where σ_s is given by formula (3.13).

Suppose that on one side of the layer, where $\tau = 0$, parallel radiation is falling with a given spectral distribution. Let this be a Planck distribution, equal to $B_\lambda(T)$, where T is the effective temperature of the photospheric radiation of the star.

Our problem is to determine the intensity of the radiation of frequency ν leaving

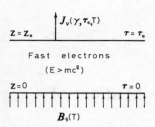

Fig. 4.1. The problem of radiation transfer through a medium of fast electrons (one-dimensional problem).

the layer of fast electrons, as a function of the total optical thickness τ_0 of the layer, the energy of the fast electrons γ, and the effective temperature T of the Planck radiation, i.e. we want to find the form of the function $J_\nu(\tau_0, \gamma, T)$.

The equation for energy transfer at frequency ν has the form

$$\cos \theta \, \frac{dJ_\nu}{dz} = - n\sigma_s J_\nu + \varepsilon_\nu \, , \qquad (4.2)$$

where ε_ν is the emission coefficient per unit volume at frequency ν. The condition for radiative equilibrium is, in our case:

$$4\pi\varepsilon_\nu = n\sigma_s \gamma^2 B_{\nu_0} e^{-\tau} + n\sigma_s \gamma^2 \int J_{\nu_0} d\omega \, , \qquad (4.3)$$

where the first term in the right-hand side is due to photons of the direct radiation B_{ν_0} of frequency ν_0, transformed to the frequency ν in the elementary process of scattering by fast electrons, and the second term to photons of the diffuse radiation J_{ν_0} at the same frequency ν.

Substituting (4.3) in (4.2) we find the following differential equation for the function J_ν:

$$\cos \theta \, \frac{dJ_\nu}{d\tau} = -J_\nu + \frac{\gamma^2}{4\pi} B_{\nu_0} e^{-\tau} + \frac{\gamma^2}{4\pi} \int J_{\nu_0} d\omega \, , \qquad (4.4)$$

where we must put instead of ν_0 the quantity $\nu_0 = \dfrac{\nu}{\gamma^2}$, and B_{ν_0} is the Planck function, which for this case can be written as:

$$B_{\nu_0}(T) = \frac{2h}{c^2} \, \frac{\nu_0{}^3}{e^{h\nu_0/kT}-1} = \frac{2h}{c^2} \frac{1}{\gamma^6} \, \frac{\nu^3}{e^{h\nu/\gamma^2 kT}-1} \, . \qquad (4.5)$$

The solution of equation (4.4) gives the intensity of the radiation $J_\nu(\tau_0, \gamma, T)$ leaving the layer of fast electrons.

Unfortunately the third term in the right-hand side of the equation makes the solution very difficult. However, if τ_0 is not too large, second-order scattering can be neglected. Besides, to obtain a qualitative picture we restrict ourselves to the one-dimensional problem, i.e. in (4.4) we take $\cos \theta = 1$. Then integration of (4.4) is easy and we find for the intensity of the radiation leaving the fast electron layer:

$$J_\nu(\tau_0, \gamma, T) = B_\nu(T)e^{-\tau_0} + \frac{\gamma^2}{4\pi} B_{\nu_0}(T)\tau_0 e^{-\tau_0} \, , \qquad (4.6)$$

where the condition $J_\nu(0) = B_\nu(T)$ for $\tau = 0$ has been used. This solution will be called the "first approximation." The relation (4.6) gives at the same time the spectral distribution of the radiation leaving a layer of fast electrons of small optical thickness; it was derived for the first time in [1,2].

It is more convenient to write (4.6) as follows:

$$J_\nu(\tau, \gamma, T) = B_\nu(T) C_\nu(\tau, \gamma, T) , \qquad (4.7)$$

where $C_\nu(\tau, \gamma, T)$ is a quantity equal to

$$C_\nu(\tau, \gamma, T) = \left(1 + \frac{1}{4\pi} \frac{1}{\gamma^4} \frac{e^x - 1}{e^{x/\gamma^2} - 1} \tau \right) e^{-\tau} , \qquad (4.8)$$

where

$$x = h\nu/kT .$$

The function C_ν has a simple physical meaning: it shows how many times the intensity of the radiation leaving the layer of fast electrons is larger (when $C_\nu > 1$) or smaller (when $C_\nu < 1$) than the intensity of the radiation $B_\nu(T)$ at the inner boundary of the layer.

Note that the function C_ν is connected to the flux of excess radiation i, which is an observed quantity (cf. §5, Ch. 1), as follows:

$$i = C_\nu - 1 . \qquad (4.9)$$

Analysis of formula (4.8) shows that for sufficiently low temperatures of the Planck radiation, the function C_ν is of the order of unity in visual wavelengths (5000-6000 Å), it is larger than one in photographic wavelengths (4000-5000 Å) and is considerably larger than one in ultraviolet wavelengths (3000-4000 Å). At the same time C_ν is smaller than one in infrared wavelengths ($\lambda > 10,000$ Å). Thus the intensity of the radiation leaving the layer is enhanced at photographic wavelengths, is very much higher in ultraviolet wavelengths, and decreases somewhat in the infrared spectral region. This last intensity decrease is due to the transformation of long wavelength photons to the short wavelength region of the spectrum. Since the energy of the fast electrons--the primary source of the additional energy given to the long wavelength photons--is non-thermal in character, we can speak of the continuous emission by the star of non-thermal character in the photographic and ultraviolet regions of the spectrum.

3. APPLICATION TO STELLAR PHOTOSPHERES

We now consider the actual problem of interaction of fast electrons with photospheric radiation. We assume that the layer of fast electrons with effective optical thickness lies either directly above the photosphere, as shown in Fig. 4.2, or at a certain distance from it. The difference between these two cases is only quantitative. In the second case, for instance, the radiation falling on the inner boundary of this layer is weakened by the dilution coefficient, which depends on the relative distance of the layer from the surface of the star. The spectral distribution of the intensity of the radiation leaving the layer of fast electrons will

be the same, whether this layer lies
directly above the photosphere or at
some distance from it, on the assumption that in both cases the photosphere
is equally well screened by the fast
electrons, i.e. that the effective
optical thickness of the medium is the
same in both cases.

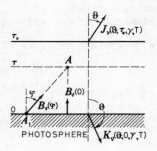

The equation for the energy transfer
is again written in the form (4.2.).
However, in the formation of the
equation for radiative equilibrium
it should be kept in mind that direct
contribution of the radiation at frequency ν_0, later transformed to frequency ν, will be collected from the

Fig. 4.2. The problem of radiation
transfer through a medium of fast electrons (real photosphere).

entire area of the photosphere of which each point A_1 is seen from the point A
under different angles ϕ. The value of the intensity of the radiation incident
on the inner boundary of the fast-electron layer will not be constant as in (4.3),
but will be a function $B_\nu(\phi)$. This is determined by the law of the distribution
of the intensity of the photospheric radiation over the stellar disk. Therefore,
instead of (4.3) the equation of radiative equilibrium will be:

$$4\pi\varepsilon_\nu = \gamma^2 n\sigma_s 2\pi \int_0^{\pi/2} B_{\nu_0}(\phi)e^{-\tau \sec \phi} \sin\phi d\phi + \gamma^2 n\sigma_s \int_{4\pi} J_{\nu_0} d\omega \ . \tag{4.10}$$

The form of the function $B_\nu(\phi)$, i.e. the law of limb darkening for stars of late
spectral classes is not well known. However, taking into account that in the first
term at the right hand side of equation (4.10) after the integral the exponential
factor is the most important part, we may represent this function with sufficient
accuracy in the form:

$$B_\nu(\phi) = B_\nu(0) \cos \phi \ . \tag{4.11}$$

Then we have from (4.10)

$$\frac{\varepsilon_\nu}{n\sigma_s} = \frac{\gamma^2}{2} B_{\nu_0}(0)E_3(\tau) + \frac{\gamma^2}{4\pi} \int J_{\nu_0} d\omega \ , \tag{4.12}$$

where

$$E_3(\tau) = \int_1^\infty e^{-\tau x} \frac{dx}{x^3} \ . \tag{4.13}$$

Substituting (4.12) in (4.2.) we find

$$\cos \theta \ \frac{dJ_\nu}{d\tau} = -J_\nu + \frac{\gamma^2}{2} B_{\nu_0}(0)E_3(\tau) + \frac{\gamma^2}{4\pi} \int J_{\nu_0} d\omega \ . \tag{4.14}$$

Neglecting the diffusion term in this equation and restricting the solution to small values of τ, we find for the intensity $J_\nu(\theta)$ of the outward radiation ($\theta \leqslant \frac{\pi}{2}$):

$$\cos\theta \frac{dJ_\nu}{d\tau} = -J_\nu + \frac{\gamma^2}{2} B_{\nu_0}(0)E_3(\tau) . \tag{4.15}$$

In the same way we have for the intensity $K_\nu(\theta)$ of the inward radiation ($\theta \geqslant \frac{\pi}{2}$):

$$-\cos\theta \frac{dK_\nu}{d\tau} = -K_\nu + \frac{\gamma^2}{2} B_{\nu_0}(0)E_3(\tau) . \tag{4.16}$$

These equations have exact solutions:

$$J_\nu(\theta, \tau) = B_\nu(0)\cos\theta e^{-\tau \sec\theta} + \frac{\gamma^2}{2} B_{\nu_0}(0)f_1(\tau, \theta) ; \tag{4.17}$$

$$K_\nu(\theta, \tau) = \frac{\gamma^2}{2} B_{\nu_0}(0)f_2(\tau, \theta) , \tag{4.18}$$

where

$$f_1(\tau, \theta) = \int_0^\tau E_3(t)e^{-(\tau-t)\sec\theta} \sec\theta \, dt; \tag{4.19}$$

$$f_2(\tau, \theta) = \int_\tau^{\tau_0} E_3(t)e^{-(t-\tau)\sec\theta} \sec\theta \, dt. \tag{4.20}$$

In the derivation of (4.17) and (4.18) the following boundary conditions have been used:

$$J_\nu(0, \theta) = B_\nu(0)\cos\theta \quad \text{for } \tau = 0; \tag{4.21}$$

$$K_\nu(\tau_0, \theta) = 0 \qquad \text{for } \tau = \tau_0. \tag{4.22}$$

We are interested in the fluxes of radiation H_ν at the boundaries of the layer. We have for arbitrary τ:

$$H_\nu(\tau) = 2\pi \left[\int_0^{\pi/2} J_\nu(\tau,\theta)\cos\theta\sin\theta d\theta + \int_{\pi/2}^\pi K_\nu(\tau, \theta)\cos\theta\sin\theta d\theta \right] = \tag{4.23}$$

$$2\pi \left\{ \frac{1}{3}B_\nu(0)E_4(\tau) + \frac{\gamma^2}{2} B_{\nu_0}(0)[F_1(\tau) - F_2(\tau)] \right\}$$

where $\nu_0 = \nu/\gamma^2$ and

$$E_4(\tau) = 3 \int_1^\infty e^{-\tau x} \frac{dx}{x^4} \tag{4.24}$$

$$F_1(\tau) = \int_0^1 f_1(\tau, y) \frac{dy}{y^3} \tag{4.25}$$

$$F_2(\tau) = \int_1^\infty f_2(\tau, y) \frac{dy}{y^3} \tag{4.26}$$

and $f_1(\tau, y)$ and $f_2(\tau, y)$ are the functions (4.19) and (4.20) with $\sec\theta = y$. From (4.23) we find an expression for the stream of outward radiation at the inner boundary of the fast electron layer (remembering that in our nomenclature the stream of photospheric radiation is equal to $\frac{2\pi}{3} B_\nu(0)$, when $\tau_0 = 0$):

$$H_\nu(\tau) = B_\nu(0) C_\nu(\tau, \gamma, T), \tag{4.27}$$

where

$$C_\nu(\tau, \gamma, T) = E_4(\tau) + \frac{3}{2\gamma^4} \frac{e^x - 1}{e^{x/\gamma^2} - 1} F_1(\tau) . \tag{4.28}$$

Analogously we find the inward radiation at the inner boundary of the fast electron layer, i.e. for $\tau = 0$

$$H_\nu(0) = B_\nu(0) G_\nu(\tau, \gamma, T) , \tag{4.29}$$

where

$$G_\nu(\tau, \gamma, T) = \frac{3}{2\gamma^4} \frac{e^x - 1}{e^{x/\gamma^2} - 1} F_2(\tau) . \tag{4.30}$$

The relation (4.27), analogous to (4.7), gives the spectral distribution of the intensity of the photospheric radiation leaving the layer of fast electrons as a function of the energy of the fast electrons γ, the effective optical depth τ, and the Planck temperature of the photosphere T. For convenience of the computations we give in Table 4.1 numerical values of the corresponding functions for a number of values of τ. Table 4.2 gives numerical values of the function $C_\nu(\tau, \gamma, T)$ for $\tau = 1, 0.1, 0.01, 0.001, 0.0001$ and various values of the temperature of the star, from T = 2500° (class M6) to T = 5500° (class G5).

For the solution of the transfer equation (4.14) the diffusion term in it was omitted. Therefore the formulae (4.27) and (4.29) cannot be applied for values of τ of the order of and larger than unity. For the case $\tau = 1$ the values found for C_ν from (4.28) in the short wavelength region may be of the order of one half the true value. Formula (4.28) gives the more accurate results for $\tau < 1$.

The function C_ν has a number of interesting properties. Some of them are: a) in the short wavelength region the second term in (4.28) is the most important one, and therefore we can write with sufficient accuracy:

TABLE 4.1 Numerical Values of the Functions $E_3(\tau)$, $E_4(\tau)$, $F_1(\tau)$, $F_2(\tau)$

τ	$E_3(\tau)$	$E_4(\tau)$	$F_1(\tau)$	$F_2(\tau)$
0	0.5000	1.0000	0	0
0.0001	0.4949	0.9988	0.000045	0.00005
0.001	0.4941	0.9975	0.00047	0.0005
0.002	0.4932	0.9960	0.00094	0.0012
0.004	0.4914	0.9931	0.00188	0.0021
0.006	0.4896	0.9901	0.0028	0.0030
0.008	0.4879	0.9872	0.0037	0.0040
0.01	0.4861	0.9843	0.0046	0.0049
0.02	0.4774	0.9698	0.0091	0.0130
0.04	0.4607	0.9417	0.0169	0.0207
0.06	0.4448	0.9145	0.0242	0.0278
0.08	0.4297	0.8883	0.0298	0.0343
0.10	0.4152	0.8629	0.0369	0.0403
0.2	0.3516	0.7483	0.0602	0.0811
0.4	0.2573	0.5674	0.0827	0.1086
0.6	0.1915	0.4339	0.0872	0.1228
0.8	0.1443	0.3339	0.0842	0.1306
1.0	0.1097	0.2582	0.0769	0.1349

TABLE 4.2 Numerical Values of the Function $C_\nu(\tau, \gamma, T)$ for $\gamma^2 = 10$

λ Å	τ				
	1	0.1	0.01	0.001	0.0001
			T = 2500° (M6)		
2000	2.196×10^8	1.054×10^8	1.313×10^7	1.342×10^6	1.284×10^5
2500	1.027×10^6	4.929×10^5	6.145×10^4	6.279×10^3	6.011×10^2
3000	4.301×10^4	2.064×10^4	2.573×10^3	2.629×10^2	26.152
4000	641.5	308.55	33.341	4.9165	1.3740
5000	53.54	26.435	4.172	1.323	1.0300
6000	10.81	5.9292	1.6159	1.0620	1.0042
7000	3.6387	2.4850	1.1865	1.0820	1.0007
8000	1.7220	1.5754	1.0719	1.0064	0.9997
9000	1.0309	1.2337	1.0305	1.0022	0.9992
10000	0.7261	0.9157	0.9908	0.9983	0.9990
20000	0.3159	0.8905	0.9878	0.9978	0.9989
30000	0.2899	0.8791	0.9862	0.9977	0.9988

TABLE 4.2 (continued)

λ Å	τ				
	1	0.1	0.01	0.001	0.0001

T = 2800° (M5)

2000	1.386×10^8	6.653×10^6	8.294×10^5	8.474×10^4	8.114×10^3
2500	1.438×10^5	6.970×10^4	8.604×10^3	8.791×10^2	84.172
3000	6995	3357	418.48	43.75	5.0929
4000	168.09	81.397	11.0238	2.0233	1.1006
5000	18.9679	9.8407	2.1035	1.1118	1.0098
6000	4.7280	3.0077	1.2517	1.0248	1.0015
7000	1.9019	1.6516	1.0826	1.0075	0.9990
8000	1.0466	1.2402	1.0315	1.0023	0.9985
9000	0.7091	1.0793	1.0113	1.0002	0.9983
10000	0.5494	1.0026	1.0017	0.9993	0.9982
20000	0.3051	0.8854	0.9871	0.9978	0.9980
30000	0.2863	0.8763	0.9861	0.9977	0.9980

T = 3600° (M0)

2000	8.74×10^4	4.19×10^4	5.23×10^3	5.34×10^2	51.2
2500	2589	1243	154.9	16.82	2.514
3000	255	122.4	16.24	2.556	1.1480
4000	15.03	7.953	1.8682	1.0878	1.0074
5000	3.3112	2.3278	1.1669	1.0161	1.0005
6000	1.2140	1.3215	1.0415	1.0033	0.9993
7000	0.7101	1.0799	1.0063	1.0002	0.9990
8000	0.5202	0.9886	1.0000	0.9991	0.9989
9000	0.4316	0.9461	0.9946	0.9985	0.9989
10000	0.3838	0.9232	0.9895	0.9983	0.9989
20000	0.2917	0.8789	0.9863	0.9977	0.9989
30000	0.2808	0.8738	0.9853	0.9976	0.9989

T = 4200° (K5)

2000	7100	3400	426	44.4	5.35
2500	354	171	22.3	3.18	1.22
3000	49.9	24.7	3.98	1.30	1.03
4000	4.75	3.02	1.25	1.02	1.00
5000	1.37	1.40	1.05	1.00	0.999
6000	0.71	1.08	1.01	1.00	0.999
7000	0.50	0.98	0.998	0.999	0.998
8000	0.412	0.937	0.993	0.998	0.998
9000	0.367	0.915	0.990	0.998	0.998
10000	0.342	0.903	0.989	0.998	0.998
15000	0.297	0.882	0.986	0.997	0.998
20000	0.286	0.876	0.985	0.997	0.998
30000	0.287	0.872	0.985	0.997	0.998

TABLE 4.2 (continued)

λ	τ				
Å	1	0.1	0.01	0.001	0.0001
			T = 4900° (K0)		
2000	830	399	51.0	6.10	1.51
2500	65.9	32.4	4.94	1.40	1.04
3000	12.7	6.84	1.73	1.07	1.01
4000	1.91	1.65	1.08	1.01	0.999
5000	0.77	1.11	1.01	1.00	0.999
6000	0.50	0.979	0.998	0.999	0.998
7000	0.403	0.932	0.993	0.998	0.998
8000	0.357	0.910	0.990	0.998	0.998
9000	0.333	0.899	0.988	0.997	0.998
10000	0.318	0.891	0.987	0.997	0.998
15000	0.290	0.878	0.986	0.997	0.998
20000	0.282	0.874	0.985	0.997	0.998
30000	0.276	0.871	0.985	0.997	0.998
			T = 5500° (G5)		
2000	207	100	13.5	2.27	1.13
2500	22.3	11.5	2.31	1.13	1.01
3000	5.36	3.31	1.29	1.03	1.00
4000	1.12	1.28	1.04	1.00	0.999
5000	0.57	1.01	1.00	0.999	0.999
6000	0.42	0.94	0.994	0.998	0.998
7000	0.362	0.913	0.990	0.998	0.998
8000	0.333	0.900	0.988	0.997	0.998
9000	0.317	0.891	0.987	0.997	0.998
10000	0.307	0.886	0.987	0.997	0.993
15000	0.286	0.876	0.986	0.997	0.998
20000	0.280	0.873	0.985	0.997	0.998
30000	0.275	0.871	0.985	0.997	0.998

$$C_\nu(\tau, \gamma, T) \underset{\sim}{\sim} \frac{3}{2\gamma^4} \frac{e^x - 1}{e^{x/\gamma^2} - 1} F_1(x) \ . \tag{4.31}$$

b) For small values of τ (≤ 0.01) we have

$$E_4(\tau) \underset{\sim}{\sim} 1; \quad F_1(\tau) \underset{\sim}{\sim} \frac{\tau}{2} \tag{4.32}$$

In this case formula (4.28) simplifies to:

$$C_\nu(\tau, \gamma, T) = 1 + \frac{3}{2\mu^2} \frac{e^x - 1}{e^{x/\gamma^2} - 1} \tau \ . \tag{4.33}$$

c) For a given Planck temperature the factor of $F_1(\tau)$ in (4.28) increases rapidly towards the short wavelengths; then, after reaching a maximum at a certain wavelength (for a certain value x_0), determined by the relation

$$(\gamma^2 - 1)e^{x_0}(1 + \gamma^2)/\gamma^2 - \gamma^2 e^{x_0} + e^{x_0/\gamma^2} = 0 \ , \tag{4.34}$$

it decreases fast, asymptotically approaching zero.

d) The function $F_1(\tau)$ has in the interval $\tau = 0$ to 1 a very weak maximum at $\tau \sim 0.6$ (cf. Table 4.2). On account of the approximate character of our solution for $\tau \sim 1$, the position found for the maximum cannot be considered to be accurate. From physical considerations a rapid decrease of the function $F_1(\tau)$ with increasing τ should be expected, for $\tau > 1$.

e) Since C_ν is of dimension zero we have $C_\nu(\tau, \gamma, T) = C_\lambda(\tau, \gamma, T)$. Then one can write, analogously to (4.27)

$$H_\lambda(\tau) = B_\lambda(T)C_\nu(\tau, \gamma, T) \ . \tag{4.35}$$

4.   THE CASE OF AN EXPONENTIAL LAW FOR THE ENERGY DISTRIBUTION OF THE ELECTRONS

In the above discussion monoenergetic electrons were considered, where all electrons in the layer or shell have the same energy equal to γ. It is interesting to consider other possible energy spectra of the fast electrons as well. For instance the spectrum where the concentration of the fast electrons N_e depends exponentially on their energy E, which is mostly used in cosmic-ray physics:

$$\frac{dN_e}{dE} = KE^{-\alpha} \ , \tag{4.36}$$

or, with the energy γ of dimension zero

$$\frac{dN_e}{d\gamma} = K(mc^2)^{1-\alpha}\gamma^{-\alpha} \ . \tag{4.37}$$

This spectrum has a boundary at the low-energy side, for $\gamma = \gamma_m$, say. At the high-energy side there is no such boundary ($\gamma \to \infty$). Evidently not all electrons in such an energy spectrum can take part in the processes of transformation of long wavelength photons into short wavelength ones. Electrons for which $\gamma < 1$ can only give ordinary Thomson scattering, without change of the frequency of the photon. The problem is therefore dealing with two processes, where part of the electrons, with $\gamma > 1$ give the inverse Compton effect and part, for which $\gamma < 1$, neutral scattering only. However, in order to avoid the introduction of new parameters we consider only the solution of the problem for $\gamma_{pr} = 1$. The expression for the optical depth, equal for all wavelengths is

$$d\tau = \sigma_s dz \int\limits_{\gamma_m}^{\infty} dN_e = \sigma_s K \frac{(mc^2)^{1-\alpha}}{\alpha - 1} \frac{dz}{\gamma_m^{\alpha-1}} \; .$$

Hence

$$\tau = \sigma_s K \frac{(mc^2)^{1-\alpha}}{\alpha - 1} \gamma_m^{1-\alpha} z \; . \tag{4.38}$$

For the spatial radiation coefficient we have (cf. details in [1])

$$\varepsilon_\nu = K \frac{\sigma_s}{8\pi} e^{-\tau}(mc^2)^{1-\alpha} x^{-\frac{\alpha-3}{2}} \int\limits_{o}^{x} B_u(T) u^{\frac{\alpha-5}{2}} du \; , \tag{4.39}$$

where $x = h\nu/kT$; $u = h\nu_o/kT$, and $B_u(T)$ is the Planck function, ν_o having changed into u.

Substituting (4.39) in the transfer equation (4.2) and solving this for our simplest scheme (see Fig. 4.1) we find for the intensity of the radiation leaving the layer of fast electrons

$$J_\nu(\alpha, \tau, T) = B_\nu(T) A_\nu(\alpha, \tau, T) \; , \tag{4.40}$$

where

$$A(\alpha, \tau, T) = \left\{ 1 + \delta\tau(e^x-1)x^{-\frac{\alpha+3}{2}} \int\limits_{o}^{x} \frac{u^{\frac{\alpha+1}{2}} du}{e^u-1} \right\} e^{-\tau} \; ; \tag{4.41}$$

$$\delta = \frac{\alpha-1}{4\pi} \gamma_m^{\alpha-1} \; .$$

An analysis of equation (4.41) shows that for cool stars $A_\nu > 1$ in the region of photographic and ultraviolet radiation, and $A_\nu < 1$ in the infrared. This means that for an energy spectrum of the fast electrons of the form $N_e \sim \gamma^{-\alpha}$ the brightness of the star will increase in U and B radiation.

However, as shown by quantitative comparisons, fast electrons with energy spectrum $\gamma^{-\alpha}$ cannot produce the very low color indices often observed in strong flares. An energy spectrum of the form $\gamma^{-\alpha}$ cannot explain the observed radio-radiation from flare stars (cf. Chapter 13). Thus the exponential electron energy spectrum is found to be less suitable for the case of flare stars.

5. FAST ELECTRONS WITH A GAUSSIAN DISTRIBUTION

For the sake of completeness we also consider the case where the energy spectrum of the fast electrons follows a Gaussian distribution:

$$n_e(\gamma) = n_o \frac{2}{\sqrt{\pi}\sigma} e^{-\left(\frac{\gamma-\gamma_o}{\sigma}\right)^2} \tag{4.42}$$

where, as before, $\gamma = E/mc^2$, and σ is the dispersion expressed in γ, n_0 is the total number of fast electrons per unit volume with energy from zero to infinity, γ_0 is the average energy of the electrons.

The spectrum (4.42) is interesting because quite often the energy spectrum of the electrons resulting from the decay of unstable atomic nuclei (β-decay) is a curve which rather resembles a normal distribution. Therefore consideration of a spectrum of type (4.42) may shed some light on the possible mechanism of generation of fast electrons in the outer regions of stellar atmospheres.

The spectrum (4.42) lies between the two extreme cases considered above (Fig. 4.3). For very small values of the dispersion ($\sigma \to 0$) the spectrum (4.42) approaches the case of mono-energetic electrons ($\gamma =$ const.), but for very large values of the dispersion ($\sigma \to \infty$), $n_e(\gamma)$ depends weakly on γ and finally it approaches an exponential law ($\gamma^{-\alpha}$).

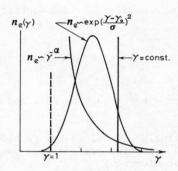

In solving the equation for the energy transfer through a medium of fast electrons with a spectrum of type (4.42) one should remember that only electrons with $\gamma > 1$ take part in the processes of Thomson scattering related with the transformation of long wavelength photons into short wavelength ones. Hence we have for the effective optical depth of the medium

Fig. 4.3. Three types of energy spectra of fast electrons considered in the text: mono-energetic ($\gamma =$ const.); exponential ($\sim \gamma^{-\alpha}$), and Gaussian distribution.

$$\tau' = \sigma_s \int_0^z dz \int_1^\infty n(\gamma) d\gamma = \sigma_s \int_0^z n_0' \, dz \,, \tag{4.43}$$

where n_0' is the total number of fast electrons per unit volume with energy $\gamma > 1$.

Electrons with energy $\gamma < 1$ produce Thomson scattering without transformation of the frequency of the photons; they act like ordinary "thermal" electrons. The optical depth due to pure scattering is

$$\tau'' = \sigma_s \int_0^z dz \int_0^1 n(\gamma) d\gamma = \sigma_s \int_0^z n_0'' dz \,, \tag{4.44}$$

with $n_0 = n_0' + n_0''$.

Considering that in our case $\gamma_0 > 1$, we shall neglect the electrons with $\gamma < 1$ and assume that $n_0 = n_0'$, i.e. $\tau = \tau'$. The solution of the equation of energy transfer in the case of a Gaussian distribution of the energy of the electrons leads to the following results (model Fig. 4.1): for the outward flow of radiation at the outer boundary of the fast electron layer, where the total optical depth is equal to τ, we have

$$J_\nu(\tau, \gamma_0, \sigma, T) = B_\nu(T) C_x(\tau, \gamma_0, \sigma, T) \,. \tag{4.45}$$

Analogously we have for the inward radiation (directed towards the star) at the inner boundary of the layer, where $\tau = 0$:

$$H_\nu(\tau, \gamma_0, \sigma, T) = B_\nu(T)G_x(\tau, \gamma_0, \sigma, T). \qquad (4.45')$$

In these expressions

$$C_x(\tau, \gamma_0, \sigma, T) = \left[1 + \frac{\tau}{2\pi\sqrt{\pi}\sigma}(e^x - 1)\Phi_x(\gamma_0, \sigma)\right]e^{-\tau} \qquad (4.46)$$

$$G_x(\tau, \gamma_0, \sigma, T) = \frac{\tau}{2\pi\sqrt{\pi}\sigma}(e^x - 1)\Phi_x(\gamma_0, \sigma)e^{-\tau}, \qquad (4.47)$$

where $x = h\nu/kT$ and

$$\Phi_x(\gamma_0, \sigma) = \int_1^\infty \frac{e^{-\left(\frac{\gamma - \gamma_0}{\sigma}\right)^2}}{e^{x/\gamma^2} - 1} \frac{d\gamma}{\gamma^4}. \qquad (4.48)$$

Numerical values of the function $\Phi_x(\gamma_0, \sigma)$ in the range from $x = 0.5$ to $x = 20$ and for two values of the parameters γ_0 and σ are given in Table 4.3.

In the special case where $\sigma \to 0$ and thus $\gamma \to \gamma_0$, we find from (4.46)

$$C_x(\tau, \gamma_0, T) = \left[1 + \frac{1}{4\pi\gamma^4} \frac{e^x - 1}{e^{x/\gamma^2} - 1}\tau\right]e^{-\tau}. \qquad (4.49)$$

As expected this expression is the same as (4.8), which was derived for the case of mono-energetic electrons.

In the same way we find for the case of the real photosphere (cf. Fig. 4.2), instead of (4.28) and (4.30):

$$C_x(\tau, \gamma_0, \sigma, T) = E_4(\tau) + \frac{3}{\sqrt{\pi}\sigma}(e^x - 1)\Phi_x(\gamma_0, \sigma)F_1(\tau), \qquad (4.50)$$

$$G_x(\tau, \gamma_0, \sigma, T) = \frac{3}{\sqrt{\pi}\sigma}(e^x - 1)\Phi_x(\gamma_0, \sigma)F_2(\tau), \qquad (4.51)$$

where the functions $F_1(\tau)$ and $F_2(\tau)$ are given by (4.25) and (4.26). Numerical values of the functions $C_x(\tau, \gamma_0, \sigma, T)$ are given in Table 4.4 for a number of values of the wavelength and of the temperature of the star.

An analysis of equation (4.45) shows that for a Gaussian distribution of the energy of the electrons, the main parameters of the flare--color indices and amplitudes of the increase in brightness--are found just within the observational limits for flare stars. The same is true in the case of mono-energetic electrons. In this respect these two distributions are, at least for stars of late classes, equivalent, and the question whether the fast electrons have all the same energy or show a normal distribution is more or less irrelevant. However, the form of the energy spectrum of the electrons is directly related to the nature of their origin or the mechanism of their generation. Therefore it will be better to use the more general assumption

TABLE 4.3 Numerical Values of the Functions $\Phi_x(\gamma_0, \sigma)$

x	$\gamma_0 = 2$		$\gamma_0 = 3$		x	$\gamma_0 = 2$		$\gamma_0 = 3$	
	$\sigma = 1$	$\sigma = 2$	$\sigma = 1$	$\sigma = 2$		$\sigma = 1$	$\sigma = 2$	$\sigma = 1$	$\sigma = 2$
0.5	0.861	1.967	0.455	0.974	4	0.0527	0.0736	0.0410	0.0722
0.6	0.703	0.976	0.376	0.799	5	0.0349	0.0494	0.0300	0.0510
0.7	0.590	0.819	0.319	0.674	6	0.0243	0.0351	0.0229	0.0378
0.8	0.505	0.701	0.276	0.580	7	0.0178	0.0259	0.0180	0.0291
0.9	0.440	0.610	0.243	0.507	8	0.0131	0.0198	0.0145	0.0230
1.0	0.388	0.537	0.217	0.449	10	0.00765	0.0121	0.0097	0.0151
1.2	0.310	0.429	0.177	0.362	12	0.00476	0.0080	0.0069	0.0106
1.4	0.255	0.352	0.149	0.301	14	0.00308	0.0056	0.0051	0.0078
1.6	0.214	0.295	0.128	0.255	16	0.00208	0.0041	0.0038	0.0059
1.8	0.182	0.252	0.112	0.220	18	0.00142	0.0030	0.0028	0.0046
2.0	0.157	0.217	0.0987	0.192	20	0.00102	0.0023	0.0022	0.0036
3	0.0855	0.118	0.0599	0.110					

TABLE 4.4 Numerical Values of the Functions $C_x(\tau, \gamma_0, \sigma, T)$ for a Gaussian Distribution of the Fast Electrons for $\gamma_0 = 3$, $\sigma = 2$ [Equation (4.46)]

λ, Å	T = 2800° (M5)			T = 3600° (M0)			T = 4200° (K5)		
	τ			τ			τ		
	1	0.1	0.01	1	0.1	0.01	1	0.1	0.01
3000	8881	4265	531	326	157	20.4	69.5	34.1	5.1
4000	221	107	33	21.7	11.2	2.3	7.15	4.17	1.40
5000	27.4	13.9	18.3	4.7	3.0	1.25	2.08	1.74	1.09
6000	7.1	4.2	1.4	1.83	1.62	1.08	1.05	1.24	1.03
7000	2.9	2.1	1.14	1.05	1.24	1.03	0.71	1.08	1.01
10000	0.8	1.12	1.02	0.51	0.98	1.00	0.43	0.95	0.99

that the energy spectrum of the electrons is given by the normal distribution with dispersion σ, of which a particular case (for $\sigma \to 0$) is a group of electrons all with the same energy.

6.   THE SHELL OF FAST ELECTRONS AROUND THE STAR

The fast electrons above the star's photosphere may be produced as follows. First, from the nucleus of the star, matter of unknown composition is transported or ejected outwards. Fast electrons escape from this material produced by β-decay of certain nuclei; more details will be considered in Chapter 16. The nature and the rate of production of fast electrons from this matter is the same for all flares and for all flare stars. However, according to the duration of the ejection or flow of matter from the stellar interior, and also depending on the initial velocity of the ejection, the source of the fast electrons may be closer to or farther from the surface of the star.

If the source of fast electrons screens part of the star's surface (this part will be called y), then the observed flux of radiation H^*_ν can evidently be represented as the sum of the Compton (non-thermal) and normal (thermal) radiation of the star, i.e.

$$H^*_\nu = yH_\nu(\tau, \gamma, T) + (1 - y)B_\nu(T) \ . \tag{4.52}$$

However, the real picture seems to be somewhat different. In fact, the fast electrons which appear at a certain distance from the stellar surface will speed in all directions practically with the velocity of light. At the same time they will be intercepted by the dipole magnetic field of the star. Hence, for a certain time the star is rapidly surrounded by a shell of fast electrons. At the same time on account of the diffusion of the electrons this "momentary" shell will dissipate from the star with a certain velocity.

In this model we must observe a flare also if the expulsion of the primary matter occurred at the back side of the star with respect to the observer. Then the flare does not have a local character; it encompasses the whole star. Therefore everywhere in the following y will be taken equal to 1.

7.   APPLICATION TO DOUBLE STARS

The fast electron hypothesis applies very well for the excitation of an optical flare in double star systems. In fact, according to formula (3.5) the maximum shift in frequency of the photon is found for $\alpha \approx \pi$, i.e. for a frontal collision of photons with fast electrons. In the case of a single star only a small part of the fast electrons, produced at some distance from the star (the "source of fast electrons") will be directed towards the photons leaving the star (Fig. 4.4), thus leading to a frontal or nearly frontal encounter producing the Compton photons. But in the case of double systems a source of fast electrons equally powerful as in the first case can lead to a considerably stronger effect than in the case of a single star. This is illustrated in Fig. 4.5. One source of fast electrons lying between two stars will have about equal efficiency when compared to photons ejected by two separate stars A and B.

Quantitative consideration of this problem is much more complicated for the case of a double star system than for a single star. It is even hard to predict qualitatively the possible behavior of fast electrons in the magnetic field of the two

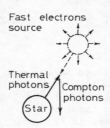

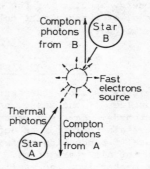

Fig. 4.4. Excitation of "Compton"
radiation (collision of thermal
photons with fast electrons) in the
case of a single star.

Fig. 4.5. Excitation of "Compton"
radiation in the case of a binary
stellar system.

stars. However, we can approximately predict the following:

a) The flare frequency must be higher for double stars than for single stars.

b) The strongest flares (absolute amount of the energy) are to be expected in
binary stellar systems.

c) In the case of binary stellar systems sometimes irregularities may appear in
the lightcurve (after the maximum of the flare) in the shape of secondary maxima.

d) A correlation could be expected between some flare parameters like: frequency,
strength, duration, etc.--and the phase of the binary stellar system.

e) The percentage of flare stars must be higher for binary stars than among single
stars. This appears to be really the case; more than half of the flare stars of
UV Cet type in the neighborhood of the Sun (Table 1.1) are double systems; for the
rest there is not enough data yet to suggest a binary nature.

With regard to the flare stars in stellar associations and young clusters, we do
not know at all whether they are binaries or not. Since they are so extremely
faint it will not be possible to decide this point by special spectroscopic observations. But from general considerations of stellar statistics it is to be expected that there will be many double systems among the flare stars in associations.

This problem should be the subject of special investigations. Considering the
data available at present we should not be too astonished if all flare stars are
found to be double systems.

8. STELLAR ATMOSPHERES CONSISTING OF A MIXTURE OF THERMAL AND FAST ELECTRONS

Let us consider the following problem. Suppose that in the atmosphere of a certain
star there are so many thermal electrons that the optical depth of the medium for
Thomson scattering is larger than one. At the same time there are fast electrons
in much smaller numbers than the thermal electrons. Then the optical depth due to
the fast electrons is much smaller than unity. Under these conditions a photon
will be scattered many times by thermal electrons without change of frequency.

The time a photon needs to pass through the medium becomes longer and thus the probability that it encounters a fast electron increases, which results by the inverse Compton effect in a change of frequency of the photon. In this way a certain number of long wavelength photons may be transformed to short wavelengths. We note that a large concentration of fast electrons is expected for the atmospheres of hot stars like type O, Wolf-Rayet stars, nuclei of planetary nebulae, etc. A relatively small admixture of fast electrons in their atmospheres may lead to strengthening the ultraviolet continuum by new photons of Compton origin. Since at the same time the maximum of the Planck radiation for these stars lies in the region $\sim$ 1000 Å, the presence of fast electrons with $\gamma^2 \sim 10$ in their atmospheres may give rise to a second maximum around 100 Å, i.e. in the region of soft x-ray radiation. This maximum will be mainly of non-thermal origin.

As applied to hot stars the concept of fast electrons can have interesting consequences. It can explain, in particular, the ultraviolet excess in their spectra [3]. There is not the slightest doubt that such an excess exists, particularly from the following facts:

1. The occurrence of emission lines of O VI in the spectra of the nuclei of some planetary nebulae [4, 5]. There are even nuclei for which the lines of O VI are stronger than the lines of He II (!). Fivefold ionization of the oxygen atoms requires very hard radiation, shorter than 100 Å, of sufficiently high intensity.

2. The existence of nuclei of planetary nebulae with positive color index [6, 7]. Evidently since they have no excess radiation in the ultraviolet, these relatively cold nuclei cannot assure the radiation of the planetary nebulae in general.

3. The strong increase--from 20,000 to 80,000 K of the excitation temperature of Wolf-Rayet stars with increasing ionization potential of the ion from the lines of which the temperature is determined [8].

Formal treatment of the problem of radiative transfer through a medium consisting of a mixture of thermal and fast electrons leads to the following equation (in the case of mono-energetic electrons):

$$\cos\theta \, \frac{dJ_\nu}{dt} = -J_\nu + \frac{1 - p}{4\pi} B_\nu e^{-\tau} + \frac{p}{4\pi} \gamma^2 B_{\nu_0} e^{-\tau} +$$

$$+ \frac{1 - p}{4\pi} \int J_\nu d\omega + \frac{p}{4\pi} \gamma^2 \int J_{\nu_0} d\omega , \qquad (4.53)$$

where

$$p = \frac{n_e}{n_e + N_e} , \qquad \nu_0 = \frac{\nu}{\gamma^2} ; \qquad (4.54)$$

n_e and N_e are the concentrations of thermal and fast electrons respectively.

Contrary to the cases considered above, here the diffuse component must be taken into account in the solution of the equation of energy transfer since $\tau > 1$ in this particular problem.

Clearly, this question has no direct relation to flare stars. However, it is interesting to study this problem in particular as applied to the very hot stars.

9. THE DEVIATION OF THE STELLAR RADIATION
FROM PLANCK'S LAW

In the derivation of the equations giving the spectral distribution of a star's radiation during a flare, Planck's law has been adopted for the spectral distribution of the star's photospheric radiation in its quiescent state. However, the actual distribution may differ from Planck's law in the optical (UBV) as well as in the infrared spectral regions.

For individual flare stars the true energy distribution during its quiescent state can be found in the UBV and infrared spectral regions. Below we give the corresponding formulae for finding the energy distribution in the star's spectrum at the time of the flare.

Let the dimensionless coefficients K_λ and K_{λ_0} characterize the degree of departure of the actual distribution from the Planck distribution at wavelengths λ and λ_0 = $\lambda\gamma^2$ respectively. Then we have, instead of (4.27) and (4.28) for the spectral distribution of the intensity of the star's radiation at the time of the flare:

$$H_\lambda (\tau) = B_\lambda (0) \left[K_\lambda E_4 (\tau) + K_{\lambda_0} \frac{3}{2\gamma^4} \frac{e^x - 1}{e^{x/\gamma^2} - 1} F_1 (\tau) \right] \qquad (4.55)$$

To find the radiation intensity during the flare, in V wavelengths (λ = 5500Å) we must know the value of the two coefficients $K_{0.55}$ and $K_{5.5}$ (with γ^2 = 10). For finding the intensities in the UBV system, we must have the numerical values of six coefficients: $K_{0.36}$, $K_{0.44}$, $K_{0.55}$, $K_{3.6}$, $K_{4.4}$ and $K_{5.5}$.

Another way for finding the radiation intensities during the flare assumes the knowledge of the actual values of the fluxes during the undisturbed stellar phase in the visual region $H_\lambda (0)$, as well as in the infrared $H_{\lambda_0} (0)$. In that case we have:

$$H_\lambda (\tau) = H_\lambda (0) E_4 (\tau) + \gamma^2 H_{\lambda_0} (0) F_1 (\tau), \qquad (4.56)$$

where $\lambda_0 = \lambda\gamma^2$, as before. In the UBV system this relation can be written in the form:

$$H_U (\tau) = H_{0.36} (0) E_4 (\tau) + \gamma^2 H_{3.6} (0) F_1 (\tau);$$

$$H_B (\tau) = H_{0.44} (0) E_4 (\tau) + \gamma^2 H_{4.4} (0) F_1 (\tau);$$

$$H_V (\tau) = H_{0.55} (0) E_4 (\tau) + \gamma^2 H_{5.5} (0) F_1 (\tau).$$

These relations can be used for determining flare amplitudes and color indices for those cases where the true energy distribution in the spectrum of the flare star is known.

REFERENCES

1. G. A. Gurzadyan, Astrofizika 1, 319 (1965).
2. G. A. Gurzadyan, Doklady Academy Nauk USSR 166, 53 (1966).
3. G. A. Gurzadyan, Symposium IAU No. 34, Planetary Nebulae. Ed. D. E.
   Osterbrock and C. R. O'Dell. Dordrecht, Holland, p. 332 (1968).
4. L. H. Aller, Ap. J. 140, 1601 (1964).
5. L. H. Aller, Symposium IAU No. 34, p. 339 (1968).
6. G. Abell, Ap. J. 144, 259 (1966).
7. M. A. Kazaryan, Commun. Byurakan Obs. 38, 25 (1967).
8. L. H. Aller, Ap. J. 97, 135 (1943).

CHAPTER 5
The Continuous Spectrum of a Flare

1. THE THEORETICAL SPECTRUM OF A STAR DURING A FLARE

In the preceding chapter we derived the energy distribution of a flare as a function of the energy of the fast electrons and the effective temperature of the star.
In the present chapter we shall analyze the theoretical, continuous stellar spectrum
during a flare.

For convenience, we consider a star of a spectral class, say M5, with effective temperature 2800 K. The energy distribution in the spectrum of such a star
in its quiescent state is given, for wavelengths from 2000 to 10,000 Å, by the
heavy line in Fig. 5.1 (the lowest curve). Let us assume that above the photosphere of such a star fast electrons (all with the same energy) have appeared in a
quantity equivalent to optical depth of unity, for processes due to Thomson scattering. The appearance of fast electrons leads—by their interaction with the
infrared photons of the normal stellar radiation (inverse Compton effect)—to an
immediate increase of the radiation intensity, i.e. to a stellar flare in the short-
wavelength region of the spectrum. The absolute increase of the intensity and the
character of the energy distribution in the star's spectrum depend on the value of
the energy γ of the electrons, and to a lesser degree on their energy spectrum (in
what follows we restrict ourselves to monoenergetic electrons). This can easily
be seen in Fig. 5.1; as γ increases the curves become higher and higher and their
maxima are shifted towards the short wavelengths. For the construction of these
curves formula (4.27) was used (H_ν being replaced by J_ν).

For a given spectral class of the star there is a limit to the increase of the
electron energy γ, above which further increase leads to a decrease of the flare
intensity in the spectral region which is accessible to observation. For stars of
class about M5 the limiting energy of the fast electrons corresponds to $\gamma^2 \sim 10$.
For values $\gamma^2 > 10$ the curves for the energy distribution are shifted so far towards
the short wavelengths that the relative increase of the flare intensity in U, B and
V radiation decreases (Fig. 5.2). Physically this is quite understandable: for
relatively small values of the energy of the fast electrons a considerable part of
the scattered photons will be concentrated in the nearby short-wavelength spectral
region. However, as the energy of the fast electrons increases further, the main
part of the scattered photons will be converted into very short wavelengths, not
accessible to observations.

In the above example given the stimulating role of fast electrons in the process
of flare generation was illustrated and at the same time the probable value of
their energy was indicated: $\gamma \sim 3$, i.e. $E \sim 1.5 \times 10^6$ eV. In the following this
estimate will be confirmed by various observations.

Figs. 5.1 and 5.2 indicate that the intensity increase in the short-wavelength part
of the spectrum is accompanied by a decrease of the intensity in the long-wavelength
(infrared) part of the spectrum.

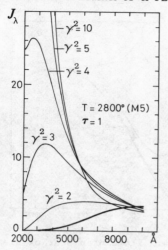

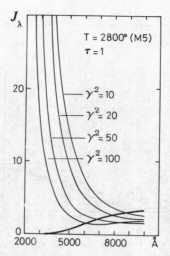

Fig. 5.1. Theoretical energy distribution J_λ in the spectrum of flares of stars of class M5 depending on the energy of the fast electrons γ.

Fig. 5.2. Theoretical energy distribution J_λ in the spectrum of flares of a star of class M5 for large values of the energy of the fast electrons γ (from E = 1.5 MeV to E = 5.1 MeV).

Finally, the stellar flare caused by the inverse Compton effect has a spectrum (Figs. 5.1 and 5.2) qualitatively quite different from the spectrum of a normal star of class M5. Whereas usually for stars of class M5 the intensity decreases steadily towards short wavelengths, in the theoretical flare spectrum the intensity increases rapidly towards the short wavelengths. In other words, during the flare the star becomes bluer.

2. THEORETICAL SPECTRA OF STARS OF DIFFERENT CLASSES DURING FLARES

The form of the theoretical spectrum of a star during a flare depends not only on the energy of the fast electrons but also on the effective temperature of the star and on the strength of the flare. By strength of the flare we shall mean the effective optical depth τ of the layers of fast electrons producing Thomson scattering; this quantity is proportional to the total energy of the fast electrons appearing during the flare. The optical depth τ enters as a parameter in all the formulae which give the theoretical flare spectra for various energy distributions of the electrons. These formulae also contain the effective temperature of the radiation of the photosphere T. Therefore it is interesting to consider the properties of theoretical star spectra during flares for various values of T and τ.

Consider first the case of monoenergetic electrons applied to real photospheres. The energy distribution in the flare spectrum is in this case given by formula (4.28). Here the values for the effective temperatures of stars of different spectral classes are approximate:

Class	M6	M5	M0	K5	K0	G5
T_{eff}°K	2500	2800	3600	4200	4900	5500

In Figs. 5.3, 5.4, and 5.5 the theoretical curves of energy distributions in the spectra of stars from M6 to K5, during flares, are given. The curves have been

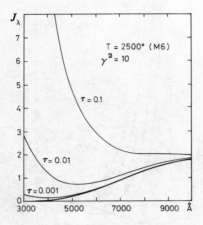

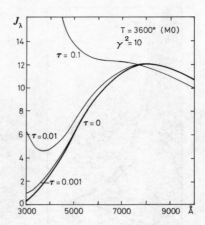

Fig. 5.3. Theoretical energy distribution in the spectrum of a star of class
M6 during a burst induced by fast electrons (γ^2 = 10) depending on the strength
of the burst τ (J_λ in arbitrary units).

Fig. 5.4. Flare of a star of class M0
(cf. Fig. 5.3).

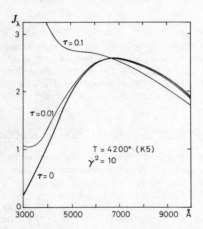

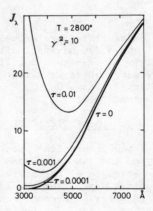

Fig. 5.5. Flare of a star of class K5
(cf. Fig. 5.3).

Fig. 5.6. Structure of the short-
wavelength part of the spectrum of an
M5 star for faint flares (τ ∿ 0.01 to
0.0001, cf. Fig. 5.3).

constructed for energies of the fast electrons corresponding to γ^2 = 10 and for
values of τ = 0.1, 0.01, 0.001. Analogous curves for stars of classes K0 and G5
are given in [1, 2].

These curves clearly show the increase of the relative radiation intensity during
a flare, in particular in the ultraviolet. All other conditions being equal this
increase has a maximum for stars of class M6 and decreases rapidly towards earlier
classes. For example, for τ = 1 the increase in intensity at 3400 Å is a factor of
6000 times for M6 stars, but for stars of classes M5, M0, K5, K0 and G5 this increase

is 1200, 66, 16, 5 and 2.5 times, respectively.

The sharp decrease of relative intensity of the flare from late to earlier spectral classes is one of the chief properties resulting from the hypothesis of fast electrons.

There is still another "temperature effect." In fact in all cases the theoretical curves of the flare spectra intersect the Planck energy curves of the undisturbed stars at some wavelength λ_0. The increase of brightness of the flare as compared with the undisturbed state of the star, occurs only in the region $\lambda < \lambda_0$. For $\lambda > \lambda_0$ a "negative flare" occurs, i.e. a lowering of the level of the continuous spectrum. This question will be considered in detail later; here we only note that the region of "zero" amplitude, λ_0, is shifted towards shorter wavelengths for increasing effective temperature of the star.

With regards to the strength of the flare τ, the larger τ is, the larger the flare amplitude at a given wavelength. However, the dependence of the amplitude on τ decreases rapidly with increasing stellar temperature. For $\tau = 0.01$ the intensity increase at $\lambda = 3000$ Å for a K0 star is a factor of only 1.7, whereas for an M6 star it is 2600 times (!) for the same value of τ. Even for $\tau = 0.0001$ the intensity at $\lambda = 3000$ Å increases 30 times for M6 stars and less than one percent for stars of class K0. Hence for the case of stars of classes M5-M6 the increase of brightness in ultraviolet radiation can make very faint flares detectable (Fig. 5.6), whereas for K5-K0 stars this is practically insignificant. Such results seem to agree with the observations (§2, Ch. 2).

3. THE SPECTRUM OF THE PURE FLARE

Let us consider the theoretical spectrum of the flare itself, i.e. the spectrum of the radiation of the flare, after subtracting the normal radiation of the star. For the energy distribution F_ν of the flare we have:

$$F_\nu = B_\nu(T)(C_\nu - 1) , \qquad\qquad (5.1)$$

where $C_\nu = C_\nu(\gamma, \tau, T)$ is given by (4.28) for the case of monoenergetic fast electrons, or (4.41) and (4.46) for an exponential or Gaussian energy spectrum of the electrons; $B_\nu(T)$ is Planck's function.

When the observational results are given in magnitudes we have

$$\Delta m_\nu = -2.5 \lg F_\nu + \text{const.} \qquad\qquad (5.2)$$

Using Table 4.2 theoretical spectra of the flare were found for a number of values of τ and three values of stellar temperatures: 2500 K (M6), 2800 K (M5) and 3600 K (M0); the results are given in Fig. 5.7, which show the dependence of m_ν on $1/\lambda$ (the numbers along the m_ν axis are arbitrary and indicate only the scale and the direction of the changes of m_ν).

In those cases where the observational results on energy distribution in the flare spectrum are given in magnitudes, but on a wavelength scale, it is desirable to have, for comparison, the theoretical spectrum of the flare on the wavelength scale as well. Such a spectrum is shown in Fig. 5.8.

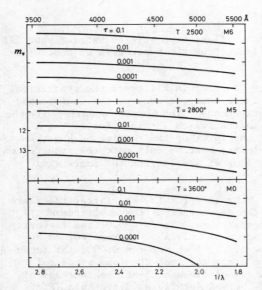

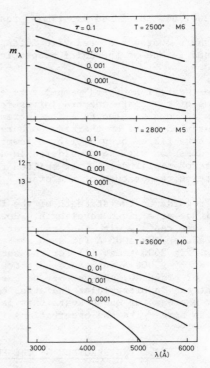

Fig. 5.7. Theoretical spectra of only
the flare due to the inverse Compton
effect of fast electrons with $\gamma^2 = 10$,
in the case of flares of stars of
classes M6, M5 and M0. The radiation
intensity is given in stellar magnitudes m_ν and on a frequency scale
$(1/\lambda)$.

Fig. 5.8. The same as Fig. 5.7, with
the intensity of the radiation given in
stellar magnitudes m_ν, and on a wavelength scale.

4. CONTINUOUS SPECTRA

To obtain a spectrogram of the star at the time of the flare is not that easy.
This explains the relatively small number of flare spectrograms, in particular of
spectrograms suitable for accurate measurements.

The first spectrograms of stars at the time of the flare were obtained accidentally.
These "accidents" occurred most often with Joy: he obtained, together with Humason,
in 1949 the first spectrogram of a flare of UV Cet [3] and afterwards the first
spectra of flares of YZ CMi, DO Cep, Ross 614, EQ Peg, V L216 Sgr [4]. In the same
accidental way spectra of flares of the stars V 645 Cen [5], DY Dra [6], Wolf 1130
[7], Wolf 359 [8], etc. were obtained.

Unfortunately, in nearly all these spectra the plates have no calibration; and
only qualitative estimates can be made, mostly of the form and strength of the
emission lines and also of the level of the continuous spectrum.

The situation changed completely when scientists began to organize the ways of obtaining spectra of the stars during the flares--by regular observational patrols
and by the construction of special equipment. Thus, e.g. at the Crimean Observatory a method was developed to obtain spectrograms at the time of flares while the
star was observed photoelectrically at the same time. With this method the moment

a flare appears the shutter of the spectrograph is opened. Such observations of
flares were made of UV Cet and AD Leo [9, 10].

There is also a method of obtaining flare spectra with the aid of a special camera
with a moving photographic plate mounted perpendicularly to the dispersion of the
spectrograph and thus recording all changes in the spectrum of the star during an
exposure. By this method Kunkel obtained a series of spectrograms of AD Leo,
EV Lac, and YZ CMi at the moment of their flares [11, 12]. Using both of the above
methods, Moffett and Bopp [13, 14] obtained excellent spectrograms during flares
of YY Gem and UV Cet.

Let us consider some results of the spectrophotometric studies of flares, restricting ourselves mainly to analysis of the continuous spectrum; the line spectrum will
be considered in Chapter 9, which is devoted entirely to the problem of excitation
of emission lines in flare stars.

The observations show that at the moment of the flare continuous emission appears,
increasing in strength towards the short-wavelength end of the spectrum and more
or less concealing the normal absorption structure of the spectrum of a late-type
dwarf star. At the same time the Balmer emission lines are strengthened considerably, and the lines themselves become much wider. However, the continuous emission
is the dominating part of the total flare radiation and the duration of this emission is of the same order as that of the flare itself (12, 15]. whereas the emission
lines disappear more slowly. From measurements by Kunkel [12] for a flare in YZ CMi
(5.XII.1965) the slope of the curve for the decrease in intensity of the Balmer
emission lines is smaller by more than a factor of two than the gradient of the
curve of the decrease of the continuous emission. Such a behavior should be taken
as a clear indication for the non-recombination origin of the continuous emission,
as compared to the emission lines which appear during the flare.

For a flare in EV Lac (11.XII.1965) a spectrogram obtained by Kunkel was studied
carefully. Curves of the energy distribution in the spectrum before and at the
moment of the flare have been con-
structed; they are shown in Fig.
5.9. In the upper half of this
figure the full-drawn line gives
the energy distribution of EV
Lac before the outburst, and
the broken line gives the
Planck distribution for T =
2800 K. A strong depression
in the continuous spectrum in
the region from 4100 Å to 4500 Å
with maximum depth at 4226 Å,
is noticed, which is caused by
absorption due to neutral
calcium.

In the lower half of Fig. 5.9
the continuous line shows the
energy distribution in the
spectrum of this star at the
time of the flare. Two things
are seen at once: the strengthening of the continuous spectrum in the short-wavelength
part of the spectrum (shorter
than 4500 Å), and the

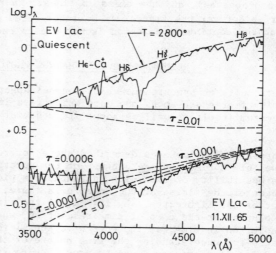

Fig. 5.9. Spectrum of EV Lac in the region 3500-
5000 Å before the flare (above) and during the
flare (below). Broken lines--theoretical spectra
of a flare star of class M5 with $\gamma^2 = 10$ and for
various strengths of the flare τ.

considerable increase of intensity of the emission lines. The increase of the
level of the continuous spectrum is seen between the strong emission lines H_δ and
H_ϵ, and also between H_γ and H_δ, i.e. at those spectral regions where there are no
other emission lines. Particular attention should be given to this fact because
some authors have tried to explain the strengthening of the continuous spectrum at
short wavelengths (up to the Balmer jump) as a result of blending of higher order
Balmer emission lines.

We can now compare the observed energy distribution in the continuous spectrum of
the flare found in Fig. 5.9 with that given by the "fast electron" theory. To do
this we must first obtain the pure continuous spectrum, without emission lines,
and Balmer and Paschen continua. This cannot be done easily; therefore we confine
ourselves to a comparison of theory with observations in a general way.

The broken lines in Fig. 5.9 show the theoretical energy distribution curves in
the continuous spectrum of a star of class M5 (T = 2800 K) corresponding to various
values of τ--from 0.01 to 0.0001. It is seen that the best agreement is reached
for $\tau \approx 0.0006$, taking into account that due to the depression at 4200 Å, and also
at 3800 and 3900 Å, the spectrum of the star in its normal state differs generally
from the Planck distribution. Further, using the curves given in Chapter 6 (§ 1)
we can find the theoretical amplitudes corresponding to this flare of EV Lac, i.e.
for the value τ = 0.0006; they are ΔU = 1.4, ΔB = 0.2 and $\Delta V \approx 0$. Next we have to
ask how large is the contribution of the emission lines in the total flare radia-
tion? This question is very important, since it finally leads to finding the mechanism or physical process which generates stellar flares. By special measurements
Moffett and Bopp [14] have established that the contribution of the emission lines
in the total flare radiation is small; in B radiation, it is from 10% to 20%.

However, this fraction is not constant and may change from one flare to the next,
even for the same star. Now, if the continuous emission was due to recombinations
we would have in general a constant ratio between the two components of the radia-
tion--the continuous radiation (identified with the series continua, and with free-
free processes), and the emission lines. Therefore the fact that this fraction
varies for various flares may be interpreted as an argument in favor of the theory
that the continuous emission is not due to recombinations.

5.   RESULTS OF MULTI-CHANNEL PHOTOMETRY

The results of synchronous multi-channel spectrophotometric measurements of stars
during flares are among the data which give important information about the flare
spectrum itself as well as about its behavior at different phases of its development.

Contrary to the methods described above for recording flare spectra, which always
give only a time-average picture of the spectrum, multi-channel photometry gives
the possibility of recording instantaneous pictures of the spectrum at different
stages of the flare--at the onset of a flare, at maximum, and at the descending
part of the lightcurve. By its nature the first method gives a high spectral
resolution at the cost of time resolution, whereas the second method gives high
resolution in time, at the cost of spectral resolution. Both these extremes are
needed for understanding the true nature of the flare; therefore both methods of
studying flare spectra must be equally encouraged.

The usual wide-band UBV spectrophotometry for constructing lightcurves of stellar
flares has been used for a long time. However, the results found by this method
for the energy distribution in the spectra of the light of the flare are not very
accurate. Unique in this respect are the spectrophotometric measurements of two

flares of EV Lac made by Kodaira et al. [16], who recorded synchronously in five wavelength regions, from 3300 Å to 6000 Å, and with a time resolution of 0.1 second. The five channels were centered at the following wavelengths: 3570, 3850, 4170, 4550, 5000 and 5500 Å; the effective widths of the bands were for all nearly the same, ~ 370 Å. Fig. 5.10 gives the lightcurves for these two flares of EV Lac (3.VIII.1975); one is of average intensity, with amplitude Δ(UV) = 1ᵐ9, and the second is very strong, with Δ(UV) = 5ᵐ9.

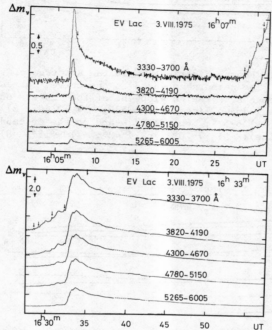

Fig. 5.10. Lightcurves obtained for two flares of EV Lac with the aid of a five-channel spectrophotometer in the wavelength range 3300–6000 Å [16].

From the results of their measurements the authors [16] constructed spectral energy curves of the additional flare radiation, subtracting the radiation of the star in its normal state. These curves are given in Figs. 5.11 and 5.12. The numbers next to the curves indicate various stages of the development of the flare, on the rising branch (full-drawn lines) as well as on the falling branch (broken lines). The indices a and b correspond with the maxima of the lightcurves.

The most characteristic property of these spectra given as a function of frequency is the clearly expressed flat form in the wavelength range considered, and also the fact that the spectrum almost does not depend on the strength of the outburst or on the stage in its development. The authors stress in particular their conclusion that the main part of the flare radiation is not due to the emission component of hydrogen, but to the continuous emission.

The measurements by Kodaira et al. appear to be sufficiently accurate and reliable to justify using them for theoretical interpretation. In particular the fact that at the outermost channel of the spectrophotometry (band center at 3515 Å), no excess radiation--as compared with the other channels--was found, which might have been due to the Balmer continuum, is a clear indication that hydrogen emission does

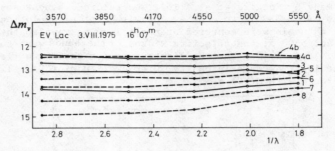

Fig. 5.11. Energy distribution in the spectrum of the flare
(16^h07^m). The numbers appended to the curves correspond to
various moments in the development of the flare: on the rising
branch (full-drawn lines): 1) 6^m53^s; 2) 6^m44^s; 3) 6^m48^s;--at
maximum: 4a) 6^m53^s; 4b) 7^m00^s;--and on the descending branch
(broken line): 5) 7^m20^s; 6) 7^m42^s; 7) 9^m15^s; 8) 11^m45^s.

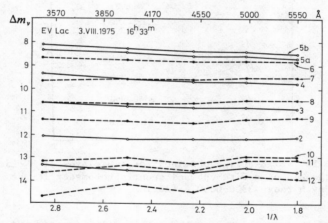

Fig. 5.12. Energy distribution in a flare spectrum (16^h33^m).
On the ascending branch: 1) 30^m00^s; 2) 31^m33^s; 3) 32^m44^s;
4) 32^m57^s. At maximum: 5a) 33^m21^s; 5b) 33^m47^s;--on the
descending branch: 6) 34^m28^s; 7) 36^m27^s; 8) 43^m17^s;
9) 53^m06^s; 10) $17^h38^m13^s$; 11) 18^h00^m; 12) $18^h55^m00^s$.

not play an important role in the generation of the additional radiation even during strong flares (Fig. 5.12). At the same time these results are in complete
agreement with the theoretical spectrum of the light of the flare derived above
(Fig. 5.7) on the basis of the fast-electron hypothesis.

In connection with this conclusion it is interesting to ask whether a flat spectrum
(on a frequency scale) for the flare is actually characteristic for the inverse
Compton effect only, or that qualitatively this form of the spectrum is also possible for other processes, among which processes of thermal character.

According to the estimate of the authors [16] the influence of individual emission
lines of hydrogen and ionized calcium on the flare spectra found by them is small--

from $0^m.01$ to $0^m.1$ for each channel of the spectrophotometer. Therefore a comparison of the theoretical continuum flare spectra (Fig. 5.7) with the observed spectra is in order.

6. FLARE SPECTRA FOR THE CASE OF A GAUSSIAN DISTRIBUTION OF FAST ELECTRONS

The theoretical flare spectrum in the case of a Gaussian distribution of the fast electrons does not differ qualitatively from the spectrum produced by mono-energetic electrons. The difference between them is quantitative only; for equal values of the energy of the fast electrons, γ_0, and the strength of the flare, τ, the curves of the energy distribution in the flare spectrum for the case of the Gaussian distribution lie slightly above the curves found for the mono-energetic electrons. In other words, the amplitudes in U and B radiation are somewhat larger in the case of the Gaussian distribution than in that of the mono-energetic electrons (cf. Ch. 6). As for the amplitudes in V radiation, the situation is somewhat different.

The stellar spectrum during the flare can also be characterized, as we have seen above, by the parameter λ_0--the region of zero amplitude. For stars of class M5-M6 this lies in the infrared ($\gtrsim 8500$ Å), independent of the energy spectrum of the fast electrons. Hence, for M5-M6 stars the type of the electron energy spectrum has no qualitative influence either on the character of the flare spectrum, or on the value of the amplitude in V radiation.

But for flares of stars of class K5 the situation is different. For both types of electron energy spectra the zone of zero amplitude lies in the photovisual spectral region--at 5400 Å and 6100 Å, respectively. Passing from one type of energy spectrum of the electrons to the other gives a qualitative change in the amplitude of the V radiation itself, namely, the amplitude is negative for mono-energetic electrons and positive for the Gaussian distribution. In the first case we have a negative flare in V radiation, in the second case a positive flare. Thus in the case of K5 stars the sign of the amplitude ΔV is sensitive to the type of energy spectrum of the fast electrons. Comparing this with observation we can thus find the probable form of the energy spectrum of the electrons. The observations give a positive value for the flare amplitude in V radiation for stars of class K5.

Thus analysis of theoretical energy distribution curves allows us to find, in the case of stars of class M5, the probable energy of the fast electrons, which is of the order of $\gamma \sim 3$. The same analysis for stars of class K5 allows us to find the probable type of the energy spectrum of the fast electrons. Clearly this spectrum is not mono-energetic, but it is described closely by a Gaussian curve.

REFERENCES

1. G. A. Gurzadyan, Astrofizika 1, 319 (1965).
2. G. A. Gurzadyan, Astrofizika 2, 217 (1966).
3. A. H. Joy, M. L. Humason, PASP 61, 133 (1949).
4. A. H. Joy, Non-Stable Stars, Ed. G. H. Herbig, Cambridge, 1957, p. 31.
5. A. D. Thackeray, M.N.R.A.S. 110, 45 (1950.
6. D. M. Popper, PASP 65, 278 (1953).
7. A. H. Joy, Ap. J. 105, 101 (1947).
8. J. L. Greenstein, H. Arp, Ap. J. Lett. 3, 149 (1969).
9. R. E. Gershberg, P. F. Chugainov, Astron. J. USSR 43, 1168 (1966).
10. R. E. Gershberg, P. F. Chugainov, Astron. J. USSR 44, 260 (1967).
11. W. E. Kunkel, Flare Stars, Doctoral Thesis, Texas, 1967.

12. W. E. Kunkel, *Ap. J.* **161**, 503 (1970).
13. T. J. Moffett, B. W. Bopp, *Ap. J. Lett.* **168**, L117 (1971).
14. B. W. Bopp, T. J. Moffett, *Ap. J.* **185**, 239 (1973).
15. P. F. Chugainov, *Izv. Crimean Obs.* **26**, 171 (1960); **28**, 150 (1962).
16. K. Kodaira, K. Ichimura, S. Nishimura, *Publ. Astron. Soc. Japan* **28**, 665 (1976).

CHAPTER 6
Brightness of Flares

1. THEORETICAL FLARE AMPLITUDES

In this chapter we determine, on the basis of the fast electron hypothesis, theoretical flare amplitudes in the photometric UBV system and compare them with observations. By flare amplitude we mean the increase in brightness of the star at the time of the flare, i.e. at the moment the fast electrons appear above its photosphere ($\tau > 0$), as compared with the brightness in the normal state, when there are no such electrons ($\tau = 0$). The numerical values of the amplitudes of the flare ΔU, ΔB, ΔV in the UBV system are found by means of the following relations:

$$\Delta U = m_U(0) - m_U(\tau) = 2.5 \log \frac{U}{U_o} ;$$

$$\Delta B = m_B(0) - m_B(\tau) = 2.5 \log \frac{B}{B_o} ; \qquad (6.1)$$

$$\Delta V = m_V(0) - m_V(\tau) = 2.5 \log \frac{V}{V_o} ,$$

where

$$U = \int J_\lambda(\tau,\gamma,T)U_\lambda d\lambda; \quad U_o = \int B_\lambda(T)U_\lambda d\lambda ;$$

$$B = \int J_\lambda(\tau,\gamma,T)B_\lambda d\lambda; \quad B_o = \int B_\lambda(T)B_\lambda d\lambda ; \qquad (6.2)$$

$$V = \int J_\lambda(\tau,\gamma,T)V_\lambda d\lambda; \quad V_o = \int B_\lambda(T)V_\lambda d\lambda .$$

In these relations the numerical values of $J_\nu(\tau,\gamma,T)$ as a function of the parameters τ, γ and T are taken from the preceding chapter, $B_\lambda(T)$ is the Planck function for the given effective temperature of the star and U_λ, B_λ and V_λ are the relative sensitivities of the ultraviolet, photographic and photovisual spectral regions in the UBV system. In the computations below the numerical values of these coefficients are taken from Johnson and Morgan [1]. Table 6.1 shows the numerical values for U_λ, B_λ and V_λ used here in arbitrary scale.

The numerical values of ΔU, ΔB and ΔV have been determined for different cases of the energy spectrum of the fast electrons.

MONOENERGETIC ELECTRONS. Table 6.2 gives the theoretical flare amplitudes in U, B, V radiation computed for the simplest case--the one-dimensional problem (formula 4.8) and T = 2800 K. The amplitudes have been computed for values of γ^2 from 2 to 50 and for two values of the optical thickness $\tau = 1$ and $\tau = 0.1$. From

63

TABLE 6.1 Values Used for the Sensitivity Coefficients in
the UBV System

$\lambda \overset{\circ}{A}$	U_λ	B_λ	V_λ	$\lambda \overset{\circ}{A}$	U_λ	B_λ	V_λ
3000	1.20			5000		2.65	2.10
3200	5.00			5200		1.25	5.55
3400	7.55			5400		0.40	5.85
3600	8.10			5600			4.85
3800	5.90	0.80		5800			3.50
4000	0.70	7.00		6000			2.30
4200		7.55		6200			1.40
4400		7.00		6400			0.55
4600		5.85		6600			0.20
4800		4.25					

TABLE 6.2 Theoretical Flare Amplitudes ΔU, ΔB, ΔV in the
One-Dimensional Problem (T = 2800 K)

γ^2	$\tau = 1$			$\tau = 0.1$		
	ΔU	ΔB	ΔV	ΔU	ΔB	ΔV
2	2.5	1.0	0.1	1.3	0.4	0.1
3	4.3	2.2	0.7	2.8	1.0	0.3
5	5.4	2.8	1.0	3.8	1.5	0.4
10	6.9	2.9	0.9	4.2	1.6	0.4
20	5.4	2.5	0.5	3.9	1.3	0.2
50	4.7	1.8	-0.05	3.2	0.8	0.04

the data in Table 6.2 a number of interesting conclusions may be drawn.

1. In all cases, independent of the energy of the fast electrons and their
effective number, i.e. independent of the strength of the flare, the condition

$$\Delta U > \Delta B > \Delta V$$

is strictly fulfilled. This result can also be stated in a different way: the
radiation leaving the fast electron layer is bluer than the radiation falling on
the inner boundary of the layer.

2. For rather small values of the energy of the fast electrons--$\gamma \sim 2-3$ (E $\sim$ 1 to
1.5×10^6 eV)--the theoretical amplitude reaches several stellar magnitudes, up to
3^m in photographic light, and up to 5-6^m in ultraviolet light.

3. For the values of T and τ given the maximum amplitude is found for $\gamma^2 \sim 10$.
For $\gamma^2 > 10$ the flare amplitude decreases with increasing γ.

The conclusion that the maximum amplitude is reached for $\gamma^2 \sim 10$ holds for a large
range of values of T, and for all values of τ (< 1). Thus even in this simple
one-dimensional problem the most important properties of the interaction of fast

electrons with the stellar photosphere are evident.

ACTUAL PHOTOSPHERES. In this case the radiation falls on the inner boundary of the fast electron layer from various angles (cf. Fig. 4.2). The intensity of the radiation leaving this layer is determined by formula (4.27) and the values of the coefficients $C_\lambda(\tau,\gamma,T)$ are given in Table 4.2. With these data and with the aid of formulae (6.1) theoretical flare amplitudes could be derived for a wide range of optical depths, from $\tau = 1$ to $\tau = 0.0001$ and for spectral classes from M6 to G5. It is assumed that the layer consists of monoenergetic electrons with $\gamma^2 = 10$. The results are given in Table 6.3.

TABLE 6.3 Theoretical Amplitudes ΔU, ΔB, ΔV Depending on the Strength of the Flare (τ) for Stars of Classes M6-G5 Monoenergetic Electrons with $\gamma^2 = 10$

T	Flare Amplit.	τ				
		1	0.1	0.01	0.001	0.0001
2500°	ΔU	$8^m\!.8$	$8^m\!.0$	$5^m\!.7$	$3^m\!.3$	$1^m\!.2$
(M6)	ΔB	5.6	4.8	2.6	0.8	0.1
	ΔV	3.2	2.5	0.8	0.12	0.01
2800°	ΔU	7.2	6.4	4.2	1.9	0.4
(M5)	ΔB	4.4	3.6	1.6	0.32	0.04
	ΔV	2.3	1.6	0.4	0.05	0
3600°	ΔU	4.2	3.5	1.5	0.3	0.03
(M0)	ΔB	2.1	1.5	0.4	0.04	0
	ΔV	0.6	0.5	0.08	0	0
4200°	ΔU	2.8	2.1	0.6	0.08	0.01
(K5)	ΔB	1.1	0.8	0.1	0.01	0
	ΔV	−0.1	0.2	0.03	0	0
4900°	ΔU	1.6	1.1	0.2	0.02	0
(K0)	ΔB	0.2	0.3	0.05	0	0
	ΔV	−0.6	0.02	0	0	0
5500°	ΔU	0.8	0.6	0.1	0.01	0
(G5)	ΔB	−0.2	0.14	0.02	0	0
	ΔV	−0.8	−0.04	0	0	0

The data in this table confirm the conclusions drawn above. In particular the condition $\Delta U > \Delta B > \Delta V$ is strictly fulfilled for all spectral classes of the stars and for arbitrary values of the strength of the flare. This may be seen clearly from Fig. 6.1, constructed from the data in Table 6.3 for stars of classes M5 and M6.

In particular the regularity should be noticed, that the amplitude of the flare increases for all radiations, from hot to cold stars. For instance, for $\tau = 1$ we have: $\Delta U = 0.8$ for stars of class G5; $\Delta U = 1.6$ for K0 stars, etc., and for M6 stars the theoretical amplitude in U radiation reaches the highest value--

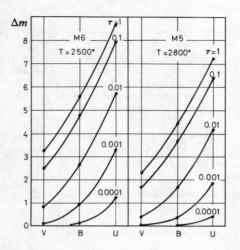

Fig. 6.1. Flares of a star of subclasses M5 and M6. The curves are theoretical values of flare amplitudes, Δm, in U, B, V radiation for various flare strengths τ. The flare is induced by inverse Compton effect of monoenergetic electrons with $\gamma^2 = 10$.

$\Delta U = 8.8$ (!).

The data of Table 6.3 can be used to determine the strength of the flare, i.e. the numerical value of τ if the values of the flare amplitudes in U, B and V radiation are known from observations. To facilitate this operation curves of ΔU, ΔB and ΔV versus τ constructed for M6 and M5 stars are given in Figs. 6.2 and 6.3. From these curves it follows that the flare amplitudes have their maxima approximately for $\tau = 0.6$. However, we can attach no significance to this since our solution of the transfer equation is inaccurate for $\tau \sim 1$.

Returning to a general estimate of the numerical values of ΔU, ΔB, ΔV given in Table 6.3 we should stress that they are in general somewhat too low, due to the following two reasons:

a) Planck's law has been used for the energy distribution in the continuous spectrum of the undisturbed star. However, the actual energy distribution in late-type stars is strongly disturbed by numerous absorption lines and bands. As a result the actual intensity of the continuous radiation may be, in U radiation, say, lower by

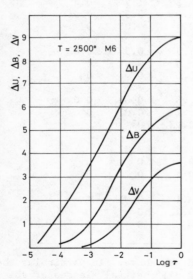

Fig. 6.2. Flare of a star of subclass M6. Theoretical dependences of the amplitudes ΔU, ΔB and ΔV on the strength of the flare τ. Monoenergetic electrons with $\gamma^2 = 10$.

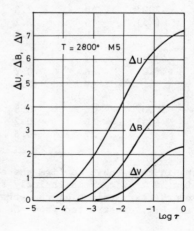

Fig. 6.3. Flare of a star of class M5. Cf. Fig. 6.2.

a factor of two or more than the Planck intensity. However, the observed amplitudes are found as the ratio of the flux of radiation from the star during the flare to the real (not Planckian) flux of radiation from the undisturbed star in a given spectral range. Therefore the limiting flare amplitude of a star of class M5-M6 in U radiation may be an entire stellar magnitude larger than the values given in Table 6.3. Then the theoretically possible maximum value of the flare amplitude in U radiation for M6 stars will be of the order $9\overset{m}{.}5$ to 10^m.

b) In the solution of the equation of transfer we have neglected the diffusion term, which is allowed only for small values of τ. For $\tau \sim 1$ the intensity of the diffuse component of the radiation becomes comparable with the intensity of the direct part of the radiation. As a result the intensity of the radiation leaving the layer will be considerably higher--more than twice than that which we found before when the diffusion component was ignored.

Thus the fast electron hypothesis predicts a flare with amplitude up to 10^m in U radiation for stars of class M6.

GAUSSIAN ELECTRON DISTRIBUTION. Table 6.4 gives numerical values of the theoretical flare amplitudes in U, B and V radiation for the case of a Gaussian energy distribution of the fast electrons (formulae 4.45 and 4.50). The computations have been

TABLE 6.4 Theoretical Amplitudes ΔU, ΔB, ΔV for Different Values of Flare Strength (τ) for Stars of Classes M6-G5. (Gaussian Distribution of the Electrons with $\gamma_0 = 3$ and $\sigma = 2$).

T	Flare Amplit.	τ				
		1	0.1	0.01	0.001	0.0001
2500°	ΔU	$9\overset{m}{.}0$	$8\overset{m}{.}2$	$6\overset{m}{.}0$	$3\overset{m}{.}6$	$1\overset{m}{.}4$
(M6)	ΔB	5.9	5.1	2.9	1.0	0.14
	ΔV	3.6	2.9	1.0	0.17	0.02
2800°	ΔU	7.4	6.6	4.4	2.1	0.5
(M5)	ΔB	4.7	3.9	1.9	0.4	0.05
	ΔV	2.7	2.0	0.6	0.07	0.006
3600°	ΔU	4.6	3.8	1.8	0.38	0.043
(M0)	ΔB	2.6	1.9	0.5	0.064	0.006
	ΔV	1.0	0.8	0.13	0.013	0
4200°	ΔU	3.2	2.5	0.8	0.12	0.011
(K5)	ΔB	1.5	1.1	0.2	0.023	0.001
	ΔV	0.3	0.4	0.05	0.005	0
4900°	ΔU	2.0	1.4	0.3	0.04	0.003
(K0)	ΔB	0.7	0.5	0.1	0.01	0.001
	ΔV	-0.17	0.15	0.02	0.001	0
5500°	ΔU	1.3	0.9	0.17	0.02	0.001
(G5)	ΔB	0.2	0.3	0.04	0.004	0
	ΔV	-0.45	0.06	0.01	0	0

made for the real photosphere and for one set of values of the parameters for the Gaussian distribution: $\gamma_0 = 3$ and $\sigma = 2$. Computations for other possible values of γ_0 and σ give about the same results since the amplitude depends only slightly on the parameters γ_0 and σ.

From a comparison of the data in Tables 6.3 and 6.4 the following conclusions may be drawn:

a) The theoretical flare amplitudes ΔU, ΔB and ΔV are for the case of the Gaussian distribution always somewhat larger than for the case of monoenergetic electrons.

b) The flare amplitudes for stars of classes M0-M6 are not very sensitive to the energy spectrum assumed for the fast electrons. Therefore it is convenient for M0-M6 stars to use the formulae derived for the case of the monoenergetic electrons.

c) The flare amplitudes of stars of classes K5-G5 differ for the Gaussian distribution significantly, and sometimes even considerably, from the amplitudes found for the monoenergetic electrons. The Gaussian distribution is closer to reality than the monoenergetic one. Therefore for stars of class K5-G5 we must use the formulae derived for the case of the Gaussian electron distribution.

2. COMPARISON WITH THE OBSERVATIONS

The comparison of theoretical flare amplitudes with their observed values can be done in different ways. For instance, it is possible to compare not only the amplitudes with each other, but also to consider the various parameters of the star and of the flare. This approach allows us to find some physical parameters of the flare. Such an analysis is applied to the flare stars near the Sun (stars of UV Cet type) as well as to those in stellar associations and young clusters.

Tables 6.5 and 6.6 give the distributions of the numbers of flares with respect to amplitude in B and U radiation for the Orion association and the Pleiades cluster and NGC 2264 (from data up to 1970). First, the maximum number of cases of flares corresponds to a value of τ of the order of 0.008 in B and 0.002 in U radiation, i.e. considerably smaller than unity. Second, the maximum observed flare amplitudes, in B as well as in U radiation, lie within the limits of the values expected theoretically. The theoretical flare amplitude in B radiation is equal to 7^m6 for $\tau = 0.1$. At the same time the maximum amplitude in B radiation for the recorded flares in associations is smaller than 6^m, and is probably of the order of 5^m5. The theoretical amplitude in U radiation is equal to 10^m0 for $\tau = 0.1$. There is one case of a flare in Orion (the star Haro 177), and at least four cases in the Pleiades [3], with maximum observed amplitude of 8^m0 to 8^m5 in U radiation.

TABLE 6.5 Amplitude Distribution of Number of Flares in B Wavelengths

Association	ΔB						
	0-1	1-2	2-3	3-4	4-5	5-6	6-7
Orion	28	63	54	25	15	4	--
NGC 2264	2	10	2	--	--	--	--
$\tau =$	0.0016	0.008	0.025	0.08	0.25	(0.3)	

TABLE 6.6   Amplitude Distribution of Number of Flares in U Wavelengths

Associa-tion	ΔU								
	0-1	1-2	2-3	3-4	4-5	5-6	6-7	7-8	8-9
Orion	20	26	36	20	5	2	1	--	1
Pleiades	26	28	36	21	10	14	5	2	4
$\tau =$	0.00012	0.00063	0.002	0.006	0.012	0.032	0.10	0.25	

Thus the observed flare amplitudes in B and U radiation of stars in stellar associations and clusters are within the limits of the values expected theoretically, assuming the fast electron hypothesis.

3.   STARS OF UV CET TYPE

Whereas the observations for flare stars in associations are more or less homogeneous, the data about flare stars in the neighborhood of the Sun, i.e. the stars of UV Cet type, are highly inhomogeneous. This is due to differences in the conditions and methods of observation, individual properties of the telescopes and recording equipment. Quite often instrumental errors occur. These differences play a role in the U,B,V measurements, and the derived colors of the flares. Andrews [4], who has studied this problem, comes to the conclusion that on account of errors of atmospheric and instrumental origin a certain fraction of the observed flares is spurious. Most of all the U band suffers from atmospheric disturbances and instrumental errors (calibration); sometimes this band, which regularly records the largest flare amplitudes (as compared to the B and V channels) gives incorrect (too low) responses [5].

Regardless of these difficulties the available observational material allows us to form a sufficiently correct picture of the details of flares of some UV Cet type stars.

Some of the UV Cet stars have been studied in great detail. Therefore it is useful to study the observational results for the most interesting ones. Data on the photoelectric parameters and radiative power of these stars [7] are collected in Table 6.7.

UV CET. This is one of the faintest (in absolute magnitude) flare stars of spectral class M5.5e [14] or M6e [15]. The effective temperature of the photospheric radiation of UV Cet appears to be lower than 2800 K   (M5), but higher than 2500 K (M6). UV Cet is a double system with brightness of the components $m_A = 12.45$ and $m_B = 12.95$. Flare activity has been found in both components. It is one of the flare stars nearest to the Sun; its distance is 2.65 parsecs. The low luminosity of each of these stars ($L_A = 5 \times 10^{-5}$ $L_\odot$, $L_B = 3 \times 10^{-5}$ $L_\odot$) is also due to their small dimensions; the radii are ~ 0.08 $R_\odot$. The total mass of the system is estimated at about 0.08 $M_\odot$ with nearly equal masses of the components. However the distance between the components is rather large; the period of orbital revolution is about 200 years. There is some indication that the distance between the components increases with time.

UV Cet is the most observed flare star. The results of spectrophotometric observations have been collected mainly in the following numbers of the IBVS

TABLE 6.7 Photovisual (V) Magnitudes, Absolute Luminosities (M_V),
Color Indices (B – V) and (U – B) and Energies Emitted in the Undisturbed
State (q(V), q(B), q(U)) for Some Flare Stars

Star	V	M_V	B – V	U – B	q(V) erg s^{-1}	q(B) erg s^{-1}	q(U) erg s^{-1}
CN Leo	13.53	16.68	2.01	1.54	$0.56 \cdot 10^{27}$	$3.15 \cdot 10^{27}$	$3.33 \cdot 10^{26}$
UV Cet	12.95	15.27	1.85	1.39	$4.88 \cdot 10^{28}$	$1.86 \cdot 10^{28}$	$2.26 \cdot 10^{27}$
Wolf 424AB	13.1	14.31	1.82	1.24	$7.59 \cdot 10^{28}$	$2.98 \cdot 10^{28}$	$4.15 \cdot 10^{27}$
YZ CMi	11.2	12.29	1.60	1.01	$4.83 \cdot 10^{29}$	$2.32 \cdot 10^{29}$	$4.00 \cdot 10^{28}$
EV Lac	10.2	11.50	1.37	0.75	$1.00 \cdot 10^{30}$	$5.94 \cdot 10^{29}$	$1.30 \cdot 10^{29}$
EQ Peg	12.4	11.38	1.56	1.05	$1.17 \cdot 10^{30}$	$5.83 \cdot 10^{29}$	$9.96 \cdot 10^{28}$
AD Leo	9.43	10.98	1.54	1.08	$1.62 \cdot 10^{30}$	$8.20 \cdot 10^{29}$	$1.32 \cdot 10^{29}$
YY Gem	9.07	8.36	1.49	1.04	$1.80 \cdot 10^{31}$	$9.59 \cdot 10^{30}$	$1.61 \cdot 10^{30}$

(Information Bulletin on Variable Stars): 210, 296, 298, 310, 315 (1968); 343,
349, 354, 404, 405, 406 (1969); 526, 604, 608 (1971); 615, 620, 736 (1972); 760
(1973); 879 (1974); 979, 1006, 1017, 1020 (1975); 1105 (1976) and also in [17].
The largest number of flares (315) of this star in U radiation was recorded by
Kunkel [6]. The largest number (114) of spectrophotometric observations of flares
of UV Cet in UBV radiation was made by Moffett [7]. A large number of lightcurves
of flares of UV Cet in photographic and UBV radiation, recorded with high time
resolution (0.5 s), was published by Cristaldi and Rodono [8].

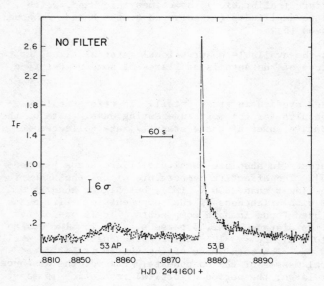

Fig. 6.4. Lightcurve of a flare of UV Cet, with
time resolution of 1 second. 53 AP is an example
of a "quiet" or "slow" flare, 53B is an example
of a "rapid" or "spike" flare [7].

Fig. 6.4 shows examples of
lightcurves of two flares of
UV Cet, one of which is
"quiet," or "slow," the
other "rapid" (spike flare)
[7]. It should be noted that
cases of flares of the second
type (spike flares) predominate. Fig. 6.5 gives the
lightcurve of another spike
flare of UV Cet with details
indicating the moments at
which the individual com-
ponents of the radiation,
emission lines, continuous
spectrum, etc., appear and
disappear [9]. One more
lightcurve of the same character is shown in Fig. 6.6 for
another flare of UV Cet. It
should be pointed out that
these last two lightcurves
are products of a high level
of experimental investigation
of stellar flares.

In Figs. 6.5 and 6.6 the two

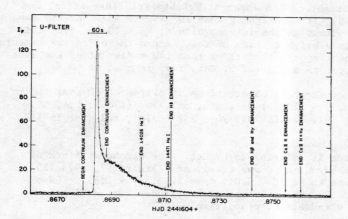

Fig. 6.5. Lightcurve of a very powerful ($\Delta U = 5.2$)
flare of UV Cet, on 17.X.72. Time resolution 1
second. Vertical arrows indicate the times of
beginning and end of the various spectral variations
of the flare. The two-component character of the
radiation of the flare (continuous spectrum and
radiation in emission lines) is seen [9].

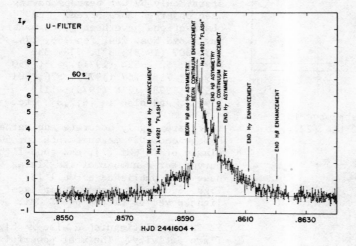

Fig. 6.6. Lightcurve of a complicated flare of UV
Cet in U radiation, observed with time constant of
1 s (cf. Fig. 6.5) [9].

most important properties of
the radiation of the flare
itself appear quite clearly.
In the first place the radiation of flares consists of
two components--the continuous spectrum and the emission
lines. In the second place,
the radiation of the flare
in the continuous spectrum
always disappears earlier
whereas the radiation in the
individual emission lines
may last for a longer time.
Even in the case that the
flare is induced only by
inverse Compton effects, the
maximum possible theoretical
amplitudes of flares of a
star of class M5.5 will be
about: $\Delta U_{max} = 8.2$, ΔB_{max}
$= 5.8$ and $\Delta V_{max} = 3.2$. What
do the observations show?
In V radiation the maximum
flare amplitude of UV Cet
appears to have been recorded
at the Abastumani Observatory
--it was equal to 2^m3 [10].
The maximum amplitude in B
radiation was recorded twice
--by Japanese [11] and
Italian astronomers [12]--as
$\sim 3^m7$. Regarding the maximum
amplitude in U radiation,
this has probably been observed by Kunkel [13] and was
approximately $5^m5 - 6^m0$. In
all cases the maximum
observed flare amplitudes of
UV Cet were found to be
within the limits of the
values expected theoreti-
cally.

One very powerful flare of
UV Cet was observed on
17.X.72 by Moffett [7], in
which the star reached a
maximum brightness, in
$t_b = 47$ s, which was 127
times the normal one, with
amplitude $\Delta U = 5.27$. The
rate of increase of the star's brightness was one of the largest, 2.7 times per
second or 1^m08 s^{-1}. The fall of the brightness after maximum lasted for
$t_a = 751$ s ≈ 12.5 minutes, and the total energy liberated during this outburst
was 1.15×10^{31} ergs (in U radiation). In other words, during the time of this
flare (~ 13 minutes) the same amount of energy was emitted as the undisturbed star
emits in about 1.5 hours.

AD LEO. This is one of the brightest (in apparent brightness) flare stars, and
is of spectral class M4.5e [16]; it is brighter than 10^m (V = 9.43). At the same
time AD Leo is one of the brightest in absolute magnitude; M_V = 10.98, i.e. this
star is at least a hundred times brighter than UV Cet. Since the effective temperatures of the two stars are about equal, it follows that the radius of AD Leo must
be one order of magnitude larger than that of UV Cet, i.e. it must be ∿0.8 $R_\odot$.

There are relatively fewer photoelectric observations of flares of AD Leo; their
results are given in IBVS Nos. 310 (1968); 334, 340, 345, 367 (1969); 534, 597
(1971); 685, 750 (1972); 790, 791, 801 (1973); 906, 932 (1974), and also in [17, 7]
etc.

As to its flare activity AD Leo is inferior to UV Cet. The strongest flare of
AD Leo appears to have been observed by Osawa et al. on February 17, 1974 [19];
the flare amplitudes were found to be ΔU = 5.01, ΔB = 2.93, ΔV = 1^m65. These
quantities proved to be lower than the theoretical amplitudes for an M5 star--
7.0, 4.5 and 2^m5 in U, B and V radiation, respectively.

Fig. 6.7 gives an example of a lightcurve of one flare of AD Leo.

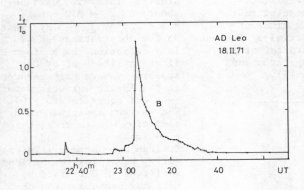

Fig. 6.7. Photoelectric lightcurve in B
radiation of one flare of AD Leo [22].

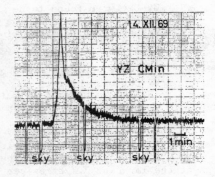

Fig. 6.8. Photoelectric recording of
one flare of YZ CMi [11].

YZ CMI. In apparent brightness
(V = 11.20) as well as in absolute
magnitude (M_V = 12.29) this star
ofclass M4.5e occupies an intermediate position between UV Cet and
AD Leo. This star has been observed
more intensively than other flare
stars, only UV Cet perhaps having
been observed more often. The
observations have been given mainly
in: IBVS Nos. 264, 265-268, 274,
305, 307 (1968); 331, 338, 339
(1969); 521, 524 (1971); 635, 750
(1972); 758, 767 (1973); 876, 901
(1974); 998, 1018 (1975); 1112
(1976), and also in [6,7,8], etc.

There are highly accurate and rather
homogeneous UBV measurements for 50
flares of YZ CMi [7], and good lightcurves for a number of its flares
have been published as well [8].
Figs. 6.8 and 6.9 show examples of
lightcurves of flares of YZ CMi.

YZ CMi is distinguished also by high
flare activity. The most powerful
flare of this star appears to have
been recorded on January 19, 1969,
when Andrews [20] observed a maximum amplitude for this star in V
radiation, equal to 1^m7; and Kunkel,
who missed the moment of the flare
maximum, estimated the amplitude
in U radiation at ∿6^m6; this corresponds, according to Fig. 6.3, with
ΔB = 3.7 and ΔV = 1.7. This flare
is also interesting because it was

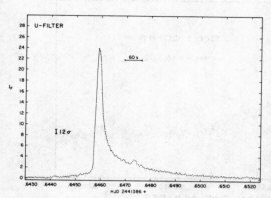

Fig. 6.9. Lightcurve of one typical
flare of YZ CMi in U radiation [9].

observed at the same time at radio frequencies (Ch. 13). Earlier, in 1965-66, two
flares of YZ CMi [6] were found with amplitudes in U radiation of $4^m_.0$ and $5^m_.0$.
These values are within the expected
theoretical limits.

EV LAC. This star of class M4.5e is only
slightly fainter than AD Leo, and therefore
its flare activity cannot be much different.
Actually, as regards the frequency of the
flares, as well as the maximum observed
amplitudes this star ranks between UV Cet
and AD Leo.

The results of spectrophotometric observations of EV Lac are given in IBVS Nos.
399, 401, 403 (1969); 600, 608 (1971); 616,
627, 672, 723, 750 (1972); 759, 802, 836
(1973); 874 (1974); and also in [7, 8, 17, 18, 21, 22], etc.

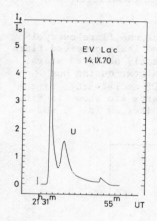

Fig. 6.10. Lightcurve in
U radiation of a flare of
EV Lac.

Fig. 6.10 shows an example of a lightcurve for one
flare of EV Lac. The maximum flare amplitudes for
this star observed at various times were found to
be $1^m_.45$ and $3^m_.2$ in V and B radiation, respectively
[17]. The maximum amplitude in U appears to have
been observed by Cristaldi and Rodono [8] during
a flare of EV Lac on 22.VIII.70, with $\Delta U \simeq 5.2$.
In all cases the amplitudes were found to be considerably lower than the theoretical values.

CN LEO = WOLF 359. The faintest among the flare
stars for which absolute luminosities are known,
is BD +4°40 48 B (object Van Biesbroeck); it has
$M_V = 19^m_.2$. Up until 1970 not a single flare of
this star had been recorded. It was included in
a list of flare stars on account of a characteristic strengthening of one spectrogram accidentally
taken by Herbig [23].

If we omit BD + 4°40 48 B, then the faintest
(absolutely) star in our list of flare stars
(Table 1.1) is CN Leo = Wolf 359 = G 406--the
fourth nearest to the Sun and the third nearest double system; it has $M_V = + 16^m_.7$.
It belongs to spectral class M6e-M8e with effective temperature about 2500 K.
This star is at least two times fainter than UV Cet and hence its radius must
be of the order of 0.05 $R_\odot$.

The star CN Leo is interesting due to its very high flare frequency--about 3 to 4
flares per hour in U radiation; this frequency is the highest among the known
flare stars. More than forty flares of CN Leo were recorded in U radiation by
Kunkel [6] in 1969 during a total observing time of less than 13 hours, and the
largest number (more than 100) of flares of this star in U, B and V radiation were
recorded by Moffett [7] in the period 1971-1972. Fig. 6.11 gives a lightcurve of
one flare of CN Leo, rather complicated and tangled in structure.

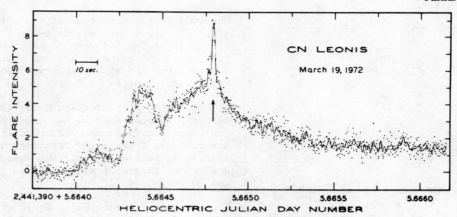

Fig. 6.11. Lightcurve with high time resolution of a complex flare of CN Leo, covering in all 3.5 minutes of time. Each point corresponds with an integration time of the photometer of 0.2 seconds, using the 208-cm telescope of the McDonald Observatory [32].

WOLF 424. This star has a very high flare frequency--about one flare every 15 minutes (cf. Table 1.3). At the same time it is one of the least observed flare stars. This star should be included in the list of constantly observed stars. Moffett [24] recorded 12 flares (in UBV radiation) of this star during one observing season (21 March 1972) of 2.62 hours with the 208-cm telescope of the McDonald Observatory. Lightcurves of four consecutive flares are shown in Fig. 6.12, together with curves of the variations of the color indices (B – V) and (U – B) of the system "star + flare" during the flare.

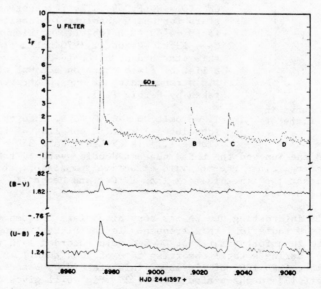

Fig. 6.12. Lightcurve (upper) of four flares of Wolf 424 in U radiation. Curves of the variations of the color indices (B – V) and (U – B) during the flare are also given, for the system "flare + quiescent star" [24].

Wolf 424 is interesting because it is found to be an ideal object, where without exception the parameters of the flare--the amplitudes in various colors, the color indices and their behavior in time, the dynamic characteristics of the lightcurves themselves, etc. were found to agree with the flare theory based on the fast electron hypothesis.

YY GEM. This is (if we do not count EQ Her) in absolute luminosity as yet the brightest star in our list of flare stars; it has $M_V = 8^m.36$. Besides, YY Gem is one of the few stars of early classes, namely M0.5.

As expected, the flare frequency of YY Gem is one of the lowest; it is more than 20 times less than that for UV Cet. On acccount of its very high absolute luminosity--nearly 1500 times (!) that of UV Cet--the total amount of radiation emitted during the flare must be very large in order that it can be observed against the background of the undisturbed radiation of the star. This can explain why there are relatively few data about flares of YY Gem. Moffett [7] for example observed only 18 flares during ∿120 hours of observation, on the average one flare per five hours. Three lightcurves in three different wavelength intervals for one very complicated flare of YY Gem are given in Fig. 6.13.

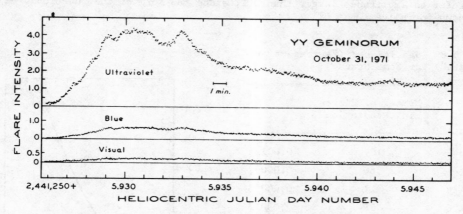

Fig. 6.13. Lightcurves of one long-lasting flare (outburst) of YY Gem taken in three wavelength ranges and with high time time resolution (1 s). In all a 30-minute recording interval is shown [32]. Example of a complex and irregular flare.

Among the remaining flare stars the following are being observed more-or-less regularly: V 1216 Sgr (IBVS Nos. 379, 602, 711, 835, 943, 968 [18], [25]), EQ Peg (very homogeneous UBV observations for 58 flares by Moffett [7]), etc.

The other flare stars have been studied even less. Among those there are some which deserve constant observations, for example V 645 Cen--the flare star closest to us; PZ Mon--belonging to class K2e, which is rare among flare stars of UV Cet type; BY Dra--the brightest in visual magnitude ($8^m.6$), etc.

4.   THE AVERAGE STRENGTH OF THE FLARES

The value of the optical depth of the fast electron medium, τ, depends on the total number of fast electrons appearing at the time of the outburst. Therefore τ may characterize the strength of the flare. Knowing the flare amplitude in some spectral range of observation we may determine the numerical value of τ by means of Table 6.4 or Figs. 6.2 and 6.3. In this connection the question arises to determine the average value $\bar{\tau}$ for all flares of a particular star.

Let $n(\Delta m)$ be the number of flares with amplitude Δm, corresponding to optical thickness $\tau(\Delta m)$. For all flares of this star, i.e. $N = \Sigma n(\Delta m)$ flares with all possible amplitudes larger than a certain fixed minimum value Δm_{min} we have an average value of the optical thickness $\bar{\tau}$, equal to

$$\bar{\tau} = \frac{\Sigma \tau(\Delta m) n(\Delta m)}{N} . \tag{6.3}$$

At the same time we have from (1.4): $n(\Delta m) = N F(\Delta m)$, where $F(\Delta m)$ is the distribution function for the flare amplitudes. Therefore we can write, instead of (6.3)

$$\bar{\tau} = \Sigma \tau(\Delta m) F(\Delta m) . \tag{6.4}$$

For some flare stars of UV Cet type the numerical values of the function $F(\Delta m)$ are given in Table 1.8, and numerical values of $\tau(\Delta m)$ in the last columns of Tables 6.5 and 6.6. With the aid of these data numerical values of $\bar{\tau}$ can be found. The results of the computations, made from U observations, are given in Table 6.8; they refer to amplitudes larger than $0\overset{m}{.}1$, since the form of the function $F(\Delta m)$ is not known for values $\Delta U < 0\overset{m}{.}1$.

TABLE 6.8 Quantities $\bar{\tau}$ ("Average Flare Strength") for Some Stars

Star	Number of flares in U radiation	$\bar{\tau}$
CN Leo	107	0.00124
UV Cet	360	0.00097
Wolf 424	11	0.00056
40 Eri C	38	0.00017
Ross 614	35	0.00017
YZ CMi	71	0.00054
EV Lac	19	0.00054
EQ Peg	27	0.00021
AD Leo	50	0.00029
AT Mic	24	0.00016
Wolf 630	119	0.00013
AU Mic	31	0.00012
YY Gem	17	0.00030

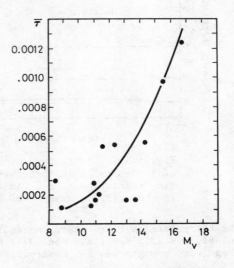

Fig. 6.14. Dependence of average strength of the flare, $\bar{\tau}$, on absolute luminosity of stars of UV Cet type (cf. Table 6.8).

From the data in Table 6.8 it follows that for only two stars--CN Leo and UV Cet --$\bar{\tau}$ is of the order of 0.001; for most of the flare stars $\bar{\tau}$ is considerably smaller than this value and of the order of 0.0001. Moreover $\bar{\tau}$ is seen to depend on the absolute magnitude M_v of the star, namely $\bar{\tau}$ is large for absolutely faint stars and decreases quite rapidly towards the absolutely bright stars (Fig. 6.14).

5. RESULTS OF THREE-COLOR OBSERVATIONS OF FLARES

As we have seen above, the numerical values of the observed flare amplitudes in U, B and V radiation are within the limits of the values expected theoretically. In all cases the inequality $\Delta U > \Delta B > \Delta V$, which also follows from the fast electron hypothesis, is strictly fulfilled. However the dependence between ΔU, ΔB and ΔV is quite specific and follows from the very nature of the hypothesis of fast electrons. Therefore comparison of the observed values ΔU, ΔB, ΔV (for a given flare) with the theoretical dependence $\Delta U \sim \Delta B \sim \Delta V \sim f(\tau)$ (for a given star) may serve as convincing verification of the theory.

In the early 1970s many flares in U,B,V were recorded reliably for a number of stars of UV Cet type. As examples, Figs. 6.15 to 6.19 give lightcurves recorded

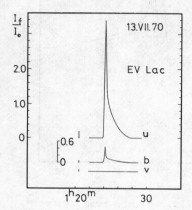

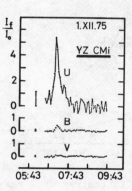

Fig. 6.15. Lightcurves of a flare of EV Lac in U,B,V radiation. Case of absence of flare in V radiation [8].

Fig. 6.16. Lightcurves of flare of YZ CMi in U,B,V radiation. Case of absence of flare in V radiation, and faint flare in B radiation [26].

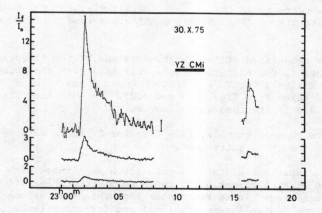

Fig. 6.17. Lightcurves of a flare of YZ CMi in U,B,V. Case of flare being recorded in all three colors [26].

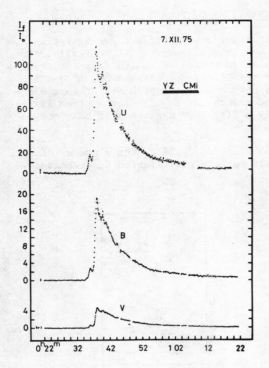

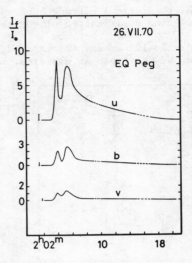

Fig. 6.18. Lightcurves of a very strong
flare of YZ CMi in UBV (U = 5.2,
B = 3.2, V = 1.95) [26].

Fig. 6.19. Lightcurves of two
successive flares of EQ Peg in UBV
[8].

during flares of EV Lac, YZ CMi and EQ Peg [7,8,26]. Cases occur where the flare
is strong in U radiation, noticeable in B and completely absent in V radiation
(Figs. 6.15, 6.16), cases where the flare is hardly visible in V radiation (Fig.
6.17), or it is quite strong (Fig. 6.18), and finally cases of two consecutive
flares (Fig. 6.19).

Fig. 6.20 shows the comparison of the observed values of the flare amplitudes of
YZ CMi and V 1216 Sgr in UBV radiation (dots, circles) with the theoretical
function (full-drawn lines) for T_{eff} = 2800 K and a given value of τ (indicated
in the diagrams and taken from Figs. 6.3 and 8.2). We note that the observed
points lie quite close to the theoretical curves.

In Fig. 6.21 we have another example, this time for EV Lac. Here too the agreement
of observation with theory is quite good.

The results of the comparison of the observed flare amplitudes with theory for
AD Leo are shown in Figs. 6.22 and 6.23, where mainly the observations by Osawa
et al. [19] have been used. It is remarkable that the observations agree well
with theory for strong flares (bremsstrahlung) and for weak flares (inverse
Compton effect). As an example two extreme flares of AD Leo may be mentioned--
the powerful flare just mentioned (17.II.74), and a faint flare on 13.II.74 (Fig.
6.22); the strengths of these two flares differ from each other by nearly three
orders of magnitude.

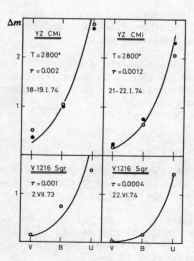

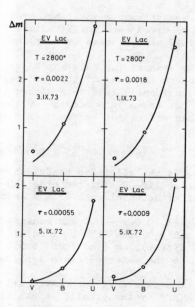

Fig. 6.20. Comparison of observed amplitudes (dots and circles) of a number of flares of YZ CMi and V 1216 Sgr with the theoretical curve $\Delta m \sim f(U,B,V)$ for T_{eff} = 2800 K and the values of flare strength, τ, indicated in the graphs.

Fig. 6.21. Comparison of observed amplitudes of flares of EV Lac with theory (cf. Fig. 6.20).

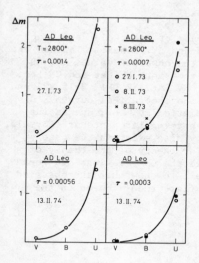

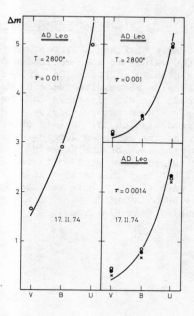

Fig. 6.22. Comparison of observed amplitudes of flares of AD Leo with theory (cf. Fig. 6.20).

Fig. 6.23. Comparison of observed amplitudes of flares of AD Leo with theory.

In these comparisons a small but systematic deviation of the theoretical results
from the observed ones for V radiation is evident, the theoretically predicted
values of ΔV always seems smaller than the observed amplitude of ΔV.

One of the reasons for this may be the possible influence of the emission line $H\alpha$
which lies in the band of the V filter. Enhancement of the H_α line during the
flare may evidently lead to a higher effective flare amplitude   in V wavelengths
which is not taken into account by the theory. The difference $\Delta m_\alpha = \Delta V_{obs} - \Delta V_{theory}$
due to this effect can be found from the following relation:

$$\Delta m_\alpha = 2.5 \log\left(1 + \eta \; \frac{W_\alpha}{\Delta\lambda_V} \; 10^{-0.4\Delta V}\right), \tag{6.5}$$

where $\Delta\lambda_V$ is the effective bandwidth of the V filter, W_α the equivalent width of
the emission line H_α (of the order 50-100 Å), and η is the relative sensitivity of
the V filter at wavelength 6563 Å as compared to the sensitivity at its maximum.

Taking $W_\alpha/\Delta\lambda_V \approx 0.2 - 0.1$ and considering that usually $\Delta V < 1$ for most of the
flares, and $\eta \approx 0.04$ (the H_α line lies in the very "tail" of the sensitivity curve
of the V filter), we find from (6.5): $\Delta m_\alpha \approx 0^m.01$. The contribution of the H_α
emission to the observed flare amplitudes in V wavelengths is small.

However, the fact that for nearly all flares and for all flare stars the observed
values ΔV are systematically larger than the theoretical values leads us to the
following question: Is there a local excess of radiation in normal conditions
(outside flares) in those stars at wavelengths ~ 5.5 μ, as compared with Planck's
distribution? Evidently, special infrared observations with the aim of explaining
the structure of the spectra of flare stars in the region 5-6 μ must be considered
very necessary.

6.   INTERNAL AGREEMENT OF FLARE AMPLITUDES

The hypothesis of fast electrons predicts the existence of a certain dependence
between the values of the amplitudes ΔU and ΔB or ΔU and ΔV. These relations are
shown graphically in Fig. 6.24 (full-drawn lines). The observed values ΔU, ΔB
and ΔV are also shown, from data by Moffett [7] for the stars UV Cet and CN Leo.
In Fig. 6.25 the same is shown for the stars AD Leo, EV Lac, YZ CMi, Wolf 424 and
V 1216 Sgr. The observed points are quite close to the theoretical curves. A
systematic deviation $\Delta V \sim 0.1$ is observed for the relation $\Delta U \sim \Delta V$, due to the
influence of the emission line H_α, which has not been taken into account (formula
6.5). Dependences of the type $\Delta U \sim \Delta B$ or $\Delta U \sim \Delta V$ may also exist in the case
of other theories for stellar flares (e.g. the nebular theory or the theory of a
hot gas). It is possible that in these cases as well the observations are found
to agree with theory. Here we have only one purpose: to show that the observed
relations $\Delta U \sim \Delta B$ and $\Delta U \sim \Delta V$ do not contradict the flare theory based on the
hypothesis of fast electrons.

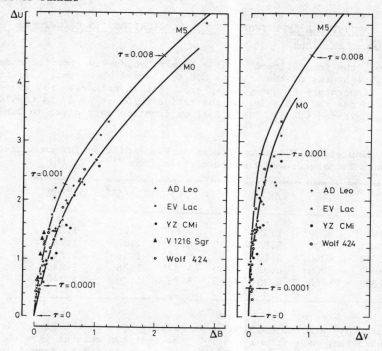

Fig. 6.24.   Comparison of observed values of amplitudes ΔU ∿ ΔB and
ΔU ∿ ΔV for flares of UV Cet and AD Leo.   Full-drawn line--theory.

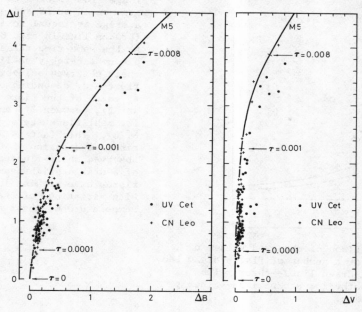

Fig. 6.25.   Comparison of observed values of amplitudes ΔU ∿ ΔB and
ΔU ∿ ΔV for a number of flare stars.   Full-drawn lines--theory.

7. THEORETICAL AMPLITUDES OF THE FLUCTUATIONS IN BRIGHTNESS FOR THERMAL PROCESSES

Strictly speaking, the inequality $\Delta U > \Delta B > \Delta V$ also holds true for the case that the increase in brightness of the star is due to thermal phenomena, i.e. to an increase of the temperature of the photosphere. But in this case the above inequality is a mild one and the value of the ratios $\Delta U/\Delta B$ or $\Delta B/\Delta V$ is considerably smaller than the observed values. This follows from the data given in Table 6.9,

TABLE 6.9 Theoretical Flare Amplitudes in UBV Radiation for an Increase of the Planck Temperature of the Photosphere from T_1 to T_2 (K)

T_1	T_2	ΔU^*	ΔB^*	ΔV^*	$\Delta U^*/\Delta B^*$	$\Delta B^*/\Delta V^*$
2800°	3000°	1.03	0.81	0.66	1.265	1.24
2800	3500	2.90	2.10	1.70	1.263	1.24
2800	4000	5.10	4.83	3.88	1.26	1.243
2800	5000	6.87	5.45	4.37	1.26	1.245
2800	6000	8.35	6.63	5.32	1.26	1.248

where ΔU^*, ΔB^* and ΔV^* are the amplitudes in brightness when the Planck temperature of the photosphere increases from T_1 to T_2. The last two columns give the numerical values of the ratios $\Delta U^*/\Delta B^*$ and $\Delta B^*/\Delta V^*$; they appear not to depend strongly on the temperature of the photosphere (at least in the range from 2800 K to 6000 K), i.e. on ΔB or ΔV and on the average are equal to 1.25 (broken line in Fig. 6.26). At the same time these ratios are considerably smaller than their observed values, which fluctuate, depending on the strength of the flare (on ΔU or ΔB), between 1.7 and 10 for real flare stars (Fig. 6.26). This fact is an additional indication that the observed sharp fluctuations of a star's brightness during flares have nothing in common with variations of the Planck temperature of the star.

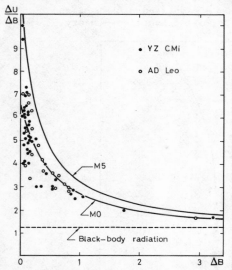

Fig. 6.26. Comparison of observed values of $\Delta U/\Delta B$ versus ΔB for a number of flares of AD Leo and YZ CMi. Full-drawn lines--theory with fast-electron hypothesis; broken line-- theory for Planck radiation.

8. THE DEPENDENCE OF FLARE AMPLITUDE ON STELLAR SPECTRAL CLASS

The flare phenomenon occurs in stars of late spectral class--later than K5. Any theory which tries to find the cause for stellar flares must take into account this firmly established fact.

Earlier, in section 1 of Chapter 6, a number of computations were made to determine theoretical flare amplitudes in UBV radiation for various values of the effective temperature of the star, i.e. essentially for various spectral classes (Tables 6.3, 6.4). A rapid glance at these results shows the flare amplitude to decrease from late-type stars (M6) to early classes (G5). These results can conveniently be represented in graphs showing flare amplitude versus effective temperature or spectral class of the star. This is given in Fig. 6.27 (Gaussian electron distribution with $\gamma_o = 3$ and $\sigma = 2$).

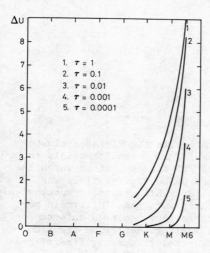

Fig. 6.27 should be considered as one of the important results of the theory of flares which can be deduced from the fast electron hypothesis. According to this theory, even in U radiation flares cannot be found in stars of classes O, B, A, F. Faint flares with amplitudes of the order 0^m5 or less may be discovered in G0 stars. But the real domain of flare stars starts at classes G5-K0. After that the flare amplitude increases strongly towards later classes, reaching 8^m to 9^m in U radiation for class M5-M6.

Let us now consider the observational results. The total number of flare stars known up to 1970 in stellar associations and aggregates is about 600, and in the neighborhood of the Sun about 50. The spectral classes of the latter are well known; they all lie between K2 and M6. Unfortunately there are very few data about spectral classes of flare stars in associations. In the Orion association, for instance, spectral classes are known for only 21 out of 254 flare stars, or about 8% only. In this respect, the situation is somewhat better in the other associations and aggregates: e.g. in the

Fig. 6.27. Theoretical relations between flare amplitude in U radiation and spectral class of flare star for various flare strengths τ, in the case of the fast-electron hypothesis.

Pleiades spectral classes are known for nearly 40% of the total number of flare stars recorded up until 1970. Spectral classes are known for many flare stars in the clusters Praesepe and Coma Berenices, which do not have many members. As a result the total number of flare stars with known spectral classes is more than a hundred. All these data are shown in Fig. 6.28, giving the observed flare amplitudes versus the star's spectral class. The theoretical curve, taken from the previous figure for $\tau = 1$ is also shown.

The conclusions to be drawn from comparison of Figs. 6.27 and 6.28 are evident: the agreement between the theoretical amplitude-spectrum relation and the observations is good. The observational material used is sufficiently large to exclude the influence of accidental errors.

The theoretical curve given in Fig. 6.28 follows from the very nature of the fast

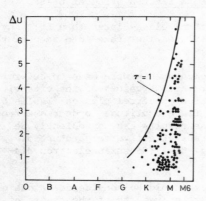

electron hypothesis. Neither the form of
the energy spectrum of the fast electrons
nor the value of the energy of these electrons has any influence at all on the
character of this relation. Therefore the
good agreement between observational and
theoretical "amplitude-spectrum" functions
in this case is of great value.

Fig. 6.28. Observed relation between
flare amplitude in U rays and spectral
class of flare star (from data in
associations).

9.   INTERPRETATION OF A FLARE OF H II 1306

During a program of colorimetric observations of stars in the Pleiades cluster
Johnson and Mitchell observed a flare of the star H II 1306 simultaneously in U,
B, V [27]. The flare reached its maximum in about 2.5 minutes, after which the
decrease began, which lasted for somewhat more than an hour. The flare was
recorded with one spectrophotometer, the light-filters which transmit successively
the U, B and V bands were alternated. Therefore there were gaps in certain parts
of the observations. In two cases, in U and V, these gaps occurred unfortunately
near the maxima of the curves, which makes it difficult to determine the true
values of the amplitudes in these radiations. Nevertheless one may find with a
sufficient degree of accuracy from these curves: $\Delta U \approx 3.7$, $\Delta B \approx 1.7$ and
$\Delta V \approx 0.65$.

In normal conditions the star H II 1306 has the following color indices (with
$m_V = 13.39$): $B - V = + 1^m35$ and $U - B = + 1^m18$. At the time of flare maximum
$B - V = + 0^m50$ and $U - B = - 1^m07$. From this data we can try to find the energetic parameters of fast electrons corresponding to the observed amplitudes of
this flare.

First we must know the effective temperature of the star H II 1306 in normal
conditions. Herbig [28] estimates the spectral class of this star at dK5 (e),
apparently from the structure of the absorption lines. The corresponding temperature is T = 4200 K. Moreover the relatively high value of B - V in normal conditions for this star is + 1^m35, corresponding to a spectral class M0 - M1 rather
than K5. This means that the effective temperature of H II 1306 must be considerably lower than 4200 K. A further argument for this assumption may be found
in the fact that, from measurements by Mumford [29] another flare star, EV Lac,
has in the normal state $B - \dot{V} = + 1^m38$, nearly the same as the value of B - V for
H II 1306. At the same time EV Lac belongs to a spectral class dM4.5e with
effective temperature about 2800 K.

Thus there are signs of abnormality in the energy distribution in the continuous
spectrum of the star H II 1306 even in its quiescent state. This distribution

corresponds to an effective temperature of the order of 4200 K, if we judge from the star's spectral class, and an effective temperature of about 2800 K if we consider its color. This circumstance should be remembered each time we compare theory with observations.

Computations of the theoretical amplitudes of the increase in brightness for a flare of a K5 star (T = 4200 K) were made for the following two models:

Model I--monoenergetic electrons with $\tau = 1$ and $\gamma^2 = 10$.

Model II--electrons with normal distribution with $\gamma_0 = 3$, $\sigma = 2$ and $\tau = 0.6$.

Moreover, a third model was computed--a star of class M5 (T = 2800 K) and monoenergetic electrons with $\gamma^2 = 10$ and $\tau = 0.01$.

The computed values ΔU, ΔB and ΔV found for these models are shown in Table 6.10.

TABLE 6.10   The Interpretation of Flare of Star H II 1306

| | Observations | Star of Class K5 | | Star of Class M5 |
		Model I	Model II	Model III
ΔU	$3^{m}\!.7$	$2^{m}\!.8$	$3^{m}\!.4$	$4^{m}\!.2$
ΔB	1.7	1.1	1.8	1.6
ΔV	0.65	−0.1	0.6	0.4

Comparison of these values with observational data (second column) shows that the agreement is best with model II, where the energy spectrum of the fast electrons is represented by a normal distribution law with the parameters $\gamma_0 = 3$ and $\sigma = 2$ and the strength of the flare corresponds to $\tau = 0.6$.

Thus there exist real parameters of the fast electrons with which the observed flare amplitudes of the star H II 1306 can be explained, even though the character of the energy distribution in the continuous spectrum of this star in normal conditions is not quite determined.

The flare observed in 1958 by Johnson and Mitchell of H II 1306 proved to be the strongest so far. Later (in 1963 and 1965) three more flares of this star were recorded by Haro and Chavira [29] with smaller amplitude in U, of the order $0^{m}\!.5$.

10.   UNUSUAL FLARE OF THE STAR HARO 177

On December 27, 1965, Haro observed an exceptionally strong flare of a star lying near the Large Orion Nebula. This star, Haro 177 = Parenago No. 1323, is at the limit of visibility of the Palomar Sky Atlas: its brightness in U is estimated at $19^{m}\!.7$.

Fig. 6.29 gives successive stages of the development of the flare of this star [30]. On the first plate it was not visible with 15-minutes exposure, taken through ultraviolet filter with the 26-31 inch Schmidt telescope of the Tonanzintla Observatory. On the second plate all five images appear with increasing brightness. However, the flare could not yet be said to have reached its maximum brightness. After about

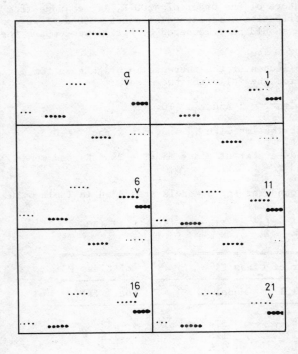

Fig. 6.29. Six successive stages in the development of a flare of star Haro 177 in Orion with flare amplitude $8\overset{m}{.}4$ in U radiation.

five minutes a third plate was obtained containing five images, again with exposure times of 15 minutes. By comparing only the second and the third plates one may conclude that about one hour passed between the first appearance of the flare and its maximum. After that the star maintained for one hour and a half nearly constant brightness at maximum, after which it began to decrease (fourth plate). Haro believes that in this case the fall in brightness may have lasted 6 to 7 hours, after which the star reached its original brightness. The amplitude of the increase in brightness was found to be $8\overset{m}{.}4$ in U radiation, i.e. during one hour the brightness increased 2300 times (!). Later, in 1976, a communication [32] appeared about the observation of flares of three stars, members of the Pleiades cluster T-18 - A106, T26 and T153 with flare amplitudes $\Delta U = 8\overset{m}{.}5$, and in one case, for the star T53 b - P2, a flare (on October 2, 1972) with amplitude even with $\Delta U > 8\overset{m}{.}5$. These are as yet the largest amplitudes ever recorded for flares of typical flare stars.

Since these observations offer a possibility of checking one or another theory of flares, such exceptional cases are of great value and certainly cannot be ignored. Unfortunately for the flare of Haro 177 we have only one observational quantity--the amplitude of the flare. We do not know, for example, the spectral class of this star. Therefore in this case, when trying to compare theory and observation we must restrict ourselves only to a comparison of the observed amplitude with its theoretical value. This comparison comes out in favor of the theory, since this predicts an amplitude in U of the order of 10^m or more (cf. §1) for M5-M6 stars.

The theoretical limiting values of the strength of a flare may be maintained for rather long times, and in any case for as long as τ does not differ from unity. This theoretical conclusion does not contradict the case of Haro 177, where the maximum brightness lasted for about one hour. In contrast the following fact related to the thermal hypothesis of the origin of flares is remarkable. For an explanation of the increase in brightness by a factor 2000 to 3000 by a thermal hypothesis the area of the star's photospheric layers must increase by the same factor in the course of one hour. This is equivalent to an increase of the radius of the star by a factor of nearly 50 during one hour. Assuming the star's radius to be equal to that of the Sun we find that the linear expansion velocity of the photospheric layers must be very large--of the order of 10 000 km s^{-1}. Earlier attention has been drawn to this fact by V. A. Ambartsumian, who came to the conclusion that in order to explain the observed rates of increase of flares of UV Cet stars by a thermal hypothesis an expansion velocity of the star's photospheric layers of the order of 50 000 km s^{-1} is required (on the condition that during the flare the temperature of the photospheric layers remains unchanged). At present we have no observational data about the expansion of the star's photospheric layers during the flare, let alone about such enormous velocities.

11. A FLARE OF A STAR OF CLASS K

The number of flare stars belonging to spectral class K is relatively small. The theoretical flare amplitudes get smaller and smaller from stars of class M to stars of class K. Moreover the values of these amplitudes for K stars are sensitive to the energy spectrum adopted for the fast electrons.

As an example, Table 6.11 gives the theoretical limiting values of the amplitudes ΔU, ΔB, ΔV for K5 stars (T = 4200 K) for a Gaussian distribution of the electrons (γ_0 = 3, σ = 2 and τ = 0.6) as well as for the case of monoenergetic electrons (γ^2 = 10 and τ = 0.6). These data are given without corrections for the deviations of the true distribution of the continuous emission in the star's spectrum from the Planck distribution for the given stellar temperature.

TABLE 6.11 Maximum Theoretical Flare Amplitudes in UBV Radiation for Stars of Class K5

	Gaussian distribution of the electrons	Monoenergetic electrons
ΔU	3^{m}4	2^{m}8
ΔB	1.8	1.1
ΔV	0.6	−0.1

In the Pleiades there are 31 flare stars of class K. Their spectral class distribution and the values of the maximum amplitudes recorded for stars of a given subclass are given in Table 6.12. Comparing these data with Table 6.11 we note that the observations agree with the theory only in the case of the Gaussian distribution.

TABLE 6.12 Maximum Observed Amplitudes of Flares in U Radiation for K Stars in the Pleiades

Spectral subclass	K3	K4	K5	K6	K7
ΔU_{max}	1^{m}5	1^{m}7	3^{m}5	3^{m}0	3^{m}4
Number of stars of given subclass	3	1	7	6	14

In Orion there are few flare stars of class K. For four among nine such stars the flare amplitude is more than 1^{m}8 (in U), for one case the maximum amplitude was found to be equal to 3^{m}2.

The data about the remaining clusters and associations are as follows. In the
Taurus Dark Cloud (TDC) there are only two stars of class K6e which have flared up
with amplitudes of 0ᵐ8 and 0ᵐ5 in U radiation. In NGC 2264 there is only one K0
star with ΔB = 0ᵐ7. In Praesepe and Coma Berenices there are no flare stars of
class K.

Thus the fast electron hypothesis can explain without additional assumptions the
observed flare amplitudes of stars belonging to class K. Moreover the comparison
of the observations with theoretical computations for K flare stars suggests that
the most probable energy spectrum of the fast electrons is close to a Gaussian.

12. A FLARE OF A STAR OF CLASS A

The flare amplitudes in ultraviolet and photographic light decrease rapidly from
late to earlier spectral classes. Let us consider now what happens in the case
of a theoretical "flare" of a star of class A0 with effective temperature T =
10 000°.

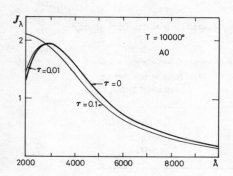

Fig. 6.30. Theoretical energy distribution for a flare of a class A0 star
with strength τ = 0.1 and τ = 0.01
(monoenergetic electrons, γ^2 = 10).

In Fig. 6.30 the curves of the theoretical
flare, corresponding to strengths τ = 0.1
and 0.01, for a star of class A0 are given.
The normal spectrum of the excited star
(τ = 0) is represented by a Planck curve
(without taking into account the absorption
of the continuous hydrogen radiation). It
follows from the figure that a flare caused
by inverse Compton effect on fast electrons
in the case of an A0 star is practically
impossible; at best the brightness may decrease in the U and B radiations, i.e. a
negative flare. However, even for τ = 0.1
the amplitude of the flare is very small,
of the order 0ᵐ1, not to mention the fact
that a flare of an A0 star with strength
τ = 0.1 either cannot happen at all or is
extremely improbable. For τ ∿ 0.01 the
flare cannot be detected (negative amplitudes of the order 0ᵐ01).

Thus the hypothesis of the inverse Compton effect with fast electrons excludes the
possibility of finding flares for stars of early spectral classes in photographic
and even ultraviolet light. In principle a flare may be discovered in such stars
in the wavelength range shorter than 2500 Å.

13. AN ULTRAVIOLET FLARE OF W UMa

During ordinary photoelectric observations of the well-known eclipsing variable
W UMa Kuhi [31] found a strange outburst of one of the components of this pair
(or of this system itself)--at a wavelength of 3300 Å with an amplitude of about
1ᵐ5 in a spectral region of 50 Å, "without a corresponding increase in brightness
at visual wavelengths". Maximum brightness was reached in about two minutes after
the beginning of the flare; afterwards the brightness decreased relatively slowly.
The total duration of the flare was about seven minutes.

The strangeness of this flare lies in the fact that both components of the system
W UMa are stars of class F8, unusually early for flare stars.

According to the fast electron hypothesis flares may be found in stars of early spectral classes in the region of short wavelengths, shorter than 3000 Å. With regard to class F, in principle flares of such stars may be found in the interval 4000–3000 Å; the whole question is how much the expected amplitude of the increase in brightness will be. The computations show that for an effective temperature of T = 6200 K, there exist no real values of γ and τ--for the case of monoenergetic electrons, and of γ_o, σ and τ--in the case of electrons with a Gaussian energy distribution, for which the observed amplitude of 1^m5 near 3300 Å can be explained. The maximum possible amplitude at 3300 Å is found equal to 0^m93 in the case of mono-energetic electrons, and 1^m1 in the case of electrons with a Gaussian distribution.

The case of the flare of W UMa leading to such a difference between theory and observations is unique. It is possible that here the effects of the absorption lines play a role, where the actual level of the continuous spectrum at 3300 Å, and within a band of 50 Å, is lower than what we expect in the case of a Planckian radiation.

It is also possible that this difference is simply due to imperfections of the theory; for a strong flare, i.e. for $\tau \sim 1$, our formulae give only a lower limit for the intensity of the Compton radiation leaving the fast electron cloud.

Finally the degree of reliability for this flare is unknown--up until now it is the only case for stars of class F8. However, the question remains open until the possibility of flares from stars of class F has been confirmed by more observations.

However, flares of stars of class F8 and even earlier are possible in principle, if the concentration of the fast electrons is very high; in such cases the additional energy appears as a result of non-thermal bremsstrahlung of fast electrons (cf. Chapter 8).

14.   THE PROBABLE VALUE OF THE ENERGY OF THE FAST ELECTRONS

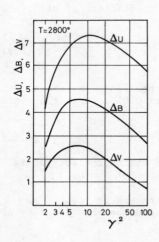

Fig. 6.31. The problem of finding the probable value of the energy of the fast electrons ("amplitude method").

The values of the flare amplitude depend in particular on the energy of the fast electrons. In this connection the problem arises of finding the probable value of the energy of the fast electrons. This may be done by various methods. One of them is to find the value of the energy of a fast electron for which the theoretical amplitude will be equal to the maximum observed amplitude at the wavelength considered. For this it is necessary to construct curves of maximum theoretical amplitudes versus electron energy γ. Such curves for stars of class M5 are shown in Fig. 6.31 (for the case of monoenergetic electrons).

This diagram shows that the maximum flare amplitude in U radiation will be obtained for $\gamma^2 \sim 10$, in B for $\gamma^2 \sim 8$, and in V for $\gamma^2 \sim 5$. This difference in the values of γ^2 corresponds only with a small interval of the electron energy--1.2 to 1.6×10^6 eV. Here we come to the conclusion that it is possible to obtain the maximum values of the observed flare amplitudes in U, B and V for $\gamma^2 \sim 10$, i.e. for energies of the fast electrons of the order 1.5×10^6 eV. The same result is also obtained in the case where the energy spectrum of the fast electrons is represented by a Gaussian curve. Another method for finding the probable value of the fast electron energy based on an analysis of the color characteristics of flares will be considered in Chapter 7.

15. ENERGY LOSSES BY FAST ELECTRONS AND BY THE STAR DURING THE FLARE

The total strength of a single flare of a star is determined by the total energy P of the fast electrons and is equal to

$$P = 4\pi R_*^2 \varepsilon \frac{\tau}{\sigma_s} ,$$ (6.6)

where ε is the energy of one fast electron, $\tau/\sigma_s = N_e$ is the number of fast electrons in a column above the star's photosphere with cross section 1 cm^2, and R_* is the star's radius.

If we take for the coefficient of Thomson scattering $\sigma_s = \gamma^2 \sigma_T = 6.65\times10^{-24}$cm^2 (cf. Ch. 3), $R \approx 0.1\ R_\odot$, and $\varepsilon = \gamma\ mc^2 = 1.5\times10^6$ eV, we find from (6.6):

$$P \approx 2\times10^{38}\ \tau\ \text{erg.}$$ (6.7)

From this amount of energy only a fraction q is transmitted to the photons in the form of Compton losses. Evidently q will be larger the longer the fast electrons are kept near the star's boundaries. Let us first determine the quantity q.

Let E_0 be the total excess energy radiated by the star in all frequencies during the entire duration of the outburst, from the moment of its appearance to the moment of total disappearance. Evidently then E_0 can be determined by integration of the lightcurves over time and for all wavelengths, that is

$$E_0 = \int\int E_\lambda(\tau)d\lambda dt .$$ (6.8)

The fraction of the total energy of the electrons, q, that is emitted in the form of radiation due to the inverse Compton effect, a kind of "coefficient of efficiency" will be

$$q = \frac{E_0}{P} = \frac{\int\int E_\lambda(\tau)d\lambda dt}{4\pi R_*^2 \varepsilon N_e} .$$ (6.9)

The value of E_0 lies on the average between 10^{30} and 10^{31} ergs [33]. At the same time the average value of τ for flare stars is of the order 0.001. With these data we find from (6.9):

$$q \approx 10^{-3}\ \text{to}\ 10^{-4},$$ (6.10)

which means that only 0.01% or 0.1% of the total energy of the fast electrons is transformed into radiated energy by the flare.

The minimum value of q--here called q_m--is determined by the energy loss of an electron in one inelastic collision with a photon. It depends on the frequency of the photon and is equal to

$$q = \frac{h\nu(\gamma^2-1)}{\gamma mc^2} \approx \gamma \frac{h\nu}{mc^2} ,$$ (6.11)

or, with $\gamma \sim 3$ and $h\nu \sim 2$ eV we find $q_m \approx 10^{-5}$ in one scattering process. Comparing this with (6.10) we find that during the flare a fast electron suffers on the average from ten to one hundred inelastic encounters with photons before leaving the star. The total energy which is lost by the electron in these encounters is of the order of one hundred electron volts--a negligibly small quantity as compared with the original energy of the electron ($\sim 10^6$ eV). Therefore we may conclude that the fast electrons leave the star with practically their entire original energy (neglecting ionization, magnetic retardation and other forms of losses which the electrons may suffer on the way).

Let us determine the whole amount of energy emitted by the star in the form of flares during the entire period of its "flare" activity, i.e. for 10^8 years, say. Assuming that the star suffers three flares per day, or 1000 flares per year with an average strength $\tau \sim 0.001$ we have

$$\text{energy emitted per year:} \quad \sim 10^{38} \text{ ergs}$$

$$\text{energy emitted in } 10^8 \text{ years:} \quad \sim 10^{46} \text{ ergs.}$$

With a bolometric luminosity of a star of class M5, equal to $L(M5)/L_\odot \sim 10^{-2}$, the total energy losses of the star during 10^8 years by normal light radiation are $\sim 10^{47}$ ergs.

These computations are very approximate since they assume a constant rate of the "flare" activity as well as a constant radiative power for the star during 10^8 years. Nevertheless the results show that during the normal "flare" activity the energy losses in the form of emission of fast electrons are of the same order as the losses in the form of normal stellar radiation.

The appearance of fast electrons in the outer regions of the stellar atmosphere is accompanied by the ejection of gaseous matter. A lower limit to the mass of this matter can be determined from the condition that the number of protons is equal to the number of fast electrons. The number of fast electrons ejected in one flare is equal to

$$N_e^* = 4\pi R_*^2 N_e \sim 10^{44} \tau. \tag{6.12}$$

The total mass of the material ejected by the star under the above conditions--1000 flares per year with a strength per flare $\tau \sim 0.001$--will be 10^{-13} $M_\odot$ per year or 10^{-5} $M_\odot$ in 10^8 years. This is many orders smaller than stellar mass losses due to various other causes in the evolution of a star.

16. THE NEGATIVE INFRARED FLARE

The fast electron hypothesis predicts the possibility of a "negative infrared flare" [34]. During the stellar flare we observe in the short wavelength region only an increase of its brightness. But if the flare is due to fast electrons, which in inelastic collisions with infrared photons convert them into short wavelength photons, then as a result the total number of infrared photons of the normal photospheric radiation of the star must decrease. This decrease occurs strictly synchronously with the increase of the stream of short wavelength photons and hence we can speak of a negative infrared flare.

Quantitatively the phenomenon of the negative infrared flare may be characterized by at least two parameters: a) the amplitude of the negative flare, and b) the critical wavelength where the decrease of the stellar radiation begins.

The amplitude of the negative infrared flare depends directly on the amplitude of the ordinary (positive) flare of the star in the short wavelength region. However, since in normal conditions the intensity of the radiation of these stars in the far infrared is very large, the relative values of the negative infrared flares will be extremely small. This may cause definite difficulties when one tries to detect them by observations.

The second parameter is the wavelength λ_0 with zero flare amplitude in the star's continuous spectrum. To the short wavelength side of λ_0 lies the region of the positive flare and to the long wavelength side that of the negative flare. Numerically λ_0 lies at the point of intersection of the two spectra.

Calculations show that for flares of stars of class M5-M6 the negative flare is to be expected in the region with $\lambda \gtrsim 8000$ Å for very strong flares, and in the region $\lambda \gtrsim 10\ 000$ Å for faint flares. In flares of stars of early class, K5-K0, the negative flare must be looked for in the photovisual region of the spectrum with $\lambda \gtrsim 5000$ Å.

With regard the amplitudes of the negative flares, these are extremely small, of the order $0^m.1$, even for strong flares (Table 6.13). This makes it nearly impossible

TABLE 6.13 Theoretical Values of Amplitudes of Negative (−) and Positive (+) Flares at 8400, 10 000 and 20 000 Å for $\tau = 0.1$

Spectral class	8400 Å	10 000 Å	20 000 Å
M6	$+0^m.4$	$+0^m.1$	$-0^m.1$
M5	+0.1	0	-0.1
M0	0	-0.1	-0.1
K5	-0.1	-0.1	-0.1
K0	-0.1	-0.1	-0.2

to discover the "negative flares" at photographic wavelengths. Furthermore, in relatively faint flares the amplitude of the negative flare hardly depends on the star's spectral class. Finally in the case of stars of class M0-K0 the amplitude of the negative flare is nearly constant in the interval from 8400 to 20 000 Å.

It seems that observations with the photoelectric method at infrared wavelengths should be considered as one possibility for discovering negative infrared flares.

The negative infrared flare is one of the important theoretical predictions of the hypothesis of fast electrons. No other physical phenomenon is known to us which gives a positive star flare in one part of the spectrum and a negative flare in another part. Therefore it is highly important to make special observations of flare stars with the purpose of discovering the negative flare effect at infrared wavelengths. Such a result could be decisive for the theory.

17. RESULTS OF OBSERVATION OF STELLAR FLARES AT INFRARED WAVELENGTHS

Already in 1959-1960 Haro made special observations at the Tonanzintla Observatory in order to study the behavior of stellar flares in the infrared spectral region. The photographic plates used by him were sensitive from 7200 to 9200 Å, with a

maximum at 8400 Å. The plates were obtained with the 26-31 inch Schmidt telescope
by the chain method, with ∿5 images per star, and with exposure times of 10-15
minutes for each image. In all 66 infrared plates were obtained, all for the Orion
association region. Careful scrutiny of these plates did not show any positive
flares in infrared wavelengths.

Considering that in Orion in U and B radiation on the average one flare is detected
every three plates (though there are also cases where flares of three stars were
discovered simultaneously on one plate), we must say that the negative results of
this search for <u>positive</u> flares at infrared wavelengths cannot be due to insufficient
observational material. Thus the fact that no ordinary positive flare has been
found in infrared light should get special importance.

However, at the time when Haro performed these infrared observations of flare stars
nothing was known about the"negative" infrared flare. Therefore an observer looking
for positive infrared flares could have overlooked negative flares, particularly
that the expected amplitudes are extremely small. Thus, Haro's plates were re-
examined in the summer of 1968. Unfortunately the poor quality of most of the plates
did not allow luminosity fluctuations of ∿ 0ᵐ5 to be detected. From this material
the possibility of a negative infrared flare cannot be proved or disproved.

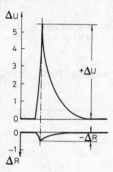

The problem of the negative infrared flare can
be solved only by photoelectric observations of
high accuracy (up to 0ᵐ01 or less). It would
be advantageous to make synchronous observations
of a star in the region of the positive flare
(ultraviolet or blue light) and in the region
of the negative flare (infrared light). The
expected results should look more or less like
Fig. 6.32.

Fig. 6.32. Lightcurves of an
ordinary--positive (above) and a
negative (below)--flare (model).
Maximum of negative flare is
reached at maximum of positive
flare.

18.   THE FREQUENCY OF NEGATIVE INFRARED FLARES

The probability of a negative infrared flare with amplitude of the order 0ᵐ1 must
be equal to the probability of a flare in U with amplitude 7ᵐ to 8ᵐ, or to the
probability of a flare in B with amplitude 4ᵐ to 6ᵐ. For example, the flare frequency in Orion with amplitude 5ᵐ to 6ᵐ in B is equal to 0.0087 flares/hour [35],
and in the Pleiades for flares in U with amplitude 7ᵐ to 8ᵐ it is 0.005 flares/hour.
From this we find for the frequencies of the negative infrared flare: in Orion one
flare per 120 hours of observations, and in the Pleiades one flare in 200 hours of
observations.

The frequency of flares in B and U radiation with amplitude larger than 0ᵐ5 in
Orion and the Pleiades is about the same and equal to 0.35 flare per hour, i.e. on
the average a flare occurs every three hours. This means that the frequency of the
negative infrared flares must be nearly 40 times less than that of the ordinary

positive flares. Therefore 160-200 hours of continuous photoelectric observations are required say on YZ CMi, in order to catch one negative infrared flare with amplitude of the order 0ᵐ1.

19. FLARE AMPLITUDES IN THE EXTREME ULTRAVIOLET

There is no doubt that sooner or later observations of flare stars will be made from outside the earth's atmosphere. One of the most important problems which can then be studied is the structure of the spectrum of the stellar flare in the far ultraviolet region, i.e. at wavelengths shorter than 3000 Å. In this connection the question arises--how large the amplitudes of stellar flares in this spectral region are, as expected from the fast electron hypothesis.

Here we restrict the problem in determining the theoretical flare amplitudes at 3000, 2500 and 2000 Å.

The flare amplitude at a given wavelength λ may be determined from the following relation:

$$\Delta m_\lambda = 2.5 \log C_\lambda(\tau, \gamma, T). \tag{6.13}$$

where the function $C_\lambda(\tau, \gamma, T)$ is determined--in the case of monoenergetic fast electrons--from (4.28) and its numerical values are given in Table 4.2. Numerical

TABLE 6.14   Theoretical Flare Amplitudes Δm_λ at Wavelengths 3000, 2500 and 2000 Å

Spectr. class	Wave-length, Å	τ				
		1	0.1	0.01	0.001	0.0001
M6	3000	11ᵐ5	10ᵐ8	8ᵐ5	6ᵐ1	3ᵐ6
	2500	15.4	14.5	12.2	9.7	7.2
	2000	21.0	20.0	17.8	15.4	12.8
M5	3000	9.5	8.8	6.6	4.1	1.8
	2500	12.8	12.2	9.8	7.3	4.9
	2000	18.0	17.0	14.9	12.3	9.8
M0	3000	6.0	5.2	3.0	1.0	0.1
	2500	8.5	7.7	5.5	3.1	1.0
	2000	12.4	11.5	9.3	6.8	4.4
K5	3000	4.2	3.5	1.5	0.3	0.0
	2500	6.3	5.6	3.4	1.2	0.2
	2000	9.5	8.9	6.6	4.1	1.8

values of flare amplitudes at these wavelengths found from (6.13) are given in Table 6.14. These amplitudes are very large, in particular for stars of class M6-M5, of the order of 10^m and more, even for small values of τ. However, this large amplitude is not due primarily to the absolute increase of the radiation during the flare, but mostly to the extremely low radiative power of the star at these short wavelengths in its undisturbed state.

In order to obtain some idea about the real increase in brightness of the star in the wavelength region shorter than 3000 Å, we can determine the relative flare brightness, i.e. the difference between the brightness of the star in the region 2000-3000 Å and the region 4000-5000 Å, during the burst.

We can write:

$$\Delta m(2000-3000) = m(2000-3000) - m(4000-5000) = 2.5 \log \frac{\int_{\lambda_1}^{\lambda_2} B_\lambda(T) C_\lambda(\tau,\gamma,T) d\lambda}{\int_{\lambda_3}^{\lambda_4} B_\lambda(T) C_\lambda(\tau,\gamma,T) d\lambda}, \qquad (6.14)$$

where λ_1 = 2000 Å, λ_2 = 3000 Å, λ_3 = 4000 Å, λ_4 = 5000 Å.

The values $\Delta m(2000-3000)$ thus found are given in Table 6.15. The sign minus or plus indicates that in the interval 2000-3000 Å the flare star is brighter or fainter by the amount given relative to its brightness in the interval 4000-5000 Å.

TABLE 6.15. Theoretical Relative Brightness Δm of a Flare Star in the Interval 2000-3000 Å as Compared with Its Brightness in the Interval 4000-5000 Å

Spectr. class	τ				
	1	0.1	0.01	0.001	0.0001
M6	-1.9	-1.9	-1.8	-1.0	+5.7
M5	-2.0	-2.0	-1.6	-0.4	+4.9
M0	-2.2	-1.9	-0.8	+1.3	+3.5
K5	-2.4	-1.7	-0.5	+1.7	+2.6
K0	-2.2	-1.3	+0.4	+1.8	+2.0

From the data in Table 6.15 it follows that typical flare stars will be brighter in the interval 2000-3000 Å by about one magnitude, compared to their brightness in photographic light.

20. THE EFFECT OF THE LUMINOSITY DECREASE OF THE STAR BEFORE THE FLARE

Flare amplitudes have a theoretical maximum at $\tau \sim 0.5$. For values of $\tau \gg 1$ the fraction of the photospheric radiation passing through the medium of fast electrons, approximately equal to

$$\frac{1}{1+\tau}, \qquad (6.15)$$

will be extremely small. In this case practically all of the radiation will be reflected back towards the photosphere since the fraction of the radiation reflected is equal to (pure scattering):

$$\frac{\tau}{1+\tau}. \qquad (6.16)$$

The fast electrons are emitted in some way from the primary matter ejected outwards from the star's interior. It is possible that an electron cloud could play the role of a dark screen, preventing the passage of normal photospheric radiation. In such cases the normal brightness of the star at all wavelengths may decrease, immediately before the flare.

A few cases of such a decrease of stellar brightness before the flare have been observed. This phenomenon occurred for instance during a number of flares of UV Cet (19.IX.65, 20.IX.65, 23.IX.65) [17], EV Lac (27.IX.71) [36], AD Leo (15.XII.73, 26,XII.73) [37], V 1216 Sgr (22.VI.74) [38], YZ CMi (2.XII.75) [39], etc. Fig. 6.33 shows an example of a lightcurve of one flare of EV Lac (9.X.73), recorded in red light ($\lambda_{eff} \approx 5400$ Å), where a real decrease of the stellar brightness immediately before the beginning of the flare is seen [40]. The value of the decrease in brightness in all cases was small--of the order of $0^{m}\!2$ to $0^{m}\!3$ (in B radiation).

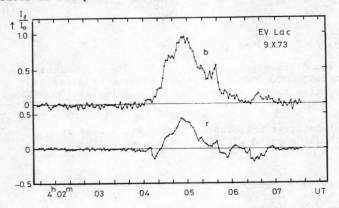

Fig. 6.33. Example of lightcurve with decrease in brightness before a flare [40] in region with λ_{peak}=5375 Å (curve r). The curve b corresponds to λ_{peak}=4300 Å.

The decrease of brightness before the flare does, how-ever, not last long; the cloud expands very rapidly, as a result of which τ de-creases. Essentially the flare starts only when τ becomes of order unity. Clearly this effect depends on the ratio between the rate of expansion of the cloud and the rate of produc-tion of new groups of fast electrons. If the process of production of fast electrons stops very quickly, then the cloud, expanding with high velocity will reach a state corresponding to $\tau \sim 0.5$ in a fraction of a second.

In certain cases even relatively faint flares may reach the theoretical limiting amplitude, but the time during which the flare is at its maximum is very short. In order to observe such flares near the theoretical maximum amplitudes one must have photometric equipment with very short time constants.

Thus the fast electron hypothesis predicts also the following two effects:

a) in certain cases the star's brightness may decrease at all wavelengths immediately before the flare;

b) most flares must show exceedingly fast increases in brightness, with amplitudes near the theoretical limit. In such cases the decrease of the star's brightness after maximum must proceed for some time at nearly the same rate as the increase before the maximum. In other words, in the most general case the lightcurve of the flare must have a very sharp and very narrow peak. This means that the average amplitude of a given series of flares will be larger, the shorter the time constant of the recording apparatus is.

Special experiments for checking these two predictions have not been made, but in the observational material available at present there are indications that these two predictions are fulfilled

21. SHORT-LIVED FLARES

In connection with the second of the above predictions the flare lightcurves obtained
by the Italian astronomers (Cristaldi, Rodono et al. [8]) are of particular interest.
Working with a spectrophotometer with a time constant nearly one order of magnitude
smaller than usual (~0.5 s) they recorded lightcurves of a large number of flares
of UV Cet and EV Lac with very high time resolution, thus showing the fine structure
of these curves. These lightcurves are shown in Figs. 6.34 to 6.37. The sharp and

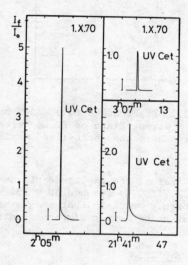

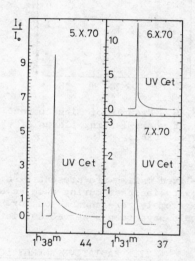

Fig. 6.34. Lightcurves of three short-
lasting flares of UV Cet in B radiation,
recorded with time resolution 0.5 s.
Vertical lines indicate measuring
errors [8].

Fig. 6.35. Lightcurves for three short-
lasting flares of UV Cet (cf. Fig.
6.34).

narrow peaks mentioned above are clearly seen. These peaks could not be found in
lightcurves obtained earlier with the aid of slower recording apparatus. These
peaks are not resolved and could even be narrower.

The large increase in the time resolution of the lightcurves has revealed qualitatively new and important properties of stellar flares. For example there exist
very shortlived flares, lasting for a few seconds, which have quite high intensities (cf. Figs. 6.34, 6.35). Furthermore the actual flare frequency is found to
be considerably higher than was thought before--at times it reaches nearly one
flare every two minutes (cf. Fig. 6.36). The better time resolution of the lightcurves also gave important information about the dynamics of the flare itself, and
finally about its nature. Particularly, the fact that the decrease of the brightness after maximum flare takes place at nearly the same rate as its increase was
due to observations with short time constants. From the lightcurves given in the
figures it follows that a decrease in brightness by a factor of six after maximum
can occur in approximately 4 to 5 seconds.

Fig. 6.37 is an eloquent illustration of the fact that very sharp and extremely

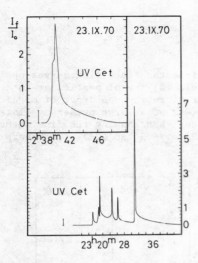

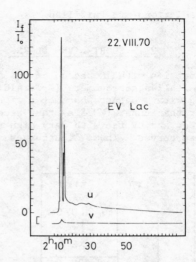

Fig. 6.36. Example of lightcurves (of a group) of short-lasting flares of UV Cet (cf. Fig. 6.34).

Fig. 6.37. Lightcurves in U and V radiation of a strong flare of EV Lac with an extremely sharp and short-lived maximum [8].

short-lived maxima are present also in very strong flares; the amplitude of the luminosity increase during such an unusual flare (22.VIII.70) reached $\Delta U \approx 5^m.2$ (!) --an extremely rare phenomenon for EV Lac. It is highly unlikely that such a maximum could have been observed earlier, with equipment using longer time constants.

22. RADIATIVE ENERGY LOSSES THROUGH THE FLARES

What is the fraction of the radiative energy emitted by the star in the form of flares? And, in general, does the absolute value of the radiative energy emitted in flares depend on the absolute luminosity of the star itself?

Let us start with the latter question--to find the absolute value of the energy E_f, emitted by the star on the average in one flare. Earlier in Chapter 1 (§7) we introduced the term "equivalent time" or "flare integral" P for one flare, which represents the time (in seconds) during which the undisturbed star with constant emission I_* (erg s^{-1}) radiates as much energy as during the entire duration of the flare. This quantity is determined from the observations by integrating the lightcurve for each flare separately. On the other hand, all the flares for a given star in a given wavelength region are characterized by a certain average value of the flare amplitude, which is found from the observations (Table 1.6). For each average amplitude, say in U radiation, $\Delta\overline{U}$ will correspond to some mean equivalent time $P(\Delta\overline{U})$. Then we may write for the energy $E_f(U)$ emitted by this star in U radiation on the average during one flare:

$$E_f(U) = P(\Delta\overline{U})I_*(U) , \qquad (6.17)$$

where $I_*(U)$ is the radiation intensity of the undisturbed star. $P(\Delta\overline{U})$ can be found empirically. The most homogeneous material for this purpose is contained in the work by Moffett [7] for a group of flare stars of UV Cet type. From this material the empirical dependence $P(\Delta\overline{U})$ on $\Delta\overline{U}$ was found (Fig. 6.38), which can be represented satisfactorily by the following relation:

$$\log P(\Delta \overline{U}) = 1.60 + 0.425 \, \Delta \overline{U} \, . \tag{6.18}$$

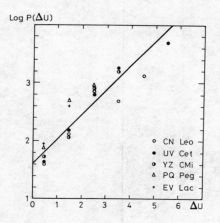

Fig. 6.38. Empirical dependence of equivalent (integral) time $P(\Delta U)$ on the absolute magnitude of flare stars M_V.

Using Table 1.6 we may find the average value of $P(\Delta U)$ for each star separately, and then the energy $E_f(U)$ emitted by each star; the results are shown in Table 6.16. The same results are represented in Fig. 6.39, which shows $\log E_f(U)$ as a function of the absolute magnitude of the star M_V. There is a very clear relation between $E_f(U)$ and M_V, namely, the energy emitted in U radiation during a flare is the larger the brighter the star. This conclusion also holds for B and V radiation. The relation found between $E_f(U)$ and M_V can be represented by the following empirical formula:

$$\log E_f(U) = 35.27 - 0.383 \, M_V \, . \tag{6.19}$$

Let us now consider the first question concerning the fraction of the energy emitted by the star during the flare. If f_U is the flare frequency (flares per hour) the number of flares in one second will be: $2.77 \times 10^{-4} f_U \, F(\Delta U)$, where $F(\Delta U)$ is the distribution function of the flare amplitudes. On the

TABLE 6.16 Average Values of the Absolute Energy Liberated by the Star in U Radiation During the Flare ($E_f(U)$) and the Fraction of the Radiant Energy Lost by the Star by Flares (J_f/I_*) in U and B bands

Star	M_V	$I_*(U)$ erg s^{-1}	$P(\Delta U)$ s	$E_f(U)$ ergs	$\dfrac{J_f(U)}{I_*(U)}$	$\dfrac{J_f(B)}{I_*(B)}$
CN Leo	16.68	$3.33 \cdot 10^{26}$	186	$6.20 \cdot 10^{28}$	0.29	0.036
UV Cet	15.27	$2.26 \cdot 10^{27}$	140	$3.16 \cdot 10^{29}$	0.31	0.044
Wolf 434	14.31	$4.15 \cdot 10^{27}$	132	$5.50 \cdot 10^{29}$	0.18	--
40 Eri C	13.73	$8.35 \cdot 10^{27}$	64	$5.32 \cdot 10^{29}$	0.035	--
Ross 614	13.08	$1.32 \cdot 10^{28}$	62	$8.20 \cdot 10^{29}$	0.060	--
YZ CMi	12.29	$4.00 \cdot 10^{28}$	100	$4.00 \cdot 10^{30}$	0.046	0.016
EV Lac	11.50	$1.30 \cdot 10^{29}$	107	$1.40 \cdot 10^{31}$	0.010	--
EQ Peg	11.33	$1.00 \cdot 10^{29}$	89	$8.90 \cdot 10^{30}$	0.014	--
AT Mic	11.09	$1.17 \cdot 10^{29}$	75	$8.75 \cdot 10^{30}$	0.014	--
AD Leo	10.98	$1.32 \cdot 10^{29}$	59	$7.80 \cdot 10^{30}$	0.012	0.002
Wolf 630	10.79	$1.45 \cdot 10^{29}$	52	$7.50 \cdot 10^{30}$	0.026	--
AU Mic	8.87	$3.57 \cdot 10^{29}$	50	$1.78 \cdot 10^{31}$	0.010	--
YY Gem	8.36	$1.61 \cdot 10^{30}$	85	$1.37 \cdot 10^{32}$	0.002	--

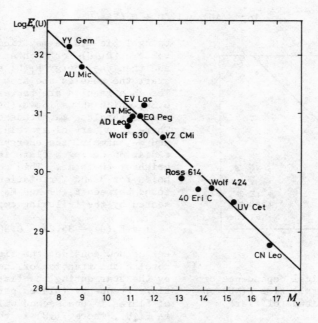

Fig. 6.39. Empirical dependence of radiant energy $E_f(U)$
lost on the average during one flare, and stellar absolute
magnitude M_V.

other hand, the energy emitted during one flare with amplitude ΔU and in the
amplitude interval $d(\Delta U)$ will be: $P(\Delta U)I_*(U)d(\Delta U)$. Thus for the fraction of the
energy $J_f(U)/I_*(U)$ radiated by the star in U radiation during the flare we have

$$\frac{J_f(U)}{I_*(U)} = 2.77 \times 10^{-4} f_U \int P(\Delta U)F(\Delta U)d(\Delta U) . \qquad (6.20)$$

An analogous expression may be written for the wavelength ranges B and V as well.
Taking the numerical values of f_U and $F(\Delta U)$ from Tables 1.3 and 1.8, and the values
$P(\Delta U)$ from (6.18) we find the numerical values of $J_f(U)/I_*(U)$, and also $J_f(B)/I_*(B)$
for a number of flare stars of UV Cet type, of which the absolute magnitudes differ
greatly (by eight stellar magnitudes). The results are given in the last two
columns of Table 6.16. Comparison of these results with the absolute magnitudes
of the stars M_V leads to the following conclusions:

a) The fraction of the radiative energy emitted by the star during the flare in U
radiation is quite large--of the order of 30%--for faint stars such as UV Cet and
CN Leo. However, with increasing absolute stellar luminosity this fraction falls
rapidly and becomes of the order of 1 to 2% for the stars YZ CMi and AD Leo, and
0.1% for YY Gem.

b) The fraction of the energy liberated by the star in B radiation is very small
--of the order of 4%--even in the case of the stars UV Cet and CN Leo. For AD Leo
this fraction is $\sim 0.1\%$.

From the two types of energy considered above the main factor to be considered
should be the relative amount of energy, since this quantity characterizes the
degree of variability of the given star. From this point of view we can conclude

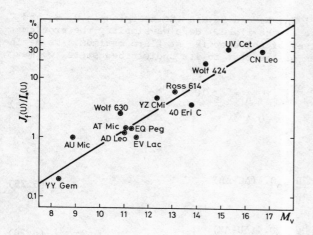

Fig. 6.40. $J_f(U)/I_*(U)$ versus M_V.

that UV Cet is more of a variable star than AD Leo.

The diagram showing $J_f(U)/I_*(U)$ (in percentage) as a function of M_V is given in Fig. 6.40. This dependence may be represented by the following empirical formula:

$$\text{Log } \frac{J_f(U)}{I_*(U)} = -4.75 + 0.26 \; M_V .$$
$$(6.20a)$$

The enormous range--three orders of magnitude--of the values of the energy radiated on the average by a star during one flare (Fig. 6.39), from a low-luminosity star (CN Leo) to a high-luminosity star (YY Gem), is quite striking. The energy liberated in one flare of YY Gem, say, is on the average 430 times as much as the radiant energy emitted in one flare of UV Cet (or 2000 times more than in the case of CN Leo). The amplitudes of flares of UV Cet and CN Leo would have been as high as 6^m5 and 8^m5 (in U radiation) if suddenly they were to flare with the same power as is usual for YY Gem. Similarly a flare of YY Gem would not be observable if it took place with the same power as for UV Cet--the amplitude of such a burst would be 0^m006. Hence the following conclusion may be drawn: in principle YY Gem may suffer frequent flares (even if these cannot be observed) with the power equivalent for the stars UV Cet or CN Leo.

23. THE ENERGY OF A FLARE

Lacy and others [33] found empirical relations between the flare energies E_U, E_B and E_V in the UBV bands respectively. These have the following form:

$$E_U = (1.20 \pm 0.08) \; E_B , \qquad\qquad (6.21)$$

$$E_U = (1.79 \pm 0.15) \; E_V . \qquad\qquad (6.22)$$

These functions were found from a group of eight flare stars of UV Cet type, with spectral classes in the range M1 - M5.5, and with absolute luminosities differing from each other by four orders of magnitude. The dispersion in flare amplitudes is up to 7^m5 in U radiation. Equations (6.21) and (6.22) seem to be valid to a large extent. To what degree do the relations (6.21) and (6.22) agree with the hypothesis of fast electrons? Can we conclude that the numerical coefficients re-lating E_U, E_B and E_V with each other do not depend on the strength of the flare, or are they averages of values with some dispersion?
Let us represent these relations in the form:

$$\frac{E_U}{E_B} = Q ; \quad \frac{E_U}{E_V} = P . \qquad\qquad (6.23)$$

Our problem is to find the quantities Q and P as a function of τ--the strength of the flare.

In the expressions (6.21) and (6.22) E_U, E_B and E_V are the energies in U, B, and V, integrated over the whole lightcurve of the flare. Here we simplify the problem by taking for E_U, E_B and E_V the values of the energies at flare maximum. We define q_U, q_B and q_V to be the luminosities of the star in its undisturbed state, we may then write:

$$E_U = q_U \, 10^{0.4 \, \Delta U}; \qquad E_B = q_B \, 10^{0.4 \, \Delta B};$$

$$E_V = q_V \, 10^{0.4 \, \Delta V} . \tag{6.24}$$

From (6.23) and (6.24) we have:

$$Q = \frac{q_U}{q_B} 10^{0.4(\Delta U - \Delta B)} , \tag{6.25}$$

$$P = \frac{q_U}{q_V} 10^{0.4(\Delta U - \Delta V)} . \tag{6.26}$$

The values of Q and P will be lower limits since we considered the energies only at the flare maximum.

The computed values of the resulting amplitudes ΔU, ΔB and ΔV, due to inverse Compton effects and nonthermal bremsstrahlung (Ch. 8) are given in Table 8.4 for a number of values of τ and for T = 2800 K. Using this table we find the following numerical values of Q and P:

τ	0.1	0.01	0.001	0.0001
Q	1.8	1.7	1.0	0.30
P	4.8	3.4	0.7	0.16

For the computations q_U/q_B was taken = 0.2 and q_U/q_V = 0.1. From Table 6.7 these quantities are between 0.105 and 0.219 for q_U/q_B, and between 0.035 and 0.110 for q_U/q_V.
These results show that the coefficients Q and P are by no means constant quantities, but vary with the strength of the flare τ. When the flare strengths vary by about three orders of magnitude, for instance, from τ = 0.0001 to τ = 0.1, the quantities Q and P change by about a factor of ten.

The lines $E_U \sim E_V$ and $E_U \sim E_B$ corresponding to different values of τ (and maybe different values of Q and P) have been drawn on the graphs of these functions constructed from observational data [33]; the results are given in Figs. 6.41 and 6.42 by broken lines.

The scattering of the points in the diagrams is explained in [33] as due to insufficiently high energy resolution of the measurements, and the relations (6.21) and (6.22) derived from these data would be no more than statistical relations. To a certain degree this is really the case. However the lines $E_U \sim E_B$ and $E_U \sim E_V$ drawn in these figures, corresponding to different values of flare strength (τ), indicate that we should expect a real scattering of the observational points within bands of a certain width.

Thus the hypothesis of fast electrons can explain naturally empirical relations of

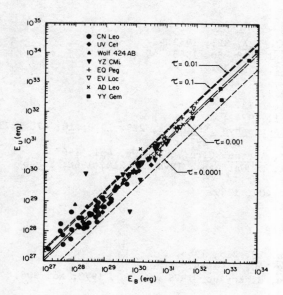

Fig. 6.41. Dependence of flare energy in U band (E_U) on flare energy in B band (E_B) for some flare stars of type UV Cet from observational data [33]. Broken lines--theoretical dependence $E_U \sim E_B$ for the case of the hypothesis of fast electrons and for various values of the strength of the flare, τ.

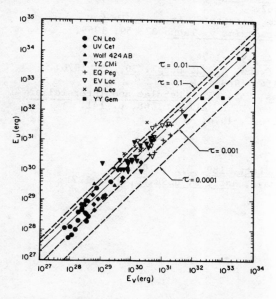

Fig. 6.42. The same as Fig. 6.41, for the relation $E_U \sim E_V$.

the form (6.21) and (6.22); in addition it reveals qualitatively new properties in these relations.

Further analysis, analogous to that just made, may lead to interesting consequences if applied to individual flare stars with given values of the parameters T, q_U/q_B and q_U/q_V, trying to encompass the most powerful as well as the faintest flares.

The results obtained allow us to find the average strength $\bar{\tau}$ of the flares for all of the stars of UV Cet type. For this it is sufficient to determine from the above table those values of $\bar{\tau}$ which correspond to the values Q = 1.20 and P = 1.79. As a result we find $\bar{\tau}$ = 0.0017 from Q and $\bar{\tau}$ = 0.0028 from P, results which are sufficiently close to each other. At the same time these results coincide with the values for $\bar{\tau}$ found above by other methods (Table 6.8) for the stars CN Leo and UV Cet.

REFERENCES

1. H. L. Johnson, W. W. Morgan, Ap. J. 117, 323, 1953.
2. T. A. Mathews, A. R. Sandage, Ap. J. 138, 30, 1963.
3. G. Haro, Bol. Inst. Tonantzintla 2, 3, 1976.
4. A. D. Andrews, PASP 80, 99, 1968.
5. G. A. Gurzadyan, IBVS No. 656, 1972.
6. W. E. Kunkel, Ap. J. Suppl. 25, 1, 1973.
7. T. J. Moffett, Ap. J. Suppl. 29, 1, 1974.
8. S. Cristaldi, M. Rodono, Astron. Astrophys. Suppl. 2, 223, 1970; 10, 47, 1973.
9. B. W. Bopp, T. J. Moffett, Ap. J. 185, 239, 1973.
10. A. D. Chuadze, T. I. Barblishwili, Astron. Circ., USSR No. 451, 1967.
11. K. Osawa, K. Ichimura, T. Noguchi, E. Watanabe, IBVS No. 210, 1968; Tokyo
    Astr. Bull. No. 192, 1969.
12. S. Cristaldi, M. Rodono, IBVS No. 404, 1969.
13. W. E. Kunkel, IBVS No. 315, 1968.
14. P. Roques, PASP 65, 19, 1953.
15. W. J. Luyten, Ap. J. 109, 532, 1949.
16. M. Petit, J. Obs. 44, 11, 1961.
17. P. F. Chugainov, Izv. Crimean Obs. 33, 215, 1965.
18. P. F. Chugainov, Izv. Crimean Obs. 40, 7, 1969.
19. K. Osawa, K. Ichimura, Y. Shimizu, H. Koyano, IBVS No. 790, 1973; No. 906,
    1974.
20. W. E. Kunkel, Nat. 222, 1129, 1969.
21. P. F. Chugainov, Izv. Crimean Obs. 26, 171, 1961.
22. P. F. Chugainov, Izv. Crimean Obs. 28, 150, 1962.
23. G. H. Herbig, PASP 68, 531, 1956.
24. T. J. Moffett, M.N.R.A.S. 164, 11, 1973.
25. A. D. Andrews, PASP 78, 542, 1966.
26. S. Cristaldi, M. Rodono, Private commun. 1976.
27. H. L. Johnson, R. I. Mitchell, Ap. J. 127, 510, 1958.
28. G. H. Herbig, Ap. J. 135, 736, 1962.
29. G. Haro, E. Chavira, ONR Symposium, Flagstaff, Arizona, 1964.
30. G. Haro, E. Chavira, Bol. Obs. Tonantzint y. Tacub 32, 59, 1969.
31. L. V. Kuhi, PASP 76, 430, 1964.
32. T. J. Moffett, Sky and Telescope 48, 94, 1974.
33. C. H. Lacy, T. J. Moffett, D. S. Evans, Ap. J. Suppl. 30, 85, 1976.
34. G. A. Gurzadyan, Astrofizika 4, 154, 1968.
35. G. Haro, Stars and Stellar Systems, Vol. VII, "Nebulae and Interstellar
    Matter," Ed. B. M. Middlehurst and L. H. Aller, 1968, p. 141.
36. D. Deming, J. C. Webber, IBVS No. 672, 1974.
37. B. B. Sanwal, IBVS No. 932, 1974.
38. G. Feix, IBVS No. 943, 1974.
39. B. N. Sanwal, IBVS No. 1180, 1976.
40. T. R. Flesh, J. P. Oliver, Ap. J. Lett. 189, L127, 1974.
41. R. E. Gershberg, N. I. Schachovskaya, Astr. J. USSR 48, 934, 1971.

CHAPTER 7
Color Indices of Flares

1. THEORETICAL COLOR INDICES

According to the fast electron hypothesis the relative increase of the intensity of the radiation during the flare is different in different regions of the spectrum. It is very large in ultraviolet light, considerable in photographic light and very small in photovisual rays. As a result the star's color changes during the flare; it becomes bluer. The variations of a flare star's color can be represented quantitatively, depending mainly on the flare parameters--its strength, the energy spectrum of the fast electrons and the star's temperature. In the UBV color system we can write

$$B - V = - 2.5 C_y + 1.04 \qquad (7.1)$$

$$U - B = + 2.5 C_u - 1.12 \qquad (7.2)$$

where

$$C_y = \log \frac{B}{V} \; ; \; C_u = \log \frac{B}{U} \qquad (7.3)$$

and U, B and V are given by the expressions (6.2). These relations contain the function $I_\lambda(\tau,\gamma,T)$, the theoretical intensity of the star's radiation at the time of the flare at a given wavelength. Taking the corresponding expressions of this function for some form of the energy spectrum of the fast electrons we can find the numerical values of (B – V) and (U – B) from (7.1) and (7.2).

The results of such computations for various cases are given below.

MONOENERGETIC ELECTRONS--ONE DIMENSIONAL PROBLEM. In Table 7.1 computed values are given of the color indices for a star of class M5 in the case of monoenergetic

TABLE 7.1 Theoretical color indices U – B and B – V for a flare of a star of class M5 depending on the electron energy γ (one-dimensional case)

γ^2	$\tau = 1$		$\tau = 0.1$		γ^2	$\tau = 1$		$\tau = 0.1$	
	U – B	B – V	U – B	B – V		U – B	B – V	U – B	B – V
2	-0^m33	$+0^m88$	$+0^m25$	$+1^m49$	10	-1^m63	-0^m18	-1^m43	$+0^m56$
3	−0.94	+0.35	−0.61	+1.04	20	−1.73	−0.19	−1.45	+0.71
5	−1.36	+0.00	−1.15	+0.66	50	−1.75	−0.03	−1.27	+1.06

electrons and for the one-dimensional problem. The values of the function $I_\lambda(\tau,\gamma,T)$ have been taken from (4.7). The computations have been made for a number of values for the electron energy from $\gamma^2 = 2$ to $\gamma^2 = 50$, and for two values of τ. The computed values of $U - B$ and $B - V$ for the undisturbed star ($\tau = 0$) are equal to: $U - B = +1^m18$ and $B - V = +1^m82$, which are somewhat different from the ordinary observed values for a star of class M5.

From the data given in Table 7.1 the following conclusions can be drawn:

a) At the time of the burst the star becomes considerably more blue. The variations of the color indices as compared with those of the undisturbed star reach 2^m8 in $U - B$ and 2^m0 in $B - V$.

b) The maximum variations of the color indices in $U - B$ as well as in $B - V$ are reached for $\gamma^2 = 10$. For $\gamma^2 > 10$ the color indices practically do not vary.

c) For the inverse Compton effect the minimum values of the color indices are: $U - B = -1^m7$ and $B - V = -0^m2$.

ACTUAL PHOTOSPHERES. This case is analyzed in more detail because it is closer to reality. The function $I_\lambda(\tau,\gamma,T)$ is replaced by the function $H_\lambda(\tau,\gamma,T)$ taken from (4.27).

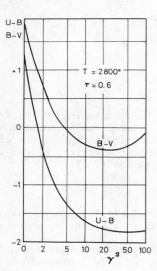

Consider first the dependence of $U - B$ and $B - V$ on γ^2 for maximum values of the flare amplitude ($\tau \simeq 0.6$). The results of the computations for an M5 star are represented in Fig. 7.1 in the form of graphs of $U - B$ and $B - V$ versus γ^2 in the interval from $\gamma^2 = 0$ to $\gamma^2 = 100$. As may be seen from this figure, the quantities $U - B$ and $B - V$ at first decrease rapidly with increasing γ^2 up to values $\gamma^2 \sim 10$; for γ^2 from 10 to 100 they do not change significantly. The limiting values are found to be: $U - B = -1^m80$, $B - V = -0^m38$, i.e. nearly the same as in the case of the one-dimensional problem. Thus the value $\gamma^2 \sim 10$ is optimal not only for explaining the observed flare amplitudes, but also for explaining the colors of the stars observed during the flare. Therefore further computations will be done only for the case of $\gamma^2 = 10$.

Fig. 7.1. Variations of color indices $U - B$ and $B - V$ as a function of γ^2. (Note: The figures in this book denote γ as μ.)

Table 7.2 shows the values of the theoretical color indices for the case of real photospheres, monoenergetic electrons ($\gamma^2 = 10$) and for various values of flare strength.

In the computations Planck's law corresponding to a given effective temperature of the star has been used for the distribution of the continuous spectrum of the undisturbed star. However due to the influence of the absorption lines the real energy distribution in the star's spectrum differs significantly from that according to Planck's law. Therefore the computed theoretical color indices may differ from their observed values. If the relative deviation is the same in all parts of the spectrum, then the difference

TABLE 7.2 Theoretical Color Indices U − B and B − V for a Flare of Stars of Classes M6–K5 Depending on Flare Strength τ (monoenergetic electrons, $\gamma^2 = 10$)

Temperature (spectrum) of the star	Color indices	τ					
		1	0.1	0.01	0.001	0.0001	0
2500° (M6)	U − B	$-1\overset{m}{.}60$	$-1\overset{m}{.}60$	$-1\overset{m}{.}52$	$-0\overset{m}{.}95$	$+0\overset{m}{.}48$	$+1\overset{m}{.}57$
	B − V	−0.29	−0.21	+0.27	+1.42	+1.99	+2.10
2800° (M5)	U − B	−1.63	−1.60	−1.38	−0.37	+0.80	+1.18
	B − V	−0.29	−0.13	+0.63	+1.55	+1.79	+1.82
3600° (M0)	U − B	−1.65 (−0.89)	−1.47 (−0.71)	−0.67 (+0.09)	+0.21 (+0.97)	+0.43 (+1.18)	+0.45 (+1.21)
	B − V	−0.23 (−0.10)	+0.24 (+0.37)	+1.01 (+1.12)	+1.26 (+1.38)	+1.29 (+1.42)	+1.30 (+1.48)
4200° (K5)	U − B	−1.62 (−0.60)	−1.25 (−0.23)	−0.38 (+0.64)	+0.03 (+1.05)	+0.09 (+1.11)	+0.10 (+1.12)
	B − V	−0.13 (+0.02)	+0.44 (+0.59)	+0.92 (+1.07)	+1.01 (+1.16)	+1.02 (+1.17)	+1.03 (+1.18)
4900° (K0)	U − B	−1.55 (−0.86)	−1.01 (−0.32)	−0.39 (+0.30)	−0.23 (+0.46)	−0.21 (+0.48)	−0.21 (+0.48)
	B − V	−0.03 (+0.00)	+0.49 (+0.52)	+0.75 (+0.78)	+0.78 (+0.81)	+0.79 (+0.82)	+0.79 (+0.82)

between theoretical and observed color indices would be insignificant. Evidently this is the case for stars of classes M6 and M5; the computed color indices for these stars in their undisturbed state (τ = 0) were found to be very close to their observed values (Table 7.3).

TABLE 7.3 Observed Color Indices for Stars of the Main Sequence from Data by Johnson and Morgan [1]

Spectrum	U − B	B − V	Spectrum	U − B	B − V
B0 V	$(-1\overset{m}{.}13)$	$(-0\overset{m}{.}32)$	F2 V	$0\overset{m}{.}00$	$+0\overset{m}{.}37$
B1 V	−1.00	−0.28	G0 V	+0.06	+0.60
B2 V	−0.86	−0.24	G0 V	+0.21	+0.68
B3 V	−0.71	−0.20	K0 V	+0.48	+0.82
B5 V	−0.56	−0.16	K5 V	+1.12	+1.18
A0 V	0	0	M1 V	+1.21	+1.48
A5 V	+0.09	+0.15	M5 V	+1.24	+1.69
F0 V	+0.02	+0.30			

But, from class M0 and earlier this difference is noticeable. Thus, for instance, for class K0 the computed color indices are: U − B = $-0\overset{m}{.}21$ and B − V = $+0\overset{m}{.}79$, whereas the observations give U − B = $+0\overset{m}{.}48$ and B − V = $+0\overset{m}{.}82$. If we want to compare the computed values U − B and B − V with the observational data, we must first

of all remove in some way these differences. Assuming that these are caused by the
deviation of the actual spectral distributions of the stars from the Planck distribution, we may apply the corresponding correction in (7.1) and (7.2) for each
stellar class separately in such a way that the computed color indices for the
undisturbed star agree with their observed values. This is equivalent to a parallel
displacement of the curves of U − B or B − V versus τ along the color index axis.
The corrected color indices found in this way are given in Table 7.2 in parentheses.

THE CASE OF A GAUSSIAN DISTRIBUTION OF FAST ELECTRONS. The computations have shown
that there is no qualitative difference between the theoretical color indices for
the case of a Gaussian distribution of the fast electrons and for the case of monoenergetic electrons. Therefore we shall use the data in Table 7.2 for all energy
spectra of the electrons.

2. THEORETICAL DEPENDENCE OF U − B ON B − V

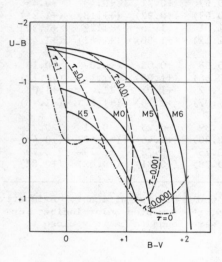

The theoretical diagram of U − B versus
B − V for the case of the fast electron
hypothesis may be constructed with the aid
of the data in Table 7.2. This diagram is
given in Fig. 7.2, where the full-drawn
lines give the functions (U − B) vs. (B − V)
for stars of classes M6, M5, M0 and K5.

All curves start practically from the main
sequence, corresponding to late-type stars;
here we have τ = 0. With increasing τ,
i.e. with increasing flare strength, the
curves run higher, gradually getting further
away from the main sequence. The upper
limiting positions of the curves correspond
with the optimum value of τ, for which the
flare amplitudes reach maximum values.
Thus moving along each of these curves we
pass from one flare to another. The
positions of the points on these curves,
corresponding to a given value of τ, are
connected by broken lines.

Fig. 7.2. Fast electron hypothesis:
theoretical diagram U − B vs. B − V for
flares of stars of classes K5 − M6
(full-drawn lines). Broken lines
equal flare strength τ.—.—, line:
main sequence.

It is remarkable that the curves corresponding to classes M6 and M5 are so far from the
main sequence; as far as 2^m5. The fast
electron hypothesis leads to a completely
new form of the diagram (U − B) vs. (B − V).
Therefore even the most approximate esti-
mates of the color indices are sufficient to determine if this radiation is due to
fast electrons or not.

3. THE DEPENDENCE OF U − B ON B − V IN THE CASE OF A HOT GAS

Some authors assume that stellar flares may be explained by radiation of an ionized
gas ejected by the star ("hot gas hypothesis"). In this connection it is interesting to compare the (U − B) vs. (B − V) diagram constructed for this case with the
color diagram for the case of fast electrons given above. The hot gas hypothesis
considers the radiation of a system consisting of a star of late spectral class
and an optically thin (or thick) layer of ionized gas.

Depending on the proportions in which these two components are present in the total radiation of the system, we have different values of $U-B$ and $B-V$. If the color indices for the purely "stellar" radiation are called $(U-B)_*$ and $(B-V)_*$, and those for the purely "gas" radiation $(U-B)_o$ and $(B-V)_o$, we have for the color indices $U-B$ and $B-V$ of the system:

$$U-B = -2.5 \log\{10^{-0.4[(U-B)_o+1.12]} + a.10^{-0.4[(U-B)_*+1.12]}\} +$$

$$+ 2.5 \log(1 + a) - 1.12; \qquad (7.4)$$

$$B-V = +2.5 \log\{10^{0.4[(B-V)_o-1.04]} + a.10^{0.4[(B-V)_*-1.04]}\} +$$

$$+ 2.5 \log(1 + a) - 1.04, \qquad (7.5)$$

where $\underline{a}$ is a parameter of the system representing the ratio of the stream of radiation from the star and that from the hot gas in the photographic region:

$$\underline{a} = \frac{\int B_\lambda(T)B_\lambda d\lambda}{\int J^o_\lambda(T_e)B_\lambda d\lambda} , \qquad (7.6)$$

B_λ is the sensitivity curve of the photometric system in B radiation. The case $\underline{a} = 0$ refers to pure gas radiation, the case $a = \infty$ to purely stellar radiation. The values of $(U-B)_*$ and $(B-V)_*$ corresponding to the colors of the "pure" star were taken from Table 7.3, and the values $(U-B)_o$ and $(B-V)_o$ corresponding to the recombination radiation of the ionized gas are given in Table 7.4 for a number of

TABLE 7.4 Theoretical Color Indices $(U-B)_o$ and $(B-V)_o$ for an Ionized Gas for Various Values of the Electron Temperature

		T_e	5000°	10 000°	20 000°	30 000°
$n_e > 10^6$ cm^{-3}		$(U-B)_o$	$-1^m.09$	$-0^m.94$	$-0^m.78$	$-0^m.51$
		$(B-V)_o$	-0.56	-0.38	-0.25	-0.19
$n_e < 10^6$ cm^{-3}		$(U-B)_o$	-0.86	-0.79	-0.25	-0.19
		$(B-V)_o$	-0.01	$+0.03$	$+0.04$	$+0.04$

values of the electron temperature, for the case of optically thick ionized gas ($n_e > 10^6$ cm^{-3}) as well as for an optically thin gas ($n_e < 10^6$ cm^{-3}) [2].

On the basis of these data theoretical color indices for the system "star + hot gas" were obtained from equations (7.4) and (7.5). The results for $n_e > 10^6$ cm^{-3} and $T_e = 10^4$ K are shown in Fig. 7.3 in the form of a theoretical diagram $U-B$ vs. $B-V$ for the system "star + hot gas"; the point $a = 0$ corresponds to purely gaseous radiation and the point $a = \infty$ to purely stellar radiation.

The family of curves constructed for $n_e < 10^6$ cm^{-3} differs very little from the curves shown in Fig. 7.3. For values of the electron temperature higher than 10 000 K all curves for the relation $U-B$ vs. $B-V$ run below those in Fig. 7.3.

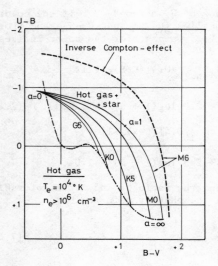

Fig. 7.3. Hot gas hypothesis: theoretical diagram U – B vs. B – V for $T_e = 10^4$ K and $n_e > 10^6$ cm^{-3} (full-drawn lines). For comparison also the theoretical U – B vs. B – V relation is shown for the case of inverse Compton effect (broken line), for a star of class M6.

One curve in Fig. 7.3 is given of the family of "Compton" curves corresponding to an M5 star, and taken from Fig. 7.2.

We see that for stars of class M5 – M6, the curves for the U – B vs. B – V relation for the case of the fast electron hypothesis are above the corresponding curves for the hot gas hypothesis. For instance, the "ceiling" of the color U – B in the case of a hot gas is equal to $-0^m.94$, whereas in the case of fast electrons it is $-1^m.6$.

With regard to stars of earlier classes (M0 – G5), the U – B vs. B – V curves simply get blended for both the hot gas hypothesis and the fast electron hypothesis (in the case of an M0 star these two types of curves practically coincide). Only for flare stars later than type M0 these two families of curves can be clearly distinguished from each other.

The situation is somewhat different for the case of the "nebular hypothesis," which is a different form of the hot gas hypothesis. According to computations by Kunkel [3] and Gershberg [4] in this case color indices with high negative values can be obtained; U – B, for example, may be up to -2^m or more.

However, this is found to be possible only under certain conditions, namely when the radiation considered is of a very high temperature gas, is transparent in the continuum, but with high optical depth in the Balmer lines. Let us note that the latter assumption--increase of the opacity in the lines--leads to a significant decrease of the radiation in the lines relative to the continuous radiation, which contradicts the observations. Rather, the observations indicate as a rule that all emission lines are strengthened during the flare. It should be stressed that here we are talking about strengthening of the emission lines with respect to the continuous radiation of the star. But the continuous radiation is increased as well. Thus the strengthening of the emission lines during the flare exceeds the strengthening of the continuous spectrum. Of course, this could not happen if it is assumed that the opacity of the medium in the Balmer lines increases strongly at the time of the flare.

But the chief deficiency of the nebular hypothesis is that it cannot explain the observed changes of the color of the star during the whole duration of the flare--from the moment of its maximum until the star has returned to its normal state. The nebular hypothesis leads to a two-model interpretation of the observed color indices. This interpretation is as follows: near flare maximum the radiation is only determined by the components of the nebular model (essentially only by the continuum), but the hot gas gets additional radiation from a hot spot in the photosphere [4].

If the contribution of the hot spot to the total radiation is so large that it exceeds by many tens and hundreds of times the normal radiation of the star, then of course the spot itself should occupy a considerable part of the star's surface.

At the same time the temperature of the hot spot must be considerably higher than the effective temperature of the star's photosphere. But as soon as the fraction of the radiation of the hot spot exceeds considerably the proper photospheric radiation of the star, the observed spectrum of the star must practically be due to the radiation of this hot spot, which should have a completely different spectral structure. However, the observations have established that the outburst of the star does not lead to a qualitative change of its spectrum.

4. COMPARISON WITH OBSERVATIONS

It is of particular interest to compare the theoretical values of the colors $U - B$ and $B - V$, derived above on the basis of the fast electron hypothesis, with the results of observations. The main point of this comparison consists in drawing in the theoretical $U - B$ vs. $B - V$ diagram the observed values of the colors of the star corresponding to its undisturbed state and at the moment of flare maximum. Evidently the star performs a "drift" in the color diagram.

TABLE 7.5 Color Indices $U - B$ and $B - V$ at Flare Maximum for a Number of Flare Stars

| Star | ΔU | Year of observation | Before the flare | | At flare maximum | |
			$U - B$	$B - V$	$U - B$	$B - V$
HII 1306	$3\overset{m}{.}7$	1957	$(+1\overset{m}{.}2)$	$(+1\overset{m}{.}4)$	$-1\overset{m}{.}07$	$+0\overset{m}{.}50$
AD Leo	1.5	1959	$+1.06$	$+1.54$	-0.14	$+1.34$
EV Lac	1.55	1967	$+1.06$	$+1.58$	-0.08	$+1.34$
EV Lac	1.00	1967	$+1.06$	$+1.58$	-0.55	$+1.27$
EV Lac	1.35	1967	$+1.06$	$+1.58$	-0.10	$+1.18$
EV Lac	3.1	1968	$+1.06$	$+1.58$	-1.08	$+0.80$
DH Car	1.0	1968	$+0.32$	$+0.90$	-0.05	$+0.44$
S 5114	4.1	1969	$+0.81$	$+1.64$	-1.34	$+0.62$

In Table 7.5 the values of $U - B$ and $B - V$ are given for some flares of a number of flare stars from earlier observational data. In Fig. 7.4 the theoretical $U - B$ vs. $B - V$ diagram constructed above is given with these observational data added to it. In the lower part of the diagram the positions of the stars in their normal states are indicated; they are nearly on the main sequence. At the time of the flare they are lifted up, performing their "drift," higher and more to the left of the diagram--the higher the larger the amplitude of the flare. For the flare, e.g. of S 5114 ($\Delta U = 4.1$) the observed value of $U - B$ reached $-1\overset{m}{.}34$.

It is remarkable that before the flare and at the moment of flare maximum nearly all stars considered are between the curves of the theoretical relation $U - B$ vs. $B - V$ corresponding to spectral classes M5-M0. This indicates once more that at the time of the flare the spectral class of the star does not suffer any essential changes. The three-color observations of flare stars by Cristaldi and Rodono [5], Osawa et al. [6], Moffett [7] and others give a large sample of material for analysis of their color characteristics. Figure 7.5 shows the results of the comparison of observations of flares of AD Leo and UV Cet according to the data [5]

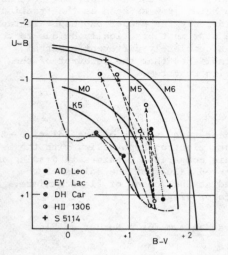

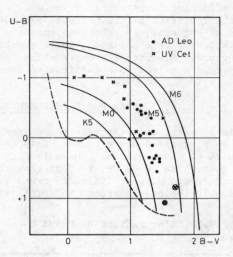

Fig. 7.4. "Drift" of a number of flare stars in the theoretical U – B vs. B – V diagram during their flares. Only the initial (before the flare) and the final (at the moment of flare maximum) positions of the stars have been indicated.

Fig. 7.5. The stars UV Cet and AD Leo in the theoretical color diagram before the flare (signs below) and at the moment of maximum of the individual flares.

and [6] with the theoretical U – B vs. B – V relation. As we see, the observations agree sufficiently well with the hypothesis of fast electrons.

Figure 7.5 is remarkable in another aspect as well: it indicates that the spectral composition of the additional radiation emitted by UV Cet and AD Leo is completely identical, although in absolute luminosity these stars differ from each other by a factor of more than a hundred.

The results of colorimetric measurements of a large number (more than 100) of flares for eight flare stars [6] are given in Fig. 7.6. It is remarkable that without exception all observed points were found to lie within the limits of the theoretical U – B vs. B – V curves.

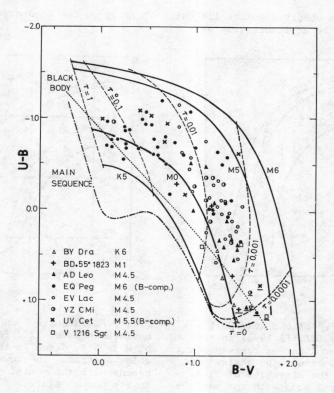

Fig. 7.6. Color indices U – B and B – V for a number of flare
stars at the moment of flare maximum. Theoretical curves--
fast electron hypothesis. Also the U – B vs. B – V relations
for stars of the main sequence and for a black body have
been drawn.

<u>5. DRIFT OF STAR IN COLOR DIAGRAM DURING FLARE</u>

In all of the above cases the color indices were determined only at the moment of
flare maximum. It is interesting to see to what degree the color variations of
the star during the whole flare period correspond with the theoretical model. Such
a case has been considered by Moffett [7], who constructed a detailed curve--given
in Fig. 7.7--of the changes in color indices during the whole time of one flare of
Wolf 424; the results show that the observed drift is in striking agreement with
the theoretical diagram.

In Fig. 7.7 the following fact draws attention: before flare maximum (point A) the
color variations follow strictly the theoretical curve corresponding to the stellar
spectral class M5. After maximum the observed points U – B and B – V are slightly
displaced towards the theoretical tracks corresponding to class M3 and even M2.
This could mean that during the flare the effective temperature of the photosphere
becomes hotter (in infrared, see §8) for a short time.

The star Wolf 424 evidently is a "model," which satisfies the requirements for comparison with theory. Additional data is shown in Fig. 7.8, where the values U – B
and B – V have been drawn from data of about 12 different flares of this star [7].

The fact that the observed points lie quite strictly on the theoretical curve must be considered to be at least striking.

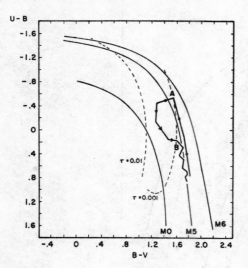

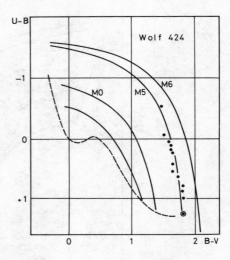

Fig. 7.7. "Drift" of the star Wolf 424 during one flare (21.III.1972) in the theoretical color diagram, corresponding with the fast electron hypothesis. The arrows indicate the direction of time. The point A corresponds with the maximum of the flare. The interval from A to B corresponds with the "rapid" part of the decrase of the lightcurve of the flare.

Fig. 7.8. Results U – B and B – V of measurements for 12 different flares of Wolf 424, drawn in the theoretical color diagram. The lower sign corresponds with the position of the star in its normal state.

6. DEPENDENCE OF COLOR ON FLARE AMPLITUDE

The colorimetric characteristics of flares can also be represented in the form of the dependence of the color indices on flare amplitude in a certain wavelength range. For example, in Fig. 7.9 the observed quantities U – B and ΔU for one flare have been drawn from data by Moffett [7] for seven flare stars of UV Cet type. Also the theoretical curves of U – B vs. ΔU have been drawn, as constructed on the basis of the fast electron hypothesis for stars of classes M5 and M6. Though the agreement of observation and theory is not bad, some deviations are apparent. First of all, the individual stars form a particular sequence of dependence of U – B on ΔU, as distinguished from the analogous sequence of the same dependence for another star, though the spectral classes of the two stars are the same. This is quite evident for the stars YZ CMi and CN Leo; the observed points for the first of these stars lie along the track for M5 and those for the second star lie along the track for M6. Still further to the left of the M5 track lie the points corre- sponding to the star EV Lac. A few points of YY Gem are lying nearly exactly on the track of M5. A certain deviation from the theoretical tracks occurs for large values of the flare amplitude ΔU (> 3). However, it should be taken into account that the theoretical curves in Fig. 7.9 correspond to a flare induced by inverse Compton effects whereas, as we shall see in the next chapter, strong flares with large values of the amplitudes are induced chiefly by non-thermal bremsstrahlung.

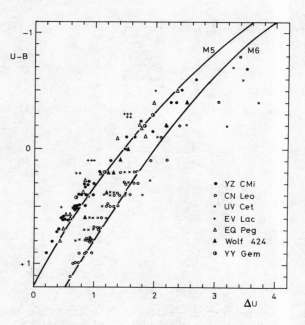

The diagrams of the type "color-amplitude" are evidently more sensitive indicators for such phenomena which cannot be found by analysis of "color-color" diagrams. In the case considered such a "new" phenomenon was a kind of stratification of the stars in the "color-amplitude" diagram. It is at present hard to indicate the cause for such a stratification. However, it is possible that the emission component of the star's radiation may play a role in this. Since the U region encompasses part of the Balmer continuum, a different degree of "emissivity" of two different stars for one and the same value of the flare amplitude leads to a difference in position of the two stars in the "color-amplitude" diagram. This question, however, requires careful analysis based on more refined observational data.

Fig. 7.9. Relation between observed values of color index U – B and amplitude ΔU at flare maximum for a number of flare stars. Also the theoretical tracks U – B vs. ΔU for stars of classes M5 and M6 based on the fast electron hypothesis have been drawn.

7. COLOR INDICES DUE ONLY TO THE FLARE

The observed color indices $U - B$ and $B - V$ at a given moment of a flare characterized by the amplitudes of the increases in brightness ΔU, ΔB and ΔV, are determined from the following relations:

$$U - B = (U - B)_* - (\Delta U - \Delta B) \qquad (7.7)$$

$$B - V = (B - V)_* - (\Delta B - \Delta V) \qquad (7.8)$$

where $(U - B)_*$ and $(B - V)_*$ are the color indices of the star in its quiescent state.

The colors $U - B$ and $B - V$ refer to the total radiation of the system, i.e. to the normal (thermal) radiation of the star plus the additional (non-thermal) radiation appearing during the flare. By means of these relations the color of the star can be determined at any phase of the development of the flare, if the amplitude values are known.

Here the color of the additional radiation, which has a non-thermal origin, is of interest. Calling the color indices of this radiation $(U - B)_f$ and $(B - V)_f$ we have:

$$(U - B)_f = U - B + (\Delta U - \Delta B) - 2.5 \log \frac{10^{0.4\Delta U} - 1}{10^{0.4\Delta B} - 1} \qquad (7.9)$$

$$(B - V)_f = B - V + (\Delta B - \Delta V) - 2.5 \log \frac{10^{0.4\Delta B} - 1}{10^{0.4\Delta V} - 1} \quad , \qquad (7.10)$$

comparing these with (7.7) and (7.8) we can write (7.9) and (7.10) in terms of the colors of the undisturbed star:

$$(U - B)_f = (U - B)_* - 2.5 \log \frac{10^{0.4\Delta U} - 1}{10^{0.4\Delta B} - 1} \qquad (7.11)$$

$$(B - V)_f = (B - V)_* - 2.5 \log \frac{10^{0.4\Delta B} - 1}{10^{0.4\Delta V} - 1} \quad . \qquad (7.12)$$

For large values of ΔU, ΔB and ΔV, we have

$$(U - B)_f \approx U - B = (U - B)_* - (\Delta U - \Delta B) \qquad (7.13)$$

$$(B - V)_f \approx B - V = (B - V)_* - (\Delta B - \Delta V) \quad , \qquad (7.14)$$

i.e. for strong flares the observed (or theoretical) colors $U - B$ and $B - V$ at the moment of maximum will be nearly the colors of the pure additional radiation $(U - B)_f$ and $(B - V)_f$. From here on the observed colors of the additional radiation are determined by means of (7.9) and (7.10) or (7.11) and (7.12). With the aid of these formulae we can determine the theoretical colors of the additional radiation from known theoretical amplitudes (cf. Table 6.4) and theoretical color indices (cf. Table 7.2).

Now we examine the observations for the colors of the additional radiation. For example we consider first what the values of $(U - B)_f$ and $(B - V)_f$ are from data by various observers [3,8,9] for a number of flare stars. These quantities are collected in Table 7.6 (for EV Lac and HII 1306 the values of $(U - B)_f$ and $(B - V)_f$

TABLE 7.6 Colors $(U - B)_f$ and $(B - V)_f$ of the Additional Radiation (Without Star) for a Number of Flares

Star	$(U - B)_f$	$(B - V)_f$	Star	$(U - B)_f$	$(B - V)_f$
YZ CMi	-1.14	(-0.35)	CN Leo	-1.17	-0.06
	-1.08	+0.12		-1.23	-0.23
	-1.08	+0.10		-1.15	+0.02
	-1.29	+0.29		-1.22	(-0.53)
	-1.00	+0.27		-1.47	-0.15
	-1.01	+0.30		-1.04	-0.25
AD Leo	-1.24	+0.21			
	-0.97	+0.01			
	-1.02	-0.02	AD Leo	-1.40	0.0
	-1.03	+0.21			
	-1.13	+0.04	EV Lac	-1.43	
	-1.26	-0.15		-1.07	--
	-1.15	0.00		-1.40	-0.20
	-0.70	-0.13	H II 1306	-1.27	-0.06
	-1.17	+0.25			

have been computed from (7.9) and (7.10) from observed colors and amplitudes).

In this table it is interesting that the colors, in particular in U and B radiation, are nearly constant, regardless of the large dispersion in the observed colors, flare strengths, absolute luminosities, and spectral classes of the stars. For instance, HII 1306 belongs to spectral class dK5e, CN Leo belongs to class dM6e, and according to Morgan's classification even to dM8e [10]. The colors given in the table for the star HII 1306 refer to a very strong flare ($\Delta U = 3\overset{m}{.}7$), whereas in the case of YZ CMi the flare amplitudes in U radiation fluctuate from 1^m to 5^m, in the case of AD Leo from $0\overset{m}{.}5$ to $2\overset{m}{.}1$, etc. Finally the star CN Leo is, in absolute magnitude, one of the faintest flare stars, and AD Leo is one of the brightest; the ratio of their luminosities is of the order of one hundred. Regardless of such differences the values of $(U-B)_f$ and $(B-V)_f$ were found to be equal on the average.

Thus we may conclude that the colors of the additional radiation are nearly the same for all flares, independent of the luminosity, spectral class of the star and flare strength. This means that the mechanism of excitation of the flares and of the generation of the non-thermal radiation is the same for all flare stars.

The problem of the resemblance or identity of the mechanisms of flares in flare stars in the neighborhood of the Sun and of flare stars in associations has been considered often. The fact that one of the flare stars in Table 7.6, HII 1306, is part of a stellar aggregate (Pleiades), and the others lie near the Sun, and that they show equal colors for the additional radiation, may serve as a convincing argument that the mechanism generating the continuous emission is common to all flare stars, independent of their age or position.

The average values of the colors of the additional radiation for all flares given in Table 7.6 are:

$$(U - B)_f = -1\overset{m}{.}20 \pm 0\overset{m}{.}18 \,,$$
$$(B - V)_f = +0\overset{m}{.}03 \pm 0\overset{m}{.}17 \,. \tag{7.15}$$

According to data by Kunkel [11] for 21 flares of the stars YZ CMi, AD Leo and CN Leo the average color indices are found to be:

$$(U - B)_f = -1\overset{m}{.}12 \pm 0\overset{m}{.}15$$
$$(B - V)_f = 0\overset{m}{.}0 \pm 0\overset{m}{.}22 \,. \tag{7.16}$$

From data of more than 400 flares of eight flare stars Moffett [12] finds the following values for the average colors:

$$(U - B)_f = -0\overset{m}{.}88 \pm 0\overset{m}{.}15$$
$$(B - V)_f = +0\overset{m}{.}34 \pm 0\overset{m}{.}44 \,. \tag{7.17}$$

About the same values were found by Cristaldi and Rodono from the results of their measurements [13].

It should be pointed out, however, that the dispersion in (7.17) is rather large. We give comparative color data for the stars UV Cet and Wolf 424:

	$(U - B)_f$	$(B - V)_f$
UV Cet	$-0\overset{m}{.}63 \pm 0\overset{m}{.}31$	$+0\overset{m}{.}25 \pm 0\overset{m}{.}55$
Wolf 424	$-1\overset{m}{.}14 \pm 0\overset{m}{.}25$	$-0\overset{m}{.}15 \pm 0\overset{m}{.}15$.

What causes this dispersion in the values of $(U-B)_f$ and $(B-V)_f$ between flares of individual stars as well as between different flares of the same star? Here we mention three possible reasons (besides that of infrared excess, see §8):

a) Ordinary photometric errors. This refers to all faint flares, and in particular to most flares in V radiation, considering that in general the amplitudes in V are either very small or even zero.

b) The influence of the Balmer emission. The fraction of the hydrogen recombination is, as has been indicated above, different not only for different stars but may also vary for different flares of the same star. Especially sensitive to this region is the B band.

c) The fact that the U, B and V measurements are not synchronous. Perhaps this effect is the most essential source of errors. The measurements give as a rule the values of the color indices at flare maximum. But quite often the duration of the maximum itself is considerably shorter than the time needed for making one cycle of measurements in U, B, V. For instance in Moffett's measurements the length of this cycle is 4 seconds. At the same time the maximum of the lightcurve for nearly all flares is so sharp that the measurements in U, B, and V wavelengths practically refer to different regions of the descending branch of the lightcurve.

In the preceding chapter we tried to find the probable value of the energy of the fast electrons responsible for the star's flare, by comparing the observed flare amplitudes with their theoretical values. The same problem is now solved by comparing the observed colors (7.15) of the additional radiation with their theoretical values ("color indices method").

The analysis shows that the theoretical color indices are not very sensitive to the effective temperature of the photospheric radiation, i.e. to the star's spectral class. They are not very sensitive either to the strength of the flare (τ). The color indices of the additional radiation appear to be very sensitive to the energies of the fast electrons. This follows, for instance, from Fig. 7.10, which shows the dependence of the theoretical quantities $(U-B)_f$ and $(B-V)_f$ on γ^2, constructed from formulae (7.9) and (7.10) and the data of Table 7.7 (star of class M5, $\tau = 0.6$).

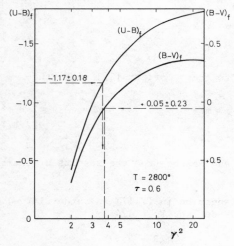

Fig. 7.10. The probable value of the energy of the fast electrons (color indices method).

Indicating in this graph the colors of the additional radiation (7.15), we can find the required value of γ^2, which is close to 4. This result is also obtained from the curve for $(U-B)_f$ as well as from the curve for $(B-V)_f$. Thus the probable value of the energy of the fast electrons determined by the "color indices method" is of the order $\gamma \sim 2$. Earlier a value $\gamma \sim 2$ to 3 was found by the "amplitude method." We cannot exclude that there exists a real dispersion in the values of γ, there are stars of T Tau type with known color index U − B which in normal circumstances equals $-1\overset{m}{.}3$ or $-1\overset{m}{.}5$; this in general is only possible for $\gamma \sim 3$. Moreover, the graph given in Fig. 7.10 is constructed for the case of monoenergetic electrons, which is an idealization of the problem. At this stage of the investigation it is better to assume that the real values of γ lie within the range 2 to 3.

TABLE 7.7 Theoretical Amplitudes ΔU, ΔB, ΔV and Color Indices U − B and B − V for the System "Flare + Star" for Different Values of the Energy of the Fast Electrons

γ^2	2	3	4	5	10	20	50	100
ΔU	4.12	5.89	6.63	6.98	7.32	7.05	6.32	5.65
ΔB	2.54	3.76	4.23	4.44	4.51	4.13	3.36	2.69
ΔV	1.41	2.17	2.44	2.53	2.40	1.96	1.21	0.81
U − B	−0.̇42	−0.̇98	−1.̇24	−1.̇38	−1.̇65	−1.̇76	−1.̇81	−1.̇81
B − V	+0.69	+0.23	+0.02	−0.09	−0.30	−0.36	−0.33	−0.07

A change of the color of the additional radiation at the time of the flare was observed [8] for four flares of EV Lac recorded simultaneously in B and V radiation. The color of the additional radiation at the moment of maximum of these flares was $(B-V)_f$ = −0.̇2 to −0.̇3. After the maximum the radiation had become redder. The recorded maximum value of $(B-V)_f \approx$ 0.̇0 to +0.̇2. However, these changes are illusive; as we shall see in the next paragraph, the colors of additional emission do not depend on flare strength τ and hence must be constant during the flare event.

In principle considerable fluctuations in the colors of the additional radiation may occur as a result of variations of the energy of the electrons themselves during the flare; however, we think this is not very probable. The data indicate that the fast electrons leave the star practically without changing their original energy.

The problem of the variations of the color of the additional radiation during the flare requires more detailed study. It is important to realize that analysis of the color indices makes it possible to explore the nature of the additional radiation liberated during a flare.

8. COLOR INDICES OF THE "PURE" FLARE AND THE FAST ELECTRON HYPOTHESIS

Here we consider the theoretical color characteristics of the "pure" flare, i.e. the radiation of the star at the time of the flare, after subtracting the radiation of the star itself. We shall see that for the pure flare the color characteristics no longer depend on the strength τ of the flare. Hence, since we do not need to consider τ as a parameter, we can include in our analysis other parameters, which were not considered before, and which may be very important and interesting.

For the intensity of the excess radiation of the flare E_ν we have from (4.27):

$$E_\nu = B_\nu(T) \frac{3}{2\gamma^4} \frac{e^x - 1}{e^{x/\gamma^2} - 1} F_1(\tau) . \qquad (7.18)$$

According to the fast electron hypothesis the intensity of the excess radiation of the flare depends on the energy distribution in the infrared spectral region, at ∼3.6 μ, ∼4.4 μ, and ∼5.5 μ. In (7.18) this distribution has been taken to follow Planck's law.

However, for flare stars one cannot be certain that their radiation in the infrared region can be represented by the Planck distribution; it will probably deviate from this. Moreover, it is not even certain that the degree of deviation will be the same at 3.6 μ, 4.4 μ and at 5.5 μ. Thus, introducing the coefficients k_1, k_2 and k_3, which characterize the degree of departure from Planck's law for these three spectral regions, we can write for the intensities of the pure flare at the centers of the U, B and V regions:

$$E_U = k_1 \frac{3}{2\gamma^4} \left[B_\nu(T) \frac{e^x - 1}{e^{x/\gamma^2} - 1} \right]_U F_1(\tau) \; ; \tag{7.19}$$

$$E_B = k_2 \frac{3}{2\gamma^4} \left[B_\nu(T) \frac{e^x - 1}{e^{x/\gamma^2} - 1} \right]_B F_1(\tau) \; ; \tag{7.20}$$

$$E_V = k_3 \frac{3}{2\gamma^4} \left[B_\nu(T) \frac{e^x - 1}{e^{x/\gamma^2} - 1} \right]_V F_1(\tau) \; . \tag{7.21}$$

The theoretical color indices of the pure flare $(B - V)_f$ and $(U - B)_f$ can be found from equations (7.1), (7.2) and the integrals (6.2). However, the computations can be simplified if we determine only the ratio of the values of these integrals at their maxima. Then (7.19), (7.20) and (7.21) give:

$$C_V = \log \left[\left(\frac{\lambda_V}{\lambda_B} \right)^5 \frac{\left[e^{x/\gamma^2} - 1 \right]_V}{\left[e^{x/\gamma^2} - 1 \right]_B} \frac{B_\lambda}{V_\lambda} \frac{k_2}{k_3} \right] \; ; \tag{7.22}$$

$$C_U = \log \left[\left(\frac{\lambda_U}{\lambda_B} \right)^5 \frac{\left[e^{x/\gamma^2} - 1 \right]_U}{\left[e^{x/\gamma^2} - 1 \right]_B} \frac{B_\lambda}{U_\lambda} \frac{k_2}{k_1} \right] \; . \tag{7.23}$$

Taking T = 2800 K for all flare stars and substituting the numerical values for U_λ, B_λ and V_λ at their maxima (Table 6.1), we find for the color indices of the pure flare according to the fast electron hypothesis ($\gamma^2 = 10$):

$$(B - V)_f = -2.5 \left[0.44 - \log \frac{k_3}{k_2} \right] + 1.04 \; , \tag{7.24}$$

$$(U - B)_f = +2.5 \left[-0.24 + \log \frac{k_2}{k_1} \right] - 1.12 \; . \tag{7.25}$$

We see that the color indices of the flare are independent of the flare strength τ. We also note that in this representation the quantities $(B - V)_f$ and $(U - B)_f$ do not depend on the three coefficients k_1, k_2 and k_3 themselves, but on their ratios k_3/k_2 and k_2/k_1, thus in fact only on two parameters.

For a Planck distribution, $k_1 = k_2 = k_3 = 1$; then (7.24) and (7.25) give:

$$(B - V)_f = -0.06 \; ;$$

$$(U - B)_f = -1.72 \; . \tag{7.26}$$

This is sufficiently close to the values found before $[(B - V)_f = -0.18$ and $(U - B)_f = -1.63$, cf., Table 7.1] for the color indices for a Planck distribution by numerical integrations.

However, for the Planck distribution the fast electron hypothesis gives only one point in the color diagram with coordinates as in (7.24). But assuming that the quantities k_3/k_2 and k_2/k_1 may differ from unity, not only from one star to another, but during different flares as well, the situation is then more complicated. Each combination k_3/k_2 and k_2/k_1 gives one point in the color diagram and the number of such combinations is unlimited. More exactly, the locus of points corresponding to a fixed value of k_2/k_1, but various values of k_3/k_2 is a line parallel to one axis. In the same way, fixed values for k_3/k_2, with various values of k_2/k_1 give another family of straight lines, parallel to the other axis. Thus we find in the color diagram $(U - B)_f$ vs. $(B - V)_f$ two sets of perpendicular lines, not a single point, as before. This result agrees well with the observations which give, even for one star, different points in the color diagram for different flares.

Furthermore, for $k_2/k_1 = 2$ and $k_3/k_2 = 2$, i.e. for relatively small departures from Planck's law in the infrared, the color characteristics of the flare change considerably; the new point has the coordinates $(B - V)_f = +0.70$ and $(U - B)_f = -0.97$. Table 7.8 shows such results for a number of values of k_3/k_2 and k_2/k_1.

TABLE 7.8   Color Indices for the "Pure" Flare, $(U - B)_f$ and $(B - V)_f$, at a Number of Values of k_2/k_1 and k_3/k_2

k_2/k_1	$(U - B)_f$	k_3/k_2	$(B - V)_f$
1	−1.72	0.6	−0.61
2	−0.97	0.8	−0.31
3	−0.52	1	−0.06
4	−0.22	2	+0.70
5	+0.03	3	+1.14

9.   COLOR "PORTRAITS" OF FLARE STARS

We now consider the observations, and see how they compare with the data in Table 7.8. The most reliable and homogeneous colorimetric data refer to ten flare stars of UV Cet type (cf. Table 7.9). For eight stars the data are shown in the color diagrams $(U - B)_f$ vs. $(B - V)_f$ in Figs. 7.11 to 7.14. In these diagrams the theoretical $(U - B)_f$ vs. $(B - V)_f$ relations have been drawn as well for a number of values of k_3/k_2 and k_2/k_1 according to Table 7.8. These diagrams show that all observed points lie within the rectangle defined by values of k_2/k_1 from 1 to 4 and of k_3/k_2 from 0.7 to 3. In other words, nearly always, for all stars and all flares the following condition holds: $k_2/k_1 > 1$, $k_3/k_2 > 1$. Hence for the group of flare stars in question,

$$k_3 > k_2 > k_1 , \qquad\qquad (7.27)$$

i.e. the departure of the real distribution of the star's radiation from the Planck distribution is largest at ~ 5.5 μ, and smallest at ~ 3.6 μ.

TABLE 7.9 Colorimetric Characteristics of the "Pure" Flare for a Group of UV Cet Type Flare Stars

Star	Sp	$\overline{(B-V)}_f$	$\overline{(U-B)}_f$	$\sigma(B-V)_f$	$\sigma(U-B)_f$	Center $\frac{k_3}{k_2}$	Center $\frac{k_2}{k_1}$	Limits $\frac{k_3}{k_2}$	Limits $\frac{k_2}{k_1}$	n	Refer.
1	2	3	4	5	6	7	8	9	10	11	12
UV Cet	M5.5	+0.31	−0.66	±0.52	±0.56	1.45	2.65	0.9−3	1.5−4	14	12
CN Leo	M6	+0.21	−0.95	0.52	0.25	1.30	2.00	0.8−3	1.5−3	20	11,12
BY Dra	M1	+0.11	−0.78	0.48	0.35	1.20	2.40	0.8−3	1.8−3.5	6	5
EQ Peg	M5.5	+0.23	−0.90	0.46	0.23	1.35	2.10	0.7−3	1.5−3.5	36	5,12
YZ CMi	M4.5	+0.24	−0.74	0.29	0.27	1.35	2.50	0.9−2	1.5−3.5	46	11−16
EV Lac	M4.5	+0.18	−1.08	0.29	0.19	1.30	1.75	1−1.6	1.5−2.1	19	5,12,16
AD Leo	M4.5	+0.24	−1.08	0.28	0.18	1.35	1.74	0.8−2	1.5−2.5	32	11,12,15
Wolf 424	M5.5	+0.14	−1.14	0.25	0.24	1.25	1.65	1−1.8	1.2−2.5	11	12,17
BD+55°1823	M1.5	+0.68	−0.48	0.18	0.18	2.00	3.10	−	−	6	5
YY Gem	M0	+0.30	−0.95	0.05	0.09	1.40	2.00	−	−	4	12

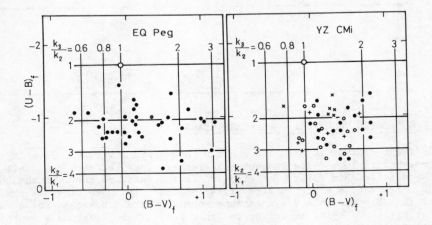

Fig. 7.11. Flare stars EQ Peg and YZ CMi in the color diagram of the "pure" flare $(U - B)_f$ vs. $(B - V)_f$. The theoretical relations $(U - B)_f$ vs. $(B - V)_f$, corresponding to different values of k_3/k_2 and k_2/k_1 are also shown.

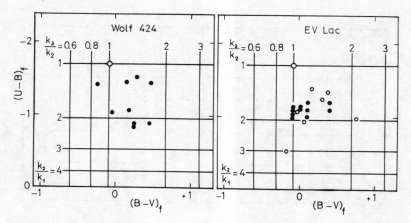

Fig. 7.12. The same as Fig. 7.11, for the flare stars Wolf 424
and EV Lac.

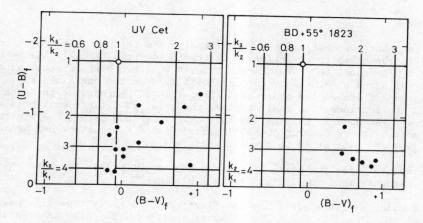

Fig. 7.13. The same as Fig. 7.11, for the flare stars UV Cet
and BD+55°1823.

Finally, the scatter of the observational points in the color diagram is mostly due
to measuring errors. However, from the fact that the dispersion in the color diagram is not the same for different stars, we can conclude that there may be a <u>real</u>
scatter as well, due to varying departures from Planck's law in the infrared.
Figure 7.11 gives the color diagrams for the stars EQ Peg and YZ CMi. In this and
all other diagrams the point $k_2/k_1 = 1$ and $k_3/k_2 = 1$ (large circle) corresponds to
the Planck distribution. For these two stars the observed points lie far from this
circle: their average position, which is about the same for the two stars, corresponds to $(B - V)_f \approx +0.23$ and $(U - B)_f \approx -0.08$ or $k_3/k_2 = 1.35$ and $k_2/k_1 = 2.10$
to 2.50 (cf. Table 7.9). The scatter of the points is about the same for the two
stars in the direction $(U - B)_f$, and considerably larger for EQ Peg in the direction
$(B - V)_f$. This must mean that the deviations from the Planck distribution in the
regions 4.4 μ and 5.5 μ are smaller for YZ CMi than for EQ Peg. Nearly the same
values are found for the scatter of the points (i.e. $\sigma(U - B)_f$ and $\sigma(B - V)_f$), as
well as for the boundary values of k_3/k_2 and k_2/k_1, for the stars Wolf 424 and EV
Lac (Fig. 7.12). For these stars the maximum values of k_3/k_2 (1.6 to 1.8) are even
smaller than 2.

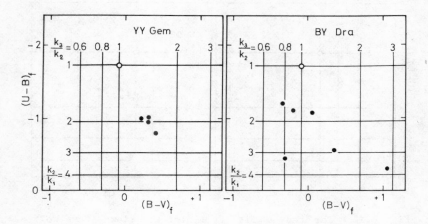

Fig. 7.14. The same as Fig. 7.11, for the flare stars YY Gem and BY Dra.

This implies that the observed scatter of the points in the color diagram can be explained if the relative deviation of the radiation from Planck's law in the region ∿5.5 μ is about 60 to 80% larger than at ∿4.4 μ. Most interesting are the following two pairs of stars--UV Cet and BD+55°1823 on the one hand (Fig. 7.13), and YY Gem and BY Dra on the other hand (Fig. 7.14). Whereas the points for BD+55°1823 are confined to a small region, those for UV Cet are scattered all over the diagram. The same is obvious for the pair YY Gem and BY Dra. Since there are only few points for YY Gem and BD+55°1823 no definite conclusions can be drawn about the stability of these stars in the infrared region.

A similar analysis for the stars CN Leo and AD Leo shows that the first has a more unstable character than the latter (Table 7.9).

The most compact diagrams are found for flare stars of early type classes, like BD +55°1823 (M1.5) and YY Gem (M0). However, another early type class star (BY Dra, M1) does not show strong compactness in its color diagram.

10. INFRARED "DIAGNOSTICS" OF FLARE STARS

Although the diagrams in Figs. 7.11 to 7.14 are not similar, they have one common feature: the average position of the points lies to the lower right of the "Planck point" with coordinates $k_2/k_1 = 1$ and $k_3/k_2 = 1$. Average positions for all flares ($\sum n = 195$) correspond with the following values for the color indices and the coefficients $\overline{k_3/k_2}$ and $\overline{k_2/k_1}$ for the "pure" flare:

$$\overline{(U - B)}_f = -0.89 ; \qquad \overline{k_2/k_1} = 2.10$$
$$\overline{(B - V)}_f = +0.20 ; \qquad \overline{k_3/k_2} = 1.35 . \tag{7.28}$$

Taking $k_1 = 1$, i.e. at 3.6 μ the true distribution does not differ from Planck's law. Then (7.28) leads to the result that the actual radiation at ∿4.4 μ is about twice that predicted from Planck's law and in the region ∿5.5 μ this factor is $(\overline{k_3/k_2}) \times (\overline{k_2/k_1}) \approx 3$.

The quantities $\overline{k_2/k_1}$ and $\overline{k_3/k_2}$ given in (7.28) represent __constant__ fractions of the deviations from Planck's law, and they are on the average nearly equal for all flare stars. Moreover there exists a __variable__ part of this deviation, which changes rather quickly, sometimes from one flare to the next.

We return to the condition (7.27). In general this condition can also be satisfied. The largest deviation is at $\sim$5.5 μ, and the smallest at $\sim$3.6 μ. Since we have no direct data on the numerical value of any of the coefficients k_1, k_2 or k_3, nothing can be said about the sign of this deviation.

The above results show that the fluctuations in radiative power of the flare stars are considerably larger in the infrared region (3 to 6 μ) than in the visual region. Any deviations from Planck's law can be represented by a variable effective temperature. Infrared excesses imply that the effective temperature in the infrared is higher. In fact, in nearly all the cases where the combined radiation of "star + flare" was considered, in the U − B vs. B − V diagram the star followed tracks corresponding to temperatures somewhat higher than the effective temperature of the star in the visual region. We can attribute these higher temperatures to the infrared region.

It should be noted that the introduction of the coefficients k_1, k_2 and k_3 in the theory has no influence on all our earlier work in the construction of the theoretical U − B vs. B − V diagrams. There, any deviations from a Planck distribution are automatically compensated by a corresponding choice of the numerical value of the parameter τ [through the function $F_1(\tau)$ in (7.18)].

For the computation of the flare amplitudes in U, B and V wavelengths three more coefficients must be introduced--k_4, k_5 and k_6, which take into account the deviations from Planck's law in the regions $\sim$0.36 μ, $\sim$0.44 μ and $\sim$0.55 μ. In §5 of Ch. 6 a remarkable fact was noted: the observed amplitudes in V are systematically larger than the values expected by the fast electron hypothesis (for a Planck distribution). This fact now finds a natural explanation, and it can also be seen as a direct indication that, in flare stars, in the region $\sim$5.5 μ, we actually have much larger radiation fluxes than expected from Planck's law with T = 2800 K.

Thus, in the most general case six coefficients are needed to determine the energy distribution in the spectrum of the pure flare--k_1, k_2 and k_3 for the infrared region, and k_4, k_5 and k_6 for the visual-photographic region; their numerical values must be found from observations. However, in the theory only ratios of these coefficients are needed, k_3/k_2 and k_2/k_1 for the determination of the color indices $(U - B)_f$ and $(B - V)_f$, and k_1/k_4, k_2/k_5 and k_3/k_6 for the amplitudes in U, B and V respectively.

In the last three sections of this chapter we have seen that the colorimetric observations do not contradict the fast electron hypothesis. Analysis of the flare characteristics from the point of view of the fast electron hypothesis has revealed a new property of these stars--the deviation of the radiative power from Planck's law in the infrared.

It is interesting that the infrared radiation of flare stars can be predicted from the colorimetric indices of the pure flare. The coefficients k_2/k_1 and k_3/k_2 can be found from the following equations, derived from (7.24) and (7.25):

$$k_2/k_1 = 4.87 \times 10^{0.4(U-B)_f} \; ; \tag{7.29}$$

$$k_3/k_2 = 1.06 \times 10^{0.4(B-V)_f} \, , \tag{7.30}$$

where $(U - B)_f$ and $(B - V)_f$ are taken from the observations. It would be interesting to apply these equations, not only to flare stars, but to T Tau type stars,
and to H_α emission stars as well, in order to find the true energy distribution in
the infrared region (3 to 6 μ) without using direct infrared observations.

REFERENCES

1. H. L. Johnson, W. W. Morgan, Ap.J. 117, 313, 1953.
2. R. E. Gershberg, Izv. Crimean Obs. 33, 206, 1965.
3. W. E. Kunkel, Flare Stars, Doctoral Thesis, Texas, 1967.
4. R. E. Gershberg, Astrofizika 3, 127, 1967.
5. S. Cristaldi, M. Rodono, Proc. IAU Symposium No. 67, Moscow, p. 75, 1975.
6. K. Osawa, K. Ishimura, Y. Simuzu, T. Okawa, K. Okida, H. Koyano, IBVS. No.
      790, 1973.
7. T. J. Moffett, Ap. J. Suppl. 28, 273, 1975.
8. P. F. Chugainov, Izv. Crimean Obs. 33, 215, 1965.
9. G. A. Gurzadyan, Bol. Obs. Tonant. y Tacub. 35, 39, 1971.
10. W. W. Morgan, Ap.J. 87, 589, 1938.
11. W. E. Kunkel, Ap. J. 161, 503, 1970.
12. T. J. Moffett, Ap.J. Suppl. 29, 1, 1974.
13. S. Cristaldi, M. Rodono, Astron. Astrophys. Suppl. 10, 47, 1973.
14. K. Osawa et al., IBVS No. 876, 906, 1974.
15. K. Ichimura et al., Tokyo Astr. Bull. No. 224, 1973.
16. S. Cristaldi, M. Rodono, IBVS No. 759, 802, 1973.
17. T. J. Moffett, M.N.R.A.S. 164, 11, 1973.

CHAPTER 8
Bremsstrahlung from Fast Electrons

1. THE PROBLEM

In the preceding chapters the problem of the excitation of flares was considered assuming that the observed strengthening at short wavelengths during the flare was due only to inverse Compton effects--inelastic collisions of fast electrons with infrared photons. The problem now is to determine how large is the production of photons by free-free transitions of fast electrons in the field of the protons (or electrons). This non-thermal bremsstrahlung can be compared with the number of photons appearing in the same spectral range as a result of inverse Compton effects.

Below it will be shown that in some flare stars the non-thermal bremsstrahlung appears only as a secondary effect and only in very strong flares, and in other flare stars it plays no role whatever in the excitation of optical flares. At the same time the non-thermal bremsstrahlung plays a highly important role in the region of very short wavelengths, the x-ray region, which we shall consider in detail in Chapter 14. Finally the non-thermal bremsstrahlung predicts the possibility of flares from dwarf stars of early spectral classes, with effective temperature of the order of 10 000 K or higher. Special discussion of the non-thermal bremsstrahlung seems necessary [1].

The energy γ of the fast electrons, which cause stellar flare, is of the order of 2 to 3. For such energies the theory of the effective cross section of the electron-proton or the electron-electron interactions leads to expressions which are highly complex for practical application [2]. It is usually thought that interpolation between the results for limiting cases--when the energy of the electrons is considerably smaller, and when the electrons are highly relativistic--is a good compromise. In our case ($\gamma \sim 3$) the electrons are not highly relativistic, but they are certainly non-thermal. Therefore for the effective cross section of the collision we may use the formula given by Joseph and Rohrlich [3], derived from the more general theory of Bethe and others [4,5,6]. This formula has the form:

$$\sigma_\nu(E)d(h\nu) = 4\alpha r_o^2 f(\nu,E)\, \frac{d(h\nu)}{h\nu}\,, \qquad (8.1)$$

where $\alpha = 1/137$, $r_o = 2.82 \times 10^{-13}$ cm and $f(\nu,E)$ is

$$f(\nu,E) = \frac{1}{E^2}\left[(E^2 + E_1^2 - \frac{2}{3}EE_1)(\ell n\, \frac{2EE_1}{mc^2 h\nu} - \frac{3}{2}) - \frac{EE_1}{9}\right]. \qquad (8.2)$$

Here E and E_1 are the energies of the electron before and after the collision with the proton. Substituting in (8.2) $E_1 = E - h\nu$, where $h\nu$ is the energy of the photon ejected as a result of the retardation of the electron, and introducing the dimensionless energy of the photon

$$\omega = \frac{h\nu}{E} = \frac{1}{\gamma} \frac{h\nu}{mc^2} , \qquad (8.3)$$

we have

$$f(\omega,\gamma) = \frac{4}{3} (1 - \omega + \frac{3}{4} \omega^2) \left[\ln(2\gamma \frac{1-\omega}{\omega}) - \frac{3}{2} \right] - \frac{1-\omega}{9} . \qquad (8.4)$$

For small frequencies of the photon, when $\omega \ll 1$, i.e. in the wavelength region of interest (longer than 3000 Å), we have

$$f(\omega,\gamma) = \frac{4}{3} (\ln \frac{2\gamma}{\omega} - \frac{3}{2}) - \frac{1}{9} . \qquad (8.5)$$

The expressions (8.1) and (8.5) will be used below for deriving the volume coefficient of the radiation of the medium due to retardation of the fast electrons.

2. INTENSITY OF NON-THERMAL BREMSSTRAHLUNG

By comparison of the observed flare parameters with their theoretical values derived on the basis of the hypothesis that the flare is due to inverse Compton effects, it is possible to find the probable form of the energy spectrum of the fast electrons. This resembles a Gaussian curve with a small dispersion (Chapter 7). Assuming for a first approximation that such electrons have all the same energy, we can write down for the volume coefficient of the retardation of radiation generated per unit time and in the energy interval of the photon from $h\nu$ to $h\nu + d(h\nu)$, the following expression:

$$\varepsilon_\nu d(h\nu) = \sigma_\nu(E) n_e n_i v h\nu d(h\nu) , \qquad (8.6)$$

where v is the velocity of the fast electrons, and n_e and n_i are the densities of electrons and protons. Transforming from frequency to wavelength we have for the volume coefficient of the bremsstrahlung per unit wavelength interval

$$\varepsilon_\lambda = 4\alpha r_o^2 n_e n_i \frac{v}{c} \gamma^2 h \left(\frac{mc^2}{h} \right)^2 F(\omega,\gamma) , \qquad (8.7)$$

where

$$F(\omega,\gamma) = \omega^2 f(\omega,\gamma) . \qquad (8.8)$$

If the cloud of fast electrons around the star occupies a volume V, then the total energy radiated by this volume per unit time and per unit wavelength interval will be

$$E_\lambda = \varepsilon_\lambda V \qquad \text{erg s}^{-1} \qquad (8.9)$$

In the case where this cloud forms a shell with outer radius $R = qR_*$ and inner radius R_*, where R_* is the radius of the star, one may write

$$V = \frac{4\pi}{3} R_*^3 (q^3 - 1) \qquad (8.10)$$

$$\tau = n_e \sigma_e R_* (q - 1) \qquad (8.11)$$

where τ is the optical depth of the medium for processes of Thomson scattering.

The additional bremsstrahlung energy E_λ is superposed on the star's normal Planck radiation corresponding to the effective temperature T. This energy is equal to

$4\pi R_*^2 B_\lambda(T)$. Therefore the stream of radiation from the system "star + shell of fast electrons" will be

$$4\pi R_*^2 J_\lambda = E_\lambda + 4\pi R_*^2 B_\lambda(T) . \qquad (8.12)$$

This relation holds as long as the self-absorption of the radiation in the shell is small, i.e. as long as $\tau < 1$.

From (8.12) we find for the brightness of the flare J_λ:

$$J_\lambda(\tau,\gamma,T) = B_\lambda(T)D_\lambda(\tau,\gamma,T) , \qquad (8.13)$$

where

$$D(\lambda,\gamma,T) = 1 + \frac{4}{3} \alpha r_o^2 n_e n_i \frac{v}{c} \left(\frac{mc^2}{h}\right)^2 R_*(q^3 - 1)\gamma^2 \frac{F(\omega,\gamma)}{B_\lambda(T)} . \qquad (8.14)$$

With $v/c = 1$, $n_e = n_i$ and n_e from (8.11), and with the numerical values of the constants we have instead of (8.14):

$$D(\tau,\gamma,T) = 1 + 0.48\times10^{40} \frac{\tau^2}{R_*} \frac{q^3-1}{(q-1)^2} \lambda^5(e^{hc/\lambda kT} - 1)\gamma^2 F(\omega,\gamma) . \qquad (8.15)$$

The dimensionless coefficient $D_\lambda(\tau,\gamma,T)$ is analogous to the coefficient $C_\lambda(\tau,\gamma,T)$ when the optical flare is induced by the inverse Compton effects; it indicates relative intensity and shows by how many times the resulting radiation exceeds the Planck radiation of the star at a given wavelength for a given flare strength. During the flare $D_\lambda > 1$ and when there is no flare, i.e. when $\tau = 0$, D_λ equals one. The quantity $(D_\lambda - 1)$ gives the ratio of the energy radiated by the shell to the star's Planck radiation.

The lightcurve of the flare given by (8.13) has a number of interesting features. First, given the assumptions D_λ is inversely proportional to the star's radius. This means that in the case of the star UV Cet with a radius equal to $0.08 R_\odot$, the relative role of the bremsstrahlung must be greater than in the case of AD Leo, the radius of which is nearly one order of magnitude larger.

It also follows from (8.15) that the value of the additional energy caused by bremsstrahlung of the fast electrons is proportional to τ^2, i.e. it depends on the optical depth of the medium much more strongly than in the case of the inverse Compton effect, where the additional radiation is proportional to τ. Hence the role of bremsstrahlung must be small for faint flares (with τ small).

Finally due to the strong dependence of the lightcurve on τ the effective duration of the flare must be considerably shorter in the case of bremsstrahlung than in the case of the inverse Compton effect.

The stars UV Cet and AD Leo are extreme cases among flare stars with regard absolute luminosity. Therefore it is useful to perform the following computations for two cases, for $R_* = 0.5\times10^{10}$ cm, and for $R_* = 5\times10^{10}$ cm. The effective temperature will be taken $T = 2800$ K in both cases. With regard the factor $(q^3-1)/(q-1)^2$ in formula (8.15), it is approximately equal to 10 (when q fluctuates between 2 and 10); and the intensity of the radiation J_λ is determined with an accuracy not worse than 40%.

With the aid of these data and formula (8.15) the numerical values of D_λ have been derived for $\gamma^2 = 10$ for two values of the star's radius, and a number of values of τ, smaller than 0.1. The results are given in Table 8.1.

TABLE 8.1 Relative Intensity of the Radiation $D_\lambda(\tau,\gamma,T)$ of the Star During the Flare, Due to Bremsstrahlung of the Fast Electrons for $R_* = 0.5\times10^{10}$ cm and $R_* = 5\times10^{10}$ cm, for a Number of Values of τ

λ, Å	ω	$F(\omega,\gamma)$	$R_* = 0.5\times10^{10}$ τ				$R_* = 5\times10^{10}$ τ			
			0.1	0.01	0.001	0.0001	0.1	0.01	0.001	0.0001
3000	2.55×10^{-6}	1.270×10^{-10}	81740	818	9.2	1.082	8174	83	1.81	1.008
4000	1.91	0.730	2728	28.3	1.27	1.003	274	3.7	1.027	1
5000	1.53	0.474	415	5.1	1.04	1.0004	42	1.4	1.004	1
6000	1.27	0.333	131	2.3	1.013	1.0001	14	1.13	1.001	1
7000	1.09	0.247	62	1.6	1.006	1	7.1	1.06	1	1
10000	0.76	0.124	21	1.2	1.002	1	3.0	1.02	1	1

If the function $J_\lambda(\tau,\gamma,T)$ is known, i.e. the law of the energy distribution in the star's spectrum during the flare, then it is easy to determine the observed parameters, in particular the color indices and the flare amplitudes in various wavelength regions.

3.   RADIATION SPECTRUM

For strong flares, as we shall see below (§5) non-thermal bremsstrahlung processes begin to predominate over the inverse Compton effect. In such a case all the main characteristics including the color indices and flare amplitudes will be determined entirely by the non-thermal bremsstrahlung.

Following the discussion above the energy distribution in the spectrum of the radiation induced by the non-thermal bremsstrahlung, in the optical wavelength region, was computed. The results for two values of the electron energy, $\gamma = 3$ (E = 1.5 MeV) and $\gamma = 15$ (E = 7.5 MeV) are given in Fig. 8.1, where the intensity is given in stellar magnitudes, m_v, on a frequency scale. The spectrum of the radiation is

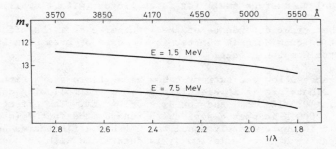

Fig. 8.1. Theoretical spectrum of the light of a flare induced by non-thermal bremsstrahlung of fast electrons with energies $\gamma = 3$ (1.5 MeV) and $\gamma = 15$ (7.5 MeV). The intensity of the radiation is given in stellar magnitudes m_v.

quite steep, and it is not very different from what we had in the case of the inverse Compton effect (Fig. 5.7). One of the flares of EV Lac, for which the spectrum of the light of the pure flare was well determined, proved to be very strong (Ch. 5, §5). According to the theory the predominating process for this flare must be non-thermal bremsstrahlung, and therefore we must have in this case a very steep spectrum. This is actually the case: the observed spectrum of this flare at maximum (Fig. 5.12) is found to be very steep.

4. COLOR INDICES

First we are interested in the numerical values of U – B and B – V for the case of pure bremsstrahlung of fast electrons. For a wide range of the electron energy these values do not depend on γ and are equal to

$$U - B = -1\overset{m}{.}33$$

$$B - V = +0\overset{m}{.}04 \ .$$

In the case of the inverse Compton effect we had: $U - B = -1\overset{m}{.}80$ and $B - V = -0\overset{m}{.}38$ (cf. Fig. 7.1). Hence radiation of Compton origin is considerably "bluer" than in the case of bremsstrahlung. A value of U – B of the order $-1\overset{m}{.}3$ for the flare radiation is observed quite often. However, one cannot conclude exactly what causes the additional radiation--the retardation of the electrons or the inverse Compton effect. But there are cases of flares where $U - B = -1\overset{m}{.}5$, which indicates that the additional radiation is of Compton origin.

Thus only in the case of small values of U – B (smaller than $-1\overset{m}{.}3$) one can speak about additional radiation of Compton origin. In the remaining cases the color characteristics are not sufficient for making an unambiguous choice between bremsstrahlung and inverse Compton effects. Additional data is clearly needed. The dependence of U – B and B – V on τ is shown in Table 8.2, and the color diagram constructed with the aid of these data is analogous to that which we had in the case of the inverse Compton effect (cf. Fig. 7.2).

TABLE 8.2   Color Indices of Flare Stars for Bremsstrahlung of Fast Electrons Depending on R_* (cm) and τ

τ	$R_* = 0.5 \times 10^{10}$		$R_* = 5 \times 10^{10}$	
	U – B	B – V	U – B	B – V
0.1	$-1\overset{m}{.}33$	$+0\overset{m}{.}05$	$-1\overset{m}{.}32$	$+0\overset{m}{.}095$
0.01	−1.23	+0.39	−0.68	+1.26
0.001	+0.50	+1.74	+0.80	+1.81
0.0001	+1.13	+1.82	+1.14	+1.82
0	+1.14	+1.82	+1.14	+1.82

5.   FLARE AMPLITUDE

Table 8.3 gives values of the theoretical flare amplitudes due to bremsstrahlung of fast electrons in U, B and V wavelengths as a function of τ and R_*.

Analysis of the data given in the last table, and also a comparison with the data of Table 6.4, leads to a number of interesting conclusions. First of all, on

TABLE 8.3 Amplitudes of Flares Due to Bremsstrahlung of Fast
Electrons in U, B and V Wavelengths

R_*, cm	Spectral range	τ			
		0.1	0.01	0.001	0.0001
0.5×10^{10}	U	$10\overset{m}{.}0$	$5\overset{m}{.}0$	$0\overset{m}{.}75$	$0\overset{m}{.}012$
	B	7.6	2.6	0.10	0
	V	5.8	1.2	0.02	0
5×10^{10}	U	7.5	2.6	0.10	0
	B	5.0	0.8	0.01	0
	V	3.3	0.2	0	0

account of the strong dependence of the bremsstrahlung with τ, it is efficient
mostly for large values of τ; for small values of τ the additional radiation is
almost totally due to inverse Compton effects. There is a certain critical value
τ_{cr}, at which bremsstrahlung and inverse Compton effect are equally important. The
quantity τ_{cr} depends on R_*, i.e. on the star's luminosity. Thus, for UV Cet
($R_* = 0.5 \times 10^{10}$ cm) $\tau_{cr} = 0.009$, for an M5 stellar type, and $\tau_{cr} = 0.06$, if it
belongs to subclass M6. In the first case the critical amplitude of the flare in
U wavelengths is equal to 4^m. Thus it can be said that all flares of UV Cet with
amplitude larger than 4^m in U radiation are caused by bremsstrahlung. When
$\Delta U < 4^m$ the flares are of Compton origin. It also follows from Table 1.8 that
flares of UV Cet with $\Delta U > 4^m$ are 2-3% of the total number of flares of this star.
This implies that about 97% of the cases of flares of UV Cet are due to inverse
Compton effects and only 3% of the flares may be due to non-thermal bremsstrahlung.

In the case of AD Leo ($R_* = 5 \times 10^{10}$ cm) the situation is different. For this star
$\tau_{cr} = 0.09$ which corresponds with $\Delta U = 6^m$. But AD Leo never shows flare amplitudes
larger than 5^m in U wavelengths (cf. Table 1.8). Hence we may conclude that for
this star bremsstrahlung plays no role at all.

All other flare stars are intermediate in luminosity between UV Cet and AD Leo.
Moreover large values of the amplitudes are extremely rare. Therefore it may be
stated that flares of UV Cet type are practically caused by inverse Compton effects.
Only for exceptionally strong flares the bremsstrahlung of fast electrons can play
a role.

The maximum amplitude observed in stars of UV Cet type is $6\overset{m}{.}6$ in U wavelengths
for YZ CMi. In stellar associations flares have been recorded with high ampli-
tudes in U radiation such as in Orion $\sim 8\overset{m}{.}4$, $8\overset{m}{.}5$, and even $> 8\overset{m}{.}5$ in the Pleiades.

Comparing these data with Table 8.3 we see that the optical depth is never more than
0.01. In other words, the total energy emitted during an exceptionally strong and
extremely rare flare never exceeds 10^{38} ergs. The overwhelming majority of the
flares correspond to cases with $\tau < 0.001$.

From the data in Tables 7.4 and 8.3 the flare amplitudes ΔU_*, ΔB_* and ΔV_* due to
bremsstrahlung and inverse Compton effects can be found together. In the case of
M5 and M6 type stars the results are given in Table 8.4 for a number of different
values of τ. For intermediate values of τ the values of the amplitudes may be
found with the aid of Fig. 8.2.

TABLE 8.4   Results for Amplitude ΔU_*, ΔB_*, ΔV_* of Flares of Stars of Classes M5 and M6, Due to Both Inverse Compton Effects and Non-thermal Bremsstrahlung

		τ = 0.1	0.01	0.001	0.0001
M5	ΔU_*	10.0	5.5	2.25	0.50
	ΔB_*	7.6	3.0	0.48	0.05
	ΔV_*	5.8	1.5	0.09	0.006
M6	ΔU_*	10.2	6.35	3.7	1.4
	ΔB_*	7.7	3.5	1.0	0.14
	ΔV_*	5.9	1.85	0.17	0.02

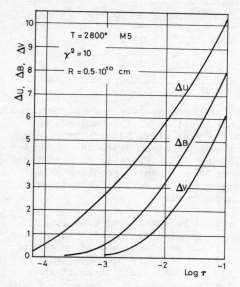

Thus the main flare characteristics-- color indices and amplitudes--do not differ qualitatively for bremsstrahlung or Compton origin of the flares. How- ever, in one aspect bremsstrahlung is quite different from the inverse Compton effect. Namely, in the infrared spectral region bremsstrahlung leads to a positive flare, as opposed to the negative flare found for the case of the inverse Compton effect in the same spectral region. The amplitudes of the positive flare are extremely small (Table 8.5) and cannot be discovered without special efforts. Since most flares correspond to cases where $\Delta U < 3^m$ ($\tau < 0.001$), then the dis- covery of the weak positive infrared flare is improbable. However, there is some probability that the negative infrared flare may be detected.

Fig. 8.2.   Resulting amplitudes of flares in UBV due to both non-thermal bremsstrahlung and inverse Compton effects.

TABLE 8.5   Theoretical Flare Amplitudes at $\lambda = 10\ 000$ Å, Caused by Bremsstrahlung of Fast Electrons ($\gamma^2 = 10$)

	τ = 0.1	0.01	0.001
$R_* = 0.5 \times 10^{10}$ cm	$3^m.25$	$0^m.19$	0^m
$R_* = 5 \times 10^{10}$ cm	1.15	0.02	0

6. APPLICATION TO HOT STARS

The hypothesis of the inverse Compton effect excludes the possibility of a positive
flare in U, B and V wavelengths in the case of hot stars, when $T \approx 10\ 000°$ (Ch. 6).
On the other hand, bremsstrahlung of fast electrons can in principle be the cause
of a positive flare in high-temperature stars. However, such a flare can reach
significant values only for large values of τ, of the order of 0.1 (Table 8.6).

TABLE 8.6 Theoretical Amplitudes of Flares Caused by Bremsstrahlung of Fast
Electrons in the Case of Hot Stars ($\gamma^2 = 10$, $R_* = 0.5 \times 10^{10}$ cm)

Flare Amplitudes	T = 10 000° τ				T = 20 000° τ			
	0.5	0.2	0.1	0.01	0.5	0.2	0.1	0.01
ΔU	3.9	1.98	0.28	0	0.68	0.15	0.03	0
ΔB	3.8	1.95	0.27	0	0.75	0.16	0.04	0
ΔV	3.8	1.94	0.24	0	0.87	0.20	0.06	0

In this connection we should mention the communications appearing now and then about
cases of flares of early-type stars. For instance, on 1 March 1964 Andrews [7]
recorded a flare of the sixth magnitude star HD 37519 of class B8 with amplitude
3^m; Page [8] has noted three flares of 66 Oph, a star of class B1 to B6, occurring
in August–September 1969 with amplitudes 1^m0, 1^m8 and 0^m8. There is a communication [9] about a flare of a B8 star in the cluster M6 with an improbably large
amplitude of 7^m (!); this flare was recorded by means of television techniques
and subsequently it was registered on a photographic plate.

These communications, especially the last one, should be considered cautiously
since it is not always possible to ascertain the circumstances in which these observations were made. However, even in cases where these are true (and if we reject
the assumption that such a flare may be due to a faint and very cool companion),
the possibility of a flare from hot stars does not in principle contradict the
results of the theory of the bremsstrahlung of fast electrons.

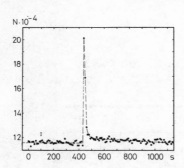

Fig. 8.3. Lightcurve of a flare of
the white dwarf G 44 – 32, recorded
in B light [10].

The communication of Warner et al. [10]
about a flare of the white dwarf G 44 - 32
is very striking. The flare was recorded
on February 7, 1970, with the 82-inch
Struve telescope equipped with a photon
counter with a photomultiplier sensitive
to blue wavelengths. The investigators
succeeded in constructing a lightcurve of
the flare—shown in Fig. 8.3—with an
integration period of 10^s. The amplitude
of the flare was equal to 0^m61, and it
lasted about one minute. The value of the
flare amplitude as well as the short
duration of the flare are in agreement
with what is expected from the mechanism of
bremsstrahlung of fast electrons in hot
dwarfs. However, additional data are
required before we conclude that actually

the hot dwarfs themselves show flares, and not their possible invisible late-type companions.

If the above results are correct, then it is possible to expect flares of those nuclei of planetary nebulae, which are white dwarfs.

REFERENCES

1. G. A. Gurzadyan, Astron. Astrophys. 20, 145, 1972.
2. H. M. Koch, J. W. Motz, Rev. Mod. Phys. 31, 920, 1959.
3. J. Joseph, F. Rohrlich, Rev. Mod. Phys. 30, 354, 1958.
4. H. A. Bethe, Proc. Cambridge Phil. Soc. 30, 524, 1930.
5. H. A. Bethe, W. Heitler, Proc. Roy. Soc. London A146, 83, 1934.
6. W. Heitler, The Quantum Theory of Radiation, Oxford: Oxford University Press, 1954.
7. A. D. Andrews, Irish Astr. J. 6, 212, 1964.
8. A. A. Page, B. Page, Sky and Telescope 40, 206, 1970.
9. G. A. Bakos, Sky and Telescope 40, 214, 1970.
10. B. Warner, G. W. van Citters, R. E. Nather, Nature 226, 67, 1970.

CHAPTER 9

Excitation of Emission Lines.
Chromospheres of Flare Stars

1. THE TWO COMPONENTS OF THE LIGHT OF THE FLARE

The photoelectric and spectroscopic observations of stellar flares have shown that they consist of two components: a) the continuous emission, and b) recombination radiation (emission lines).

At flare maximum the continuous emission predominates; it is 80-95% of the total flare radiation, the remaining 5-20% (in B radiation) is due to emission lines of hydrogen, helium, ionized calcium, etc. On the descending part of the lightcurve, the relative importance of the emission ines increases and at a certain moment the continuous emission disappears completely.
This kind of "anatomy" of the lightcurve is shown schematically in Fig. 9.1. The intersection of the two curves T_a divides the total (observed) lightcurve in two regions--C and E, in each of which one of the components predominates. Both the

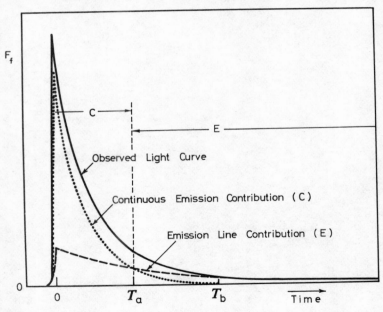

Fig. 9.1. Two-component structure of the observed lightcurve of a flare: dotted line - continuous emission (C); broken line - radiation in the emission lines (E). The point T_a corresponds to the moment of equal values of the two components, the point T_b to the moment of the disappearance of the continuous emission.

position of the point T_a and that of T_b—the moment when the continuous emission disappears—may vary from one flare to the next, even for the same star. The fact that these two regions exist indicates that the rate of the decrease of the radiation is not the same for the two components, hence there must exist certain physical conditions and processes leading to such an effect.

It should be pointed out that irrespective of the division of the lightcurve in the regions C and E, the integrated energy of the emission lines over the entire lightcurve is as small as it is at flare maximum. The energy for the excitation of the emission lines is taken from the same source as for the continuous emission itself.

The spectra of the radiation induced by the inverse Compton effect, and also by the non-thermal bremsstrahlung, are continuous and extend up to the region of the hydrogen ionization. By this short wavelength "tail" emission lines may also be excited. The point is whether the intensity of the ionizing radiation generated during ordinary flares is sufficient to ensure that the emission lines can be distinguished against the background of the stellar radiation in the visible part of the spectrum.

2. THE INTENSITY OF THE IONIZING RADIATION

In Chapter 4 the problem of the transfer of the star's photospheric radiation through the layer of fast electrons was considered. Formulae were derived for finding the spectral distribution of the intensity of the radiation leaving this layer for different models of the medium and different energy spectra of the fast electrons. These formulae are valid for the entire frequency range, including hard ultraviolet rays. For instance, in the model of the actual photosphere and for the case of monoenergetic electrons we have for the flux of radiation directed inwards towards the star:

$$H_\nu(\tau,\gamma,T_*) = B_\nu(T_*) \frac{3}{2\gamma^4} \frac{e^x - 1}{e^{x/\gamma^2} - 1} F_2(\tau) , \qquad (9.1)$$

where τ is the effective optical depth of the medium consisting of fast electrons for processes of Thomson scattering. If the value of the function $B_\nu(T_*)$ is substituted, (9.1) becomes:

$$H_\nu(\tau,\gamma,T_*) = \frac{3}{\gamma^4} \frac{h}{c^2} \left(\frac{kT_*}{h}\right)^3 \frac{x^3}{e^{x/\gamma^2} - 1} F_2(\tau) . \qquad (9.2)$$

For the case of a Gaussian energy distribution of the electrons we have

$$H_\nu(\tau,\gamma_o,\sigma,T_*) = \frac{6}{\sqrt{\pi}\,\sigma} \frac{h}{c^2} \left(\frac{kT_*}{h}\right)^3 x^3 \phi_x(\gamma_o,\sigma) F_2(\tau) . \qquad (9.3)$$

The expressions (9.2) and (9.3) are theoretical spectra of the continuous radiation. With increasing γ the maximum of these spectra is displaced towards shorter wavelengths. Figure 9.2 shows an example of such a spectrum for stars of class M5, computed from formula (9.2). It is remarkable that even for $\gamma^2 = 10$ this maximum is nearly in the region of hydrogen ionization ($\lambda \leq 912$ Å). The radiation lying in the region of wavelengths shorter than $\lambda = 912$ Å will be called L_c.

The appearance of the short wavelength "tail" in the spectrum of the Compton radiation does not give an indication of the strength of the emission lines. Clearly the ratio between the total amount of L_c radiation and the amount of energy emitted by the star in the visual region must be known. First we determine the intensity

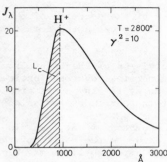

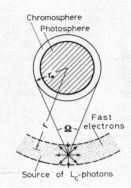

Fig. 9.2. Theoretical distribution of the energy in the region of the L_c radiation (ionizing hydrogen) for a flare of an M5 star.

Fig. 9.3. Excitation of emission lines in flare stars. The source of the ionizing L_c radiation is far from the star; the emission lines arise in the star's chromosphere.

of the L_c radiation.

The general expression for the intensity of the L_c radiation, $I_c(\tau)$, directed from the shell or layer of fast electrons towards the chromosphere of the star (Fig. 9.3) has the form:

$$I_c(\tau) = \int_{\nu_o}^{\infty} H_\nu(\tau,\gamma,T_*)d\nu \tag{9.4}$$

where ν_o is the ionization frequency of hydrogen. In the case of monoenergetic electrons we have (substituting the value of the function $H_\nu(\tau,\gamma,T_*)$ from (9.2):

$$I_c(\tau) = 3\,\frac{h}{c^2}\,\gamma^4\left(\frac{kT_*}{h}\right)^4 F_2(\tau)J_3(x_o/\gamma^2) \tag{9.5}$$

where $J_3(u)$ is

$$J_3(u) = \int_u^{\infty} \frac{z^3 dz}{e^z - 1}\;.$$

The numerical values of this function are tabulated in [1]. For temperatures which are usual for UV Cet stars (T = 2800 °K) we have $J_3(u) = 1.1 \approx 1$, so that in the following this function can be omitted.

The expression (9.5) gives the flux of ionizing L_c radiation of Compton origin directed towards the chromosphere. For the total amount of ionizing radiation of Compton origin, $E_{co}(\tau)$, we have:

$$E_{co}(\tau) = 4\pi r_*^2 I_c(\tau) \tag{9.6}$$

where r_* is the radius of the star. Substituting the value of $I_c(\tau)$ from (9.5) we find:

$$E_{co}(\tau) = 1.72\times10^{11}\,r_*^2\tau \quad \text{ergs s}^{-1} \tag{9.7}$$

Analogously we can write for the intensity of the L_c radiation induced by non-thermal bremsstrahlung of fast electrons:

$$E_{br}(\tau) = \varepsilon_c(\tau)V ,\tag{9.8}$$

where V is the effective volume around the star occupied by the fast electrons, $\varepsilon_c(\tau)$ is the volume emission coefficient:

$$\varepsilon_c(\tau) = \int_{\nu_o}^{\infty} \varepsilon_\nu(\tau)d\nu = 4\alpha r_o^2 \gamma mc^3 n_e^2 f_c(\omega_o)\tag{9.9}$$

where the value of the function $\varepsilon_c(\tau)$ from (8.6) has been substituted and where $f_c(\omega_o)$ indicates:

$$f_c(\omega_o) = \int_{\omega_o}^{\infty} f(\omega,\gamma)d\omega\tag{9.10}$$

the form of the function $f(\omega,\gamma)$ is given in (8.4).

As we saw in the preceding chapter, the short wavelength end of the radiation spectrum, in the case of non-thermal bremsstrahlung, extends to the region of X-rays (<1 Å). However, it is easy to see that the wavelength region, where the hydrogen ionization is most effective, lies between 912 Å and ∿200 Å (partly covering also the region of absorption by neutral helium). From 228 Å to about 3 Å the absorption is due exclusively to ionized helium (if n(He)/n(H) ∿ 0.1). In the region of X-rays (<3 Å) the opacity of the medium is determined chiefly by Thomson scattering on free electrons. Performing the corresponding integrations we have for the numerical value of $f_c(\omega_o)$:

Wavelength region	$f_c(\omega_o)$
900 – 200 Å	0.0007
200 – 3 Å	0.08
<3 Å	1.0

Substituting in (9.8) the values of V and n_e from (8.10) and (8.11) we find $E_{br}(\tau)$, the power of the ionizing radiation:

$$E_{br}(\tau) = 1.63\times10^{26} r_*\tau^2 f_c(\omega_o) \text{ ergs s}^{-1}\tag{9.11}$$

Substituting $f_c(\omega_o) = 0.0007$:

$$E_{br}(\tau) = 1.14\times10^{23} r_*\tau^2 \text{ ergs s}^{-1}\tag{9.12}$$

A certain part of the L_c radiation is transformed, by fluorescence, into radiation in a given emission line. Therefore, depending on which of the two components-- $E_{co}(\tau)$ or $E_{br}(\tau)$--predominates in a given case, we shall have emission lines of Compton origin or of bremsstrahlung origin. Let us give the results of the computations for UV Cet, for which $r_* = 0.08 R_\odot = 0.56\times10^{10}$ cm. From (9.7) and (9.12): $E_{co}(\tau) = 0.54\times10^{31} \tau$ ergs s^{-1} and $E_{br}(\tau) = 0.64\times10^{33} \tau^2$ ergs s^{-1}. The amounts found by means of these relations for the total amount of the ionizing radiation, as a function of flare strength τ, are given in the second and third columns of Table 9.1.

The dependence of the strength of the L_c radiation on τ is stronger in the case of the non-thermal bremsstrahlung (∿τ^2) and weaker in the case of the inverse Compton effect (∿τ). Therefore the curves of E(τ) versus τ will intersect for a certain value τ' (Fig. 9.4), separating the regions where the ionizing radiation of one or the other type predominates. Thus in the case considered, of the star UV Cet, this

TABLE 9.1 Results of Computations for $E_{co}(\tau)$ and $E_{br}(\tau)$ in the Case of UV Cet

τ	$E_{co}(\tau)$ ergs s^{-1}	$E_{br}(\tau)$ ergs s^{-1}
0.0001	5.4×10^{26}	0.6×10^{25}
0.001	5.4×10^{27}	0.6×10^{27}
0.01	5.4×10^{28}	6.0×10^{28}
0.05	0.3×10^{30}	1.6×10^{30}

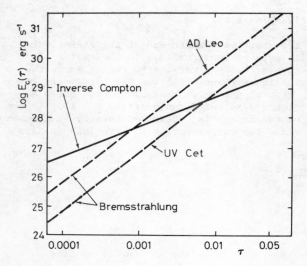

Fig. 9.4. The dependence of the power of the radiation ionizing hydrogen due to inverse Compton effect and non-thermal bremsstrahlung on the strength of the flare τ.

intersection occurs for $\tau' \approx 0.01$, hence the emission lines appearing during faint flares ($\tau < 0.01$) will be induced by ionizing radiation of Compton origin. In the case of strong flares ($\tau > 0.01$) the intensity of the emission lines will be determined by processes of non-thermal bremsstrahlung. In the case of another star, AD Leo, the picture is different: for this star $R_* \approx 5 \times 10^{10}$ cm and the intersection mentioned (in Fig. 9.4) occurs for $\tau' \approx 0.001$. For other flare stars τ' lies between 0.01 and 0.001. Therefore when we have a flare with strength $\tau < 0.001$, the excitation of the emission lines will be due chiefly to processes of the inverse Compton effect, and for $\tau > 0.001$--to non-thermal bremsstrahlung. For flares with strength $\tau \sim 0.01$-0.001 both processes play more or less equal roles.

3. EXCITATION OF EMISSION LINES

Speaking about the generation of emission lines, or about the excitation of the chromosphere and, in general, the atmosphere of flare stars, the following should be kept in mind. During the flare the existing emission lines are strengthened and also new emission lines appear. At the same time the observations do not show any strengthening of the absorption lines. No new absorption lines appear. This seems possible only in the case that the radiation exciting the chromosphere of the star falls on it from the outside.

The layer or cloud of fast electrons lies in the general case, according to our model, either at some distance from the photosphere (Fig. 9.3), or it extends from the star's photosphere to a certain distance r. Therefore only part of the L_c radiation generated in this medium will reach the chromosphere; the remaining part will leave the star forever. If the amount of L_c radiation reaching the chromosphere is

$\varepsilon_c(\tau)$, then we have:

$$\varepsilon_c(\tau) = WE_c(\tau) \tag{9.13}$$

where W is the efficiency of L_c radiation in fluorescence processes. In a particular case a geometrical interpretation may be given to W: in that case we have to do with the ordinary dilution coefficient given by the following expression:

$$W = \frac{\Omega}{4\pi} = \frac{1}{2}\left| 1 - \sqrt{1 - (r_*/r)^2} \right| \tag{9.14}$$

r_* is the radius of the star, r the radius of the shell of fast electrons.

The L_c radiation of strength ε_c which reaches the outer layers of the star's atmosphere—the chromosphere—may lead at the moment of the flare to a sudden strengthening of the ionization in these layers; as a result emission lines appear.

Thus our model implies the generation of L_c radiation in one region, at a certain distance from the star's photosphere, and the appearance of emission lines in another region, in the star's chromosphere. During this process the region where the L_c radiation is generated—the fast electrons together—expand and diffuse leaving the star quickly, but the region where the emission lines arise, the chromosphere, remains attached to the star, thought it may suffer some variations due to the unavoidable heating under the influence of the L_c radiation itself.

Thus the ensemble of fast electrons which appear spontaneously, sending towards the chromosphere a certain number of L_c photons, is quickly exhausted. But the chromosphere receiving the impulse of the L_c energy passes at once to the highest possible state of ionization, after which the process of pure recombination begins. The total energy in the form of fluorescence radiation must be equal to or somewhat less than the total L_c energy falling on the chromosphere. The duration of recombinations is determined, in particular, by the electron density of the chromosphere.

This is the general picture of the appearance and disappearance of emission lines during a flare of a cold star. In the general case the problem of the excitation of emission lines during a flare must be considered as a non-stationary problem. However, here we restrict ourselves to a qualitative analysis. We consider an equilibrium between the radiation emitted by the chromosphere with the energy in the form of emission lines on the one hand, and the L_c radiation absorbed by it on the other.

The absorption coefficient of one atom in the lines of the fundamental series of hydrogen is usually three to four orders of magnitude larger than the coefficient of continuous absorption at frequencies of the L_c radiation. Therefore the chromosphere above the level $\tau_c = 1$ must be completely opaque in the lines of the Lyman series of hydrogen. Regarding the Balmer lines, the value of the optical thickness depends on the degree of excitation of hydrogen as well. If the population of the second hydrogen level is sufficiently high, then the medium is partly or even completely opaque in the lines of the Balmer series. It is also necessary to take into account self-absorption in the lines, i.e. to solve the equation of transfer in the emission lines. However, for us it is interesting to establish the conditions under which a certain Balmer line will be visible against the background of the continuous radiation of the star. Therefore in the first approximation the self-absorption may be neglected.

According to the theory of the formation of emission lines as a result of fluorescence a certain part of the L_c energy—called Γ_i—is transformed into energy of the emission line of the Balmer series of hydrogen H_i. Assuming that in the chromosphere there are sufficient number of hydrogen atoms absorbing all the energy which falls on them from outside, we may write for the total intensity of that line:

$$E(H_i) = \Gamma_i \varepsilon_c(\tau) = \Gamma_i WE_c(\tau) , \tag{9.15}$$

where $E_c(\tau)$ is given by (9.7) or (9.12) for the case of Compton or bremsstrahlung origin of the L_c radiation.

The observations of the intensities of the emission lines in flare stars are usually given in the form of equivalent widths W_i for some line H_i. Therefore we can write for the intensity of the emission line $E(H_i)$, taking into account the deformation of the Planck distribution of the continuous spectrum of the star due to inverse Compton effect:

$$E(H_i) = W_i C_i(\tau,\gamma,T_*) B_\lambda(T_*) . \tag{9.16}$$

Equating (9.16) and (9.15) and substituting the value of $E_c(\tau)$ from (9.7) and also taking $F_2(\tau) \approx \tau/2$, we find for the equivalent width of the emission line, when it is induced by Compton radiation:

$$\frac{W_i}{\lambda_i} = W\Gamma_i \frac{3\gamma^4}{4} \frac{e^{x_i}-1}{x_i^4} \frac{\tau}{C_i(\tau,\gamma,T_*)} . \tag{9.17}$$

In the case of a very strong line, when both the emission line and the continuous spectrum are due to non-thermal bremsstrahlung, we have,

$$\frac{W_i}{\lambda_i} = W\Gamma_i \frac{f_c(H^+)}{\omega f(\omega,\gamma)} . \tag{9.18}$$

As expected in this case the ratio W_i/λ_i does not depend on τ or on the physical parameters of the star; it depends only on the dilution of the radiation W.

We use equations (9.17) and (9.18) to find the value of the equivalent width of the emission line $H\beta$ in the spectrum of a flare star of UV Cet type. Taking $\Gamma_\beta \approx 0.05$, we have from (9.17): $W\beta/\lambda\beta = 15$ $W\tau$ for faint flares ($\tau < 0.001$), using the data of Table 15.1, $W_\beta/\lambda_\beta \approx W$ for strong flares ($\tau > 0.001$). The values found by means of these relations are given in Table 9.2.

TABLE 9.2 Equivalent Width of the Emission Line $H\beta$ in the Spectrum of a Flare Star of UV Cet Type (Theoretical Computations)

τ	W_β (Å)		Mechanism of excitation
	W = 0.1	W = 0.01	
0.0001	7	1	Inverse Compton
0.001	73	7	Inverse Compton
0.01	500	50	Bremsstrahlung

Let us now consider the observations. There are relatively few reliable observational data about equivalent widths of emission lines at the moment of a flare. From measurements of spectrograms of some flares of UV Cet and AD Leo equivalent

widths were found of 10 to 120 Å. At the time of the maxima of five flares of AD
Leo P. F. Chugainov [2] found for the equivalent width W_β of the H_β line of hydrogen,
90, 80, 70, 20 and 15 Å, respectively. The maximum values of W_β, which are reached
after flare maximum, were found to be 170, 155, 300, 250 and 90 Å, respectively.
For one flare of UV Cet of medium strength ($\Delta U = 2^m4$) Bopp and Moffett [3] found
$W_\beta = 23$ Å, and for another very strong flare of the same star ($\Delta U = 5^m2$) they found
$W_\beta \approx 40$ Å (cf. Table 9.7). Comparing all these data with the results in Table 9.2
we see that actually the observations are within the limits predicted by the theory.
Moreover, from the data of the strong flare of UV Cet ($W_\beta = 40$ Å) the probable value
of the dilution coefficient W was found to be of the order of 0.01 (last line in
Table 9.2), i.e. the cloud of fast electrons in this case extends to distances five
times the star's radius.

It is interesting to compare the power of the L_c radiation of Compton origin in
flare stars with the Planck T_* radiation of some conventional stars with effective
temperature T_0.

The total number of L_c photons of Compton origin emitted by the flare star is:

$$N_c = 4\pi r_*^2 \frac{3}{2} \gamma^2 F_2(\tau) \frac{2}{c^2} \left(\frac{kT_*}{h}\right)^3 J_2 \left(\frac{x_0}{\gamma^2}\right) , \qquad (9.19)$$

where

$$J_2(u) = \int_u^\infty \frac{z^2 dz}{e^z - 1} .$$

The total number of L_c photons of thermal origin emitted by our hypothetical star
of the same dimensions, but with an effective temperature T_0, will be:

$$N_c^0 = 4\pi r_*^2 \frac{2}{c^2} \left(\frac{kT_0}{h}\right)^3 J_2(y_0) . \qquad (9.20)$$

In these expressions $x_0 = h\nu_0/kT_*$, $y_0 = h\nu_0/kT_0$.

From (9.19) and (9.20) we have

$$\frac{N_c}{N_c^0} = \frac{3}{2} F_2(\tau)\gamma^2 \left(\frac{T_*}{T_0}\right)^3 \frac{J_2(x_0/\gamma^2)}{J_2(y_0)} . \qquad (9.21)$$

If the flare parameters γ, τ and T_* are given the ratio N_c/N_c^0 depends only on T_0--
the temperature of the hypothetical star. Since we are looking for an "equivalent"
of the flare star with regards L_c activity, we may find the required temperature T_0
from (9.21), with $N_c/N_c^0 = 1$; it is found to be between 13000 and 18000 °K for M5 to
K5 stars.

Hence the radiative power in L_c of cool stars at the moment of a flare is about
equal to the radiative power of ordinary stars with a temperature of the order of
15000 °K, i.e. B type stars. It is known that stars of class B may excite rather
strong emission lines when embedded in a gas nebula.

Thus, with regards the efficiency of excitation of emission lines of non-thermal
nature, stars of classes M-K during flares are equivalent to stars of type B, in
which the excitation of the emission lines has a thermal nature.

4. DURATION OF EMISSION LINES (OBSERVATIONS)

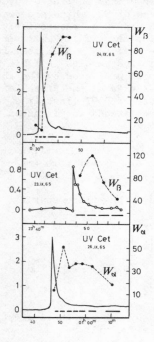

Fig. 9.5. Lightcurves (full-drawn lines) and the variations of the equivalent widths of the lines H_α and H_β (broken lines) for three flares of UV Cet.

The effective duration of the flare in the emission lines exceeds the duration of the flare in the continuous spectrum. This important phenomenon was established by synchronous photoelectric and spectrophotometric observations of individual flares. In general this was illustrated for one flare of UV Cet, and was analyzed in Ch. 6 (Fig. 6.5). In quantitative form, examples of three flares of UV Cet are shown in Fig. 9.5. Here the full-drawn lines indicate the ordinary lightcurves of the flares and the broken lines the variations of the equivalent width of the emission lines H_α or H_β during the flare; they are constructed from the results of measurements of spectrograms obtained by R. E. Gershberg at various moments during the development of the flare, indicated by short lines in the lower part of the lightcurves [4].

Judging from the curves in Fig. 9.5 the moment of maximum of the equivalent width of the emission line falls somewhat later than the moment of maximum of the flare in photovisual light. However this difference in the moments of maxima is not due primarily to the increase of the intensity of the emission line, but rather to the decrease of the flare intensity in continuous light. However, the available data are not yet sufficient to decide whether the two maxima coincide with an accuracy of more than one to two minutes of time.

Another important point is that the radiation in the emission lines continues for some time, even after the star has returned to its original brightness. The duration of this process increases the effective duration of the flare several times, or more. This phenomenon is true not only for the hydrogen lines, but also for other emission lines, among which are the lines of ionized calcium. The emission lines of Ca II appear to differ in their behavior from the hydrogen lines. The calcium lines get excited more slowly and seem to disappear more slowly [5]. Here we have a clear indication that the maximum of the intensity of the calcium lines falls later than the maximum of the flare in integrated light. The calcium lines are seen to last much longer (possibly even many hours) compared to the continuum spectrum of a flare.

The duration of the radiation of the star in the emission lines, Δt_e, measured from the moment of flare maximum until its complete extinction varies within wide limits, depending on the strength of the flare and the spectral class of the star. In most of the cases it appears to be of the order of 10-20 minutes for flare stars of UV Cet type.

The following rule is also remarkable: emission lines, belonging to elements or ions with high ionization potential of the ground state, for instance, lines of helium, always appear later than the hydrogen lines, and their radiation lasts for a relatively short time.

The characteristics of the emission lines of hydrogen and ionized calcium are also observed in flare stars which lie in stellar associations. Haro, for instance, notes two types of objects in Orion, undergoing spectral variations during the flare [6]. In one of these the intensity of the emission line H_{α} decreases after flare maximum more slowly than the integrated flare brightness; however, at the end of the flare it also disappears. In objects of the second type, characterized by a slower rate of growth of the flare, lasting 40 to 60 minutes before maximum, and 5 to 6 hours to return to the normal state, the H_{α} line is weak but remains visible for 1 to 2 days.

This phenomenon has a simple physical explanation, related with the luminescence of the medium (the chromosphere) when the processes of recombination of free electrons with protons last longer than the flare itself, i.e. when there is no longer any photoionization of the atoms. This problem will be considered in more detail in §6, 7 and 8 of the present chapter.

The energy for the excitation of the emission lines, independent of the length of time Δt_e, during which they are emitted, is derived from the fast electrons, either in the form of Compton radiation, or of bremsstrahlung. In the first case the intensity of the radiation is proportional to τ, in the second case to τ^2. This implies that the effective duration of the bremsstrahlung, Δt_{br}, must be considerably less than the effective duration of the Compton radiation Δt_{co}. The decrease of τ after maximum usually is proportional to t^{-n}, where $n \sim 1.5$ to 2 (cf. Ch. 15). Hence we find that Δt_{br} must be at least 3 to 4 times smaller than Δt_{co}.

Thus for all flares of average and higher strength the condition $\Delta t_e > \Delta t_{co} > \Delta t_{br}$ should hold. Numerically these quantities have approximately the ratios 15:3:1.

5. ELECTRON TEMPERATURE IN THE CHROMOSPHERE OF FLARE STARS

The spectral composition of the radiation of Compton origin differs essentially from the spectral composition of the Planck radiation. Hence the average value of the kinetic energy of the electron, torn away from the hydrogen atom by photoionization under the influence of Compton radiation may differ from that which we find in ordinary conditions of thermal processes in stellar atmospheres. At the same time the fact that there are practically no forbidden lines in the flare spectra indicates that one of the chief cooling mechanisms--the escape of kinetic energy out of the medium in the form of radiation of forbidden lines--does not work in the conditions of the atmospheres of flare stars. Therefore it is expected that the star's chromosphere will be hotter during a flare as compared with ordinary gaseous media exciting emission lines. In these conditions the question of finding the value of the electron temperature, or rather its upper limit, in the atmospheres of flare stars during a flare may be quite interesting [7].

In solving this problem we shall start from the assumption that the free electrons are generated by photoionization of hydrogen atoms under the influence of L_c radiation of Compton origin and lose their energy in recombination processes connected with hydrogen. There are no energy losses by excitation of forbidden lines. It is also assumed that the electrons of the medium, resulting from photoionization, have reached a Maxwellian velocity distribution. With regard to the source of the radiation of Compton origin, it is assumed, as before, that it lies outside the medium where the emission lines are excited (Fig. 9.3).

It is easily seen that qualitative considerations connected with the form of the energy spectrum of the fast electrons, as well as with their intensity (τ) do not play a role in the solution of this problem. Then we can restrict ourselves to using the expression (4.8) for computing the spectral composition of the ionizing

L_c radiation in the case of the one-dimensional problem, i.e.

$$J_\nu \sim J_x \sim \frac{x^3}{e^{x/\gamma^2} - 1} \; . \tag{9.22}$$

For the derivation of the relation between the electron temperature T_e of the medium and the energies of the fast electrons γ it is necessary to write the following two equilibrium conditions: a) the condition of stationarity: the number of atoms entering the continuum at the photoionization per unit time must be equal to the number of atoms leaving it: b) the condition of radiative equilibrium: the amount of energy spent on photoionization of hydrogen atoms per unit time must be equal to the amount of energy emitted at recombinations.

Calling the number of hydrogen atoms in the ground state per unit volume n_1, and the coefficient of continuous absorption per atom $k_{1\nu}$, we have for the number of absorptions of L_c photons per unit time:

$$n_1 \int_{\nu_o}^{\infty} k_{1\nu} \frac{J_\nu}{h\nu} \, d\nu \; , \tag{9.23}$$

where ν_o is the ionization frequency of hydrogen.

For the number of recombinations at all levels we have:

$$4\pi n^+ n_e \left(\frac{m_e}{2\pi k T_e}\right)^{3/2} \sum_{i=1}^{\infty} \int_0^{\infty} \beta_i(T_e) e^{-\frac{m_e v^2}{2kT_e}} v^3 dv \; , \tag{9.24}$$

where n^+ and n_e are the numbers of ions and free electrons per unit volume, $\beta_i(T_e)$ is the effective cross section for recombination.

Application of the condition of stationarity gives:

$$n_1 \int_{\nu_o}^{\infty} k_{1\nu} \frac{J_\nu}{h\nu} \, d\nu = 4\pi n^+ n_e \left(\frac{m_e}{2\pi k T_e}\right)^{3/2} \sum_{i=1}^{\infty} \int_0^{\infty} \beta_i(T_e) e^{-\frac{m_e v^2}{2kT_e}} v^3 dv \; . \tag{9.25}$$

In order to write down the condition of radiative equilibrium we must compute the energy absorbed during photoionization and the energy radiated upon recombination and equate them. This gives:

$$n_1 \int_{\nu_o}^{\infty} k_{1\nu} J_\nu d\nu = 4\pi n^+ n_e \left(\frac{m_e}{2\pi k T_e}\right)^{3/2} \sum_{i=1}^{\infty} \int_0^{\infty} \beta_i(T_e) h\nu \, e^{-\frac{m_e v^2}{2kT_e}} v^3 dv \; . \tag{9.26}$$

The function $\beta_i(T_e)$ entering these equations has the form:

$$\beta_i(T_e) \sim k_{i\nu} \frac{i^2 \nu^2}{v^2} \; . \tag{9.27}$$

When writing down the expressions for the absorption coefficients $k_{1\nu}$ and $k_{i\nu}$ we also take into account the influence of the negative emission:

$$k_{1\nu} \sim \frac{1}{\nu 3}\left(1 - e^{-\frac{h\nu}{kT_e}}\right) \quad ; \quad k_{i\nu} \sim \frac{1}{i^5 \nu^3}\left(1 - e^{-\frac{h\nu_i}{kT_e}}\right) . \tag{9.28}$$

Then we find from (9.25) and (9.26), using instead of $m_e v^2/2$ the quantity $h\nu - h\nu_i$:

$$\frac{\displaystyle\int_{x_o}^{\infty} x^{-1}\left(1 - e^{-x}\right)\left(e^{x'/\gamma^2} - 1\right)^{-1} dx}{\displaystyle\int_{x_o}^{\infty}\left(1 - e^{-x}\right)\left(e^{x'/\gamma^2} - 1\right)^{-1} dx} = \frac{\displaystyle\sum_{i=1}^{\infty} \frac{e^{x_i}}{i^3}\left[\int_{x_i}^{\infty} \frac{e^{-x}}{x} dx - \int_{2x_i}^{\infty} \frac{e^{-x}}{x} dx\right]}{\displaystyle\sum_{i=1}^{\infty} \frac{1}{i^3}\left(1 - \frac{1}{2} e^{-x_i}\right)} \tag{9.29}$$

where $x_o = h\nu_o/kT_e$; $x_i = h\nu_i/kT_e$; ν_i is the ionization frequency of the i-th state; $x' = xT_e/T_*$.

The only unknown in the equation (9.29) is the electron temperature T_e of the medium; it is determined unambiguously for given values of the temperature of the star T_* and the energy of the fast electrons γ. In practice the problem of finding T_e is solved as follows: first x_o is determined from (9.29) for given values of T_* and γ, and then T_e from

$$T_e = \frac{h\nu_o}{kx_o} . \tag{9.30}$$

For example we give the computations for a star of class M5 ($T_* = 2800$ K) and for a number of values of γ. The results are shown in Table 9.3, from which it follows that the theoretical value of the electron temperature of the atmosphere (chromosphere) of a cool star under the influence of L_c radiation of Compton origin is quite high--of the order of 100 000 K. It is somewhat higher than the electron temperature of the medium for synchrotron radiation [8]. For a more correct statement of the problem, however, we would also have to take into account the role of inelastic collisions of electrons with hydrogen atoms when the electron temperature is so high. This conclusion that the electron temperature in the chromospheres of flare stars must be very high--of the order of many tens of thousands of degrees seems unavoidable.

TABLE 9.3 Electron Temperature of Gaseous Medium Under the Influence of Compton Radiation

γ^2	T_e
10	154000 K
20	158000
50	175000
100	225000

6. ELECTRON DENSITY IN THE CHROMOSPHERES OF FLARE STARS

The value of the electron density n_e in the chromospheres of flare stars can be found quite reliably from the observed intensities (equivalent widths) of some emission lines in the following way. For the amount of energy ε_i radiated by an ionized gas in a given hydrogen line H_i per unit time and per unit volume we have the well-known relation (cf., for instance [1]):

$$\varepsilon_i = n_e^2 \varepsilon_i^o ,$$

(9.31)

where

$$\varepsilon_i^o = 1.777 \times 10^{-17} \frac{b_i}{T_e^{3/2}} \frac{g_{2i}}{i^3} e^{X_i} \quad \text{ergs cm}^{-3} \text{ s}^{-1}$$

(9.32)

The total energy radiated by the entire chromosphere in the given emission line will be:

$$E_i = 4\pi r_*^2 \Delta r_* \varepsilon_i \quad \text{ergs s}^{-1}$$

(9.33)

where Δr_* is the linear thickness of the chromosphere. This same quantity E_i can also be found in another way, namely from the observed value of the equivalent width of the emission line considered, W_i, and the known effective (Planck) temperature of the star T_*, i.e.

$$E_i = 4\pi r_*^2 W_i B_{\lambda_i}(T_*) .$$

(9.34)

Equating (9.33) and (9.34) and using (9.31) we find:

$$n_e = \left[\frac{W_i}{\Delta r_*} \frac{B_{\lambda_i}(T_*)}{\varepsilon_i^o} \right]^{-1/2} .$$

(9.35)

Let us apply this formula for the line $H\beta$ (i = 4). We have, for $T_e = 20\ 000°$: $b_4 = 0.448$, $X_4 = 0.491$, $g_{24} = 0.822$. Further the observations give $W(H\beta) \approx 100$ Å (cf. Fig. 9–5). Hence we have from (9.35):

$$n_e = 0.8 \times 10^{16} (\Delta r_*)^{-1/2} \text{ cm}^{-3} .$$

(9.36)

We do not know anything about the linear extent of the chromosphere of flare stars. Let us assume that it is the same as for the Sun, i.e. $\Delta r_* \sim 10\ 000$ km (lower limit) or ten times as thick (upper limit). Then we find: $n_e = 2.5 \times 10^{11}$ cm^{-3} for the first case and 0.8×10^{11} cm^{-3} for the second case. Note that on account of the relatively weak dependence of n_e on Δr_* there is no great necessity to know the exact value of the linear thickness of the chromosphere. Thus the assumed electron density in the chromospheres of flare stars must be of the order of 10^{11} cm^{-3}. Of course, this refers to the value of n_e averaged over the height of the chromosphere. Actually, since the distribution of n_e has a vertical gradient, it must be larger than 10^{11} cm^{-3}, probably up to 10^{12} cm^{-3}, at the base of the chromosphere and of the order 10^{10} cm^{-3} or lower in the transition region between chromosphere and corona. In fact Kunkel [9] could observe on spectrograms of AD Leo obtained during a flare, the emission line H_{14} of hydrogen. From this it follows that the electron density in the atmosphere (chromosphere) of this star must be less than 10^{14} cm^{-3}, in agreement with the value found above.

The known data about the solar chromosphere is as follows: According to the model of the Sun's chromosphere it is assumed that: $n_e = 6.2 \times 10^{11}$ cm^{-3} at the base of the chromosphere, $n_e = 3.6 \times 10^{10}$ cm^{-3} at a height of 10 000 km, and $n_e = 2.2 \times 10^9$ cm^{-3} at a height of 20 000 km [10]. This may give the impression that in the flare stars we have a copy of the solar chromosphere.

In §4 the assumption was made that the radiation of the star in the emission lines in the period after the flare, i.e. after the L_c radiation has stopped, may be caused by processes like recombinations in the chromosphere itself. If this is the case, then, using the known relation between the recombination time Δt_e and the electron density of the medium, which has the form [19]:

$$\Delta t_e = \frac{1}{n_e \sum\limits_1^\infty C_i(T_e) + 1} , \qquad (9.37)$$

we can find the probable value of the duration of the recombinations. Using (9.37) $n_e = 10^{11}$ cm^{-3} we find: $\Delta t_e \approx 40$ s. This value is at least 10 to 20 times smaller than that given by the observations ($\Delta t_e \approx 1000$ s). Δt_e can be larger if n_e is smaller, but in this case the intensity of the emission lines decreases very rapidly, proportionally to n_e^2, and can become completely extinct against the background of the continuous spectrum.

Thus the assumption of simple recombinations in the chromosphere cannot explain the observed durations of the emission lines in flare stars. Evidently there must exist some energy source or mechanism which maintains the radiation by the chromosphere considerably longer than that given by the relation (9.37). It seems that in the chromospheres of flare stars such a mechanism could be provided by inelastic collisions of electrons, where the energy source could be the kinetic energy of free electrons in the chromosphere itself. We have already concluded that at the moment of the stellar flare the electron temperature in the chromosphere increases quite suddenly, up to 100 000 K. As the flare fades, and no further ionizing L_c radiation reaches the chromosphere, electron collisions may for some time still maintain a sufficiently high degree of ionization to have significant influence on the rate of extinction of the emission lines.

7. DURATION OF RADIATION OF EMISSION LINES (THEORY)

The problem stated above is in fact the same as the cooling of the chromosphere itself. With decreasing T_e the intensity of the emission lines falls as well. Whereas the initial (at flare maximum) value of n_e is determined by the available ionizing radiation, during the period after the flare the quantity n_e is determined by the efficiency of the inelastic electron collisions. In that case we have for $n_e = n_e(T_e)$:

$$\frac{n_e(T_e)}{n_o} = \frac{\alpha_c(T_e)}{\alpha_c(T_e) + C(T_e)} , \qquad (9.38)$$

where n_o is the total number of hydrogen atoms per cm^3, $\alpha_c(T_e)$ is the coefficient of the probability of ionizing the hydrogen atom from the ground state by collisions, and $C(T_e)$ is the total coefficient of recombination of an electron with a proton. In the derivation of (9.38) it is assumed that $n_o = n_1 + n^+ = n_1 + n_e$, where n_1 is the density of neutral hydrogen atoms. Equation (9.38) actually gives the dependence of the intensity of an emission line on T_e. If we can find the dependence of T_e on time t, then we can get the required duration of the recombination--under the new conditions, i.e. inelastic collisions being taken into account--in the emission lines during the period after the flare.

We assume that the chromosphere is transparent for the lines and continua of all hydrogen series (including the Lyman series). Then the "leakage" of the kinetic energy of the free electrons out of the chromosphere takes place by the following cycle: ionization and excitation by electron collisions of hydrogen atoms in the ground state; recombination of electrons with protons and cascading transitions of the atoms downwards; photons produced in these transitions leave the chromosphere.

The function $T_e = T_e(t)$ can be found in the following relation, which gives the change of the kinetic energy per unit volume of the chromosphere during a time dt:

$$d(\frac{3}{2} n_e k T_e) = \left| n_1 n_e \alpha_c(T_e) h\bar{\nu}_c + n_1 n_e \sum_{i=2}^{\infty} \alpha_i(T_e) h\nu_i \right| dt \qquad (9.39)$$

where $\alpha_i(T_e)$ is the coefficient of the probability of a hydrogen atom being excited from the ground state to level i by electron collisions.

When the chromosphere is cooling this means that T_e as well as n_e are decreasing. Therefore we have from (9.39):

$$dT_e + T_e \frac{dn_e}{n_e} = n_1 \alpha_c(T_e) q(T_e) \frac{2}{3} \frac{h\bar{\nu}_c}{k} dt \qquad (9.40)$$

where

$$q(T_e) = 1 + \sum_{i=2}^{\infty} \frac{\alpha_i(T_e)}{\alpha_c(T_e)} \frac{\nu_i}{\bar{\nu}_c} . \qquad (9.41)$$

Further we have for n_1:

$$n_1 = n_o - n_e = n_o\left(1 - \frac{n_e}{n_o}\right) = n_o \frac{C(T_e)}{\alpha_c(T_e) + C(T_e)} . \qquad (9.42)$$

Substituting (9.38) and (9.42) in (9.40) and performing the integration we find for the time t during which the electron temperature of the chromosphere decreases from T_e^o to T_e:

$$t = \frac{3}{2} \frac{k}{h\bar{\nu}_c} \frac{1}{n_o} \int_{T_e^o}^{T_e} Q(T_e) dT_e , \qquad (9.43)$$

where

$$Q(T_e) = \left(\frac{1}{\alpha_c} + \frac{1}{C}\right) \left\{1 + \frac{T_e}{\alpha_c(\alpha_c + C)} \left[C \frac{d\alpha_c}{dT_e} - \alpha_c \frac{dC}{dT_e}\right]\right\} \frac{1}{q(T_e)} . \qquad (9.44)$$

In these expressions the functions $\alpha_c = \alpha_c(T_e)$ and $C = C(T_e)$ have the following form (cf., for example, [1]):

$$\alpha_c(T_e) = 0.305 \times 10^{-10} T_e^{1/2} [1 + 0.126 \times 10^{-4} T_e] \exp\left(-\frac{158300}{T_e}\right) ; \qquad (9.45)$$

$$C(T_e) = 0.206 \times 10^{-10} T_e^{-1/2} [6.415 - 0.5 \ln T_e + 0.0087 T_e^{1/3}] . \qquad (9.46)$$

The numerical values found from (9.43) for the time t in seconds, necessary for the temperature of the chromosphere to decrease from the value T_0 = 100 000 K to a given value are shown in Table 9.4 (columns 6 and 7). The table also gives numerical

TABLE 9.4   Time of Cooling of the Chromosphere, t, for Two Values of the Density of Hydrogen Atoms n_0 (Initial Electron Temperature T_e^0 = 100 000 K)

T_e	$\alpha_c(T_e)$	$C(T_e)$	$q(T_e)$	$\dfrac{n_e}{n_0}$	$t(s)$	
					$n_e = 10^{11} cm^{-3}$	$n_e = 10^{10} cm^{-3}$
1	2	3	4	5	6	7
100000 K	0.446×10^{-8}	0.693×10^{-13}	1	1	--	--
80000	0.238×10^{-8}	0.836×10^{-13}	2.3	1	7	73
60000	0.930×10^{-9}	1.055×10^{-13}	2.9	1	20	200
40000	1.740×10^{-10}	1.456×10^{-13}	4.5	0.9991	49	490
30000	0.370×10^{-10}	1.970×10^{-13}	7.0	0.9947	83	832
20000	0.196×10^{-11}	2.477×10^{-13}	16	0.890	471	4.7×10^3
15000	0.115×10^{-12}	3.066×10^{-13}	29	0.273	3180	3.2×10^4
10000	0.045×10^{-14}	4.118×10^{-13}	75	0.001	3.5×10^5	3.5×10^6

values of the functions $\alpha_c(T_e)$ and $C(T_e)$, computed from formulae (9.45) and (9.46), and also $q(T_e)$ found approximately from the data of Chamberlain [11]. The fifth column gives n_e/n_0 for various values of the electron temperature, computed from equation (9.38).

The first remarkable thing in Table 9.4 is the exceedingly fast rate of cooling of the chromosphere initially; for instance, the electron temperature falls from 100 000°K to 30 000°K in about one and a half minutes (for n_0 = 10^{11} cm^{-3}). Thus the question of how high the initial temperature is, 100 000°K or 60 000°K, has no essential significance. Later the cooling rate becomes slower and slower. For instance one hour is required to reach the value T_e = 15 000°K.

In order to clearly see how the electron density varies as the chromosphere cools, we have represented the data in columns 5 and 6 in Fig. 9.6 graphically as well, in the form of the dependence of n_e/n_0 on t (curve 2). In this figure the graph of n_e/n_0 versus t has also been drawn corresponding to the case of pure recombinations of the medium (curve 1, formula (9.37)). Comparing curves 1 and 2 with each other we see that the inelastic electron collisions actually lead to a considerable lengthening of the duration of the recombinations of the chromosphere. Thus, for instance, whereas the electron density is halved and therefore the emission lines are weakened by a factor four in 40 seconds in the case of "pure" recombinations, taking into account the electron collisions extends the same decrease of the brightness of the emission lines to ∿1800 seconds, which is more than 40 times longer (the quantities t corresponding to n_e/n_0 = 0.5 are shown in the graphs of Fig. 9.6).

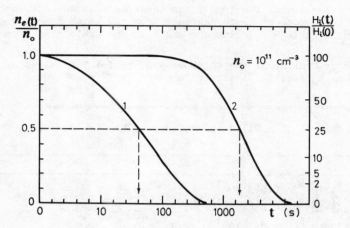

Fig. 9.6. The variations of the electron density in the
chromosphere $n_e(t)$ and intensity of the emission line $H_1(t)$
with time. The moment from which the time is counted
($t = 0$) corresponds with the moment of flare maximum.
Curve 1--"pure" recombinations, curve 2--recombinations and
electron collisions.

8.   WHAT MAINTAINS THE EMISSION LINES OF THE "QUIET" STAR?

Emission lines occur in the spectra of all or nearly all flare stars of UV Cet type.
And these are present all the time, including when the star is not flaring. During
the flare these lines are strengthened, but in the intervals between flares they
do not disappear. Evidently there must be a process maintaining the excitation of
the emission lines, though it would seem that the ionizing agent is completely absent
in the quiescent state of the cool dwarf star.

At the time of the flare the chromosphere is heated by an impulse and then the
emission lines are maintained for quite a long time. If we look at the data in
Table 9.4 we see that a three-fold decrease of the electron density or a weakening
of the emission lines by one order of magnitude occurs one hour after the moment of
the flare for $n_o = 10^{11}$ cm^{-3}, or nearly one day later for $n_o = 10^{10}$ cm^{-3}. We should
note however that stellar flares of UV Cet type take place very often--no fewer than
several flares a day.

Thus the "pumping up" of the chromosphere occurs sufficiently often so that emission
lines are maintained. This is the probable reason for the lasting presence of
emission lines in the spectra of flare stars. In those cases where the flares occur
less often than once per twenty-four hours, and where emission lines are present, it
must be assumed that the electron density in the chromosphere of such a star must be
less than 10^{10} cm^{-3}, and the star itself must be intrinsically fainter.

Moreover, relatively frequent flares or frequent "pumpings" lead to another interesting effect, namely that the electron temperature of the chromosphere is maintained at a certain level. In our case the lower limit corresponds to a temperature
of $\sim$15000 K.

What is then the structure of the chromosphere of a flare star? As it seems to us,
Fig. 9.7 can more or less give an answer to this question. At flare maximum ($t = 0$)

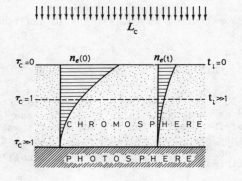

we have the highest electron density $n_e(0)$ in the <u>outer</u> layers of the chromosphere $(\tau_c = 0)$.

When the ionizing L_c radiation weakens, the number of free electrons falls exponentially. Immediately after the flare $(t > 0)$ the processes of recombination and cooling of the chromosphere are turned on; as a result n_e decreases everywhere approximately $\sim e^{-t_c}$. However, the distribution of $n_e(t)$ in the chromosphere seems to keep its initial form.

Fig. 9.7. Schematic representation of the structure of the chromosphere of a flare star. It shows the distribution of the electron density n_e with depth in the chromosphere at the moment of flare maximum $(t = 0)$ and after a time t.

9. DEGREE OF IONIZATION

If the value of the stream of L_c radiation is known, an attempt can be made to find the degree of ionization of hydrogen (and of any element) in the chromosphere of a flare star. We restrict the consideration of this problem to the surface layers of the chromosphere, i.e. where $\tau_c = 0$.

We start from the general condition of equilibrium between the number of ionizations and recombinations per unit time and per unit volume. We have:

$$n_1 \int_{\nu_0}^{\infty} k_1 \frac{WH_\nu(\tau,\gamma,T_*)}{h\nu} \, d\nu = n^+ n_e C(T_e) , \qquad (9.47)$$

where n_1, n^+, n_e are the densities of neutral and ionized hydrogen atoms and of free electrons; $C(T_e)$ is the total coefficient of recombination of electrons with protons; $k_{1\nu}$ is the coefficient of continuous absorption from the ground state per neutral hydrogen atom; $H_\nu(\tau,\gamma,T_*)$ is the flux of ionizing radiation; W is the dilution coefficient.

For the temperatures (T) which are customary for the photospheres of cool stars and for $\gamma^2 = 10$ the condition $(h\nu_0/kT_*\gamma^2) > 1$ holds. Substituting in (9.47) the expression for the function H_ν from (9.2) and $k_{1\nu} = k_0(\nu_0/\nu)^2$, we find

$$\frac{n^+}{n_1} n_0 = C_0 W \frac{T_*}{\gamma^2} \frac{F_2(\tau)}{C(T_e)} e^{-\frac{h\nu_0}{kT_*\gamma^2}} , \qquad (9.48)$$

where

$$C_0 = \frac{3kk_0\nu_0^2}{hc^2} = 4.62 \times 10^3 \text{ deg}^{-1} \text{ s}^{-1} .$$

Equation (9.48) is the ionization formula for hydrogen in the case when the radiation ionizing the chromosphere has a Compton origin.

In the case $\tau < 0.1$ we have $F_2(\tau) \approx \tau/2$. Substituting in (9.48) $\gamma^2 = 10$ and $T_* = 2800$ K we find:

$$\frac{n^+}{n_1} n_o = 2.3 \times 10^{-7} \frac{\tau W}{C(T_e)} . \tag{9.49}$$

Earlier we found for the chromosphere $n_e \approx 10^{11}$ cm^{-3}. Taking $W = 0.1$ we find the degree of ionization of the hydrogen in the chromosphere, $(n^+/n_1)_{rad}$, when the ionization is caused by radiative processes; the results for two values of τ and a number of values of the electron temperature are given in Table 9.5. The last column of the table shows, for comparison, the degree of ionization $(n^+/n_1)_{coll}$, when the ionization is due to electron collisions.

TABLE 9.5   The Degree of Ionization of Hydrogen in the Chromosphere of a Flare Star, Due to Radiation, $(n^+/n_1)_{rad}$, and Due to Electron Collisions, $(n^+/n_1)_{coll}$.

T_e	$(n^+/n_1)_{rad}$		$(n^+/n_1)_{coll}$
	W = 0.001	W = 0.01	
10000°K	55	550	0.001
15000	75	750	0.5
20000	90	900	10
30000	110	1100	200
40000	155	1550	1200

At flare maximum the electron temperature in the chromosphere must be very high, probably of the order 50 000 K. In that case the ionization will be due exclusively to electron collisions. For temperatures below 30 000 K the ionization by radiation predominates. For temperatures of 30 000 K to 40 000 K the contributions of the collisions and of the radiation in the ionization processes are about equal.

10.   THE BALMER DECREMENT OF THE EMISSION LINES

Joy [12] was the first to draw attention to the variability of the Balmer decrement of the hydrogen emission lines during the flare. The decrement becomes steeper towards shorter wavelengths and the number of lines observed in the series becomes larger, just as for stars with higher temperature.

A moving plateholder was used to obtain spectrograms of a flare of EV Lac (11.XII.65) [9]. The values found from these spectrograms for the Balmer decrement for four different instants of this flare are given in Table 9.6, where the time 3^h55^m corresponds to the moment of flare maximum.

From these data it follows that at the time of the flare the Balmer decrement is not only less steep compared to the gaseous nebulae, but in the beginning we may even have an inversion of the ratio of the line intensities. Secondly, the Balmer decrement changes during the flare, and the minimum steepness is reached at the maximum

TABLE 9.6 Balmer Decrements of Hydrogen Lines at Four Moments of the Development of One Flare of EV Lac (11.XII.65), Two Flares of AD Leo (2.III.70 and 27.IV.70), in the Chromosphere of the Sun and Gaseous Nebulae

		H_α	H_β	H_γ	H_δ	H_ϵ	H_η	H_{10}	H_{11}	$K(Ca^+)$
EV Lac:	3^h55^m	--	1	1.24	1.48	1.22	1.17	0.94	0.80	0.47
	4 00	--	1	1.04	1.16	0.92	0.63	0.64	0.47	0.59
	4 03	--	1	1.10	1.28	1.10	0.90	0.67	0.59	0.68
	4 08	--	1	1.15	1.06	0.76	0.54	0.52	0.38	0.68
AD Leo:	2.III.70	1.10	1	0.89	0.69	(0.71)	--	--	--	0.27
	27.IV.70	1.35	1	0.70	0.39	(0.84)	--	--	--	0.40
Chromosphere of the Sun		2.80	1	0.45	0.22	(0.12)	0.12	--	--	0.79
		2.72	1	0.40	0.23	(0.14)	0.17	--	--	0.69
		2.90	1	0.59	0.41	(0.27)	0.23	--	--	1.13
Gaseous nebulae (theory for T_e = 20 000°K):										
Photoionization		2.80	1	0.49	0.28	0.18	0.12	--	--	--
Electron collisions		4.80	1	0.35	0.17	0.10	--	--	--	--

of the flare. On account of the difficulties connected with the calibration of such spectrograms one cannot hope for a specific accuracy in the determination of the proper value of the decrement, but evidently there are no reasons for doubting the reality of the variations. The same table gives the values of the Balmer decrements for gaseous nebulae (theory) and the solar chromosphere (from data in [10,13]). We cannot speak about similarities of the Balmer decrements between flare stars and gaseous nebulae and the solar chromosphere. Somewhat different is the case of the Balmer decrement found at the time of one rather strong flare of EV Lac [5] which is shown in Table 9.6. In this case the inverse ratio of the line intensities has not been observed, but the decrement itself is rather steep compared with the preceding case.

The results found from the series of spectrograms obtained during the two flares of UV Cet mentioned above [3] were represented quite well. Figure 9.8 shows parts of the microphotometer tracings of these spectrograms, from which the strong increase of the intensity of the emission lines during the flare is quite evident. Table 9.7 gives the values of the Balmer decrements and the equivalent widths of the emission lines at different moments of the flare. In character the Balmer decrement does not differ from what we found above (Table 9.6) in the case of a flare of EV Lac.

Concerning the emission lines, variations of the equivalent widths of these lines may occur not only as the result of variations of the intensities of the lines themselves, but also because of variations in the level of the continuous spectrum. A sharp decrease of the level of the continuous spectrum may lead to a sharp increase of the line equivalent widths.

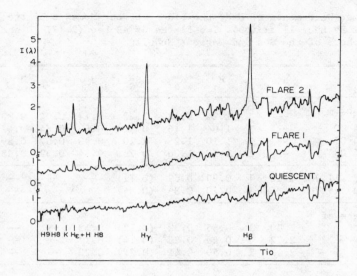

Fig. 9.8. Microphotometer tracings on an intensity scale of two flares of UV Cet (14.X.72) at the times of their maxima (flare 1 and flare 2), and also in the quiescent state of the star (lower curve). The strong enhancements of the emission lines at the time of the flare are remarkable.

TABLE 9.7 Balmer Decreements and Equivalent Widths of Emission Lines During Two Flares of UV Cet (14.X.1972)

		Time (UT)	H_β	H_γ	H_δ	K Ca 11	H_8	H_9
			Balmer decrement					
Flare	1	$08^h29^m \div 08^h31^m$	1.00	2.17	1.27	0.92	--	--
Flare	2	08 42 ÷ 08 45	1.00	1.13	1.12	0.36	0.56	0.25
		08 45 ÷ 08 48	1.00	1.52	1.35	0.85	0.60	--
		08 48 ÷ 09 01	1.00	1.15	--	--	--	--
			Equivalent widths (Å)					
Flare	1	08 29 ÷ 08 31	23.3	45.0	21.3	9.6	--	--
		Before the flare	8.3	--	--	--	--	--
Flare	2	08 42 ÷ 08 45	38.3	43.5	37.1	4.3	9.6	3.4
		08 45 ÷ 08 48	30.9	43.8	33.4	17.5	12.3	--
		08 48 ÷ 09 01	14.6	16.7	--	--	--	--
		Before the flare	6.8	8.7				

The situation is somewhat different when we consider the behavior of relative line intensities (decrement). The relative intensities of the emission lines do not depend on the level of the continuous spectrum, and therefore their values at a certain moment characterize the physical parameters of the medium (electron temperature, electron density, degree of ionization, transparency of the medium, the role of inelastic collisions, etc.). Hence we may interpret the variations of the decrement as proof of variations in the physical conditions in the chromosphere of the star for a short time compared to the duration of the flare itself.

Numerically the decrement of the hydrogen lines in flare stars greatly differs from that in the solar chromosphere. Hence it follows that the conditions and the mechanism itself for the excitation of emission lines in the chromospheres of flare stars on the one hand and in the chromosphere of the Sun on the other must differ from each other.

Finally the assumption may be made that emission lines in flare stars are not formed in their chromospheres, but simply in the gaseous clouds surrounding them, which appear at the time of the flare and are due to ejected material. However, in such a case we should observe a decrement characteristic for gaseous nebulae and this is not the case.

At present we cannot offer a quantitative theory of the production of emission lines in the atmospheres (chromospheres) of flare stars; the development of such a theory does not seem to us to be a simple problem. However, at this stage it may be useful to summarize our conclusions:

a) Chromospheres of flare stars do not represent a simple analogue of the solar chromosphere.

b) Chromospheres of flare stars cannot be identified with a completely transparent medium in the lines; this seems independent of the processes of inonization and excitation.

Photoionization on the one hand and ionization and excitation by electron collisions on the other alternately play the main role in the various stages of a flare. In addition, during a flare the intensity of the ionizing radiation is variable with time. These are more or less the conditions of excitation and transfer of the emission lines in chromospheres of flare stars.

11. ANALYSIS OF PROFILES OF EMISSION LINES

During a flare the emission lines not only become stronger, but wider as well. Joy [12] has often stressed the fact that at the time of a flare the width of the bright hydrogen lines increases. In his opinion the influence of the flare on the stellar spectrum consists in widening the hydrogen emission lines together with their becoming stronger. This appeared more clearly on a spectrogram taken accidentally with the 100-inch telescope of the Mt. Wilson Observatory during a strong flare of UV Cet on 25.IX.1948 [14]. On this spectrogram all emission lines of hydrogen, and also the weak lines of He I and 4686 He II were found to be wider than under normal conditions. Unfortunately these remarks of Joy were based on qualitative estimates only.

Quantitative data about variations in the width of emission lines at various moments during the development of flares of UV Cet and AD Leo were obtained by R. E. Gershberg [4,5]. Regardless of the coarse instrumental resolution ($\sim$3-6 $\overset{\circ}{A}$) a number of interesting conclusions can be drawn from these data. In particular, it has been established that the half-width of the emission lines first increases to a maximum and later decreases. The increase of the half-width is of the order of 1.5 to 2 times.

These facts are related to the conclusion drawn in §5, namely that at the moment of the flare the electron temperature in the star's chromosphere increases sharply, though this increase lasts a short time only. The half width of the emission line does not depend very strongly on the electron temperature of the medium, $\Delta\lambda \sim T_e^{\frac{1}{2}}$.

On the other hand, as we have seen above, the minimum value of the electron temperature in the chromosphere of a flare star is of the order 15000 K. At the moment of
the flare it may increase up to 60 000 to 100 000 K. Therefore a widening of the
emission lines by 2 to 2.5 times on the average is expected. Greenstein and Arp [15]
using the 200-inch telescope recorded a spectrogram of one flare of Wolf 359; the
microphotometer tracing of this flare, together with that of the spectrum of the
star in its quiescent state is shown in Fig. 9.9.

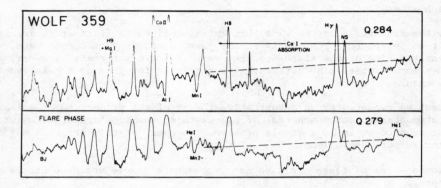

Fig. 9.9. Microphotometer tracings of spectrograms of the
star Wolf 359, obtained at the moment of a flare (below) and
in the quiescent state of the star (above). Above the
emission lines are identified, below the absorption lines.
The horizontal arrow indicates the absorption region of 4227
Ca I ∿ 400 Å wide. The strong broadening of the emission
lines at the time of the flare is remarkable.

Under normal conditions the spectrum of this star consists of the emission lines
of hydrogen and the line λ 3933 of Ca II. During the flare this spectrum undergoes a number of essential changes. First, the hydrogen lines are strengthened
and widened. The lines $\lambda\lambda$ 4713 and 4388 Å of neutral helium begin to appear. No
significant variations are observed in the structure of the absorption lines.
This holds also for the Ti O bands; during the flare they are as strong as in
normal conditions. The very wide line λ 4226 Ca I did not significantly change
either. The width of the hydrogen lines increased nearly by a factor of two, from
8 Å to 15 Å. If during the flare any continuous emission appeared, it probably
began at 3800 Å, and the faintest hydrogen line which could be seen on the spectrogram was H_{13}.

Radial velocities of the emission lines were measured at the time of the flare, as
well as later. The difference of the velocities was -23 km/s for the hydrogen
lines and 59 km/s for the K line of Ca II; these differences are small and may not
be due to outward flow or ejection of gaseous material. According to Greenstein
and Arp the negative displacements may be due to hot matter flowing out of the
star towards the observer.

Infrared observations were made of this star as well. Within the error limits no
fluctuations were observed in the brightness of the star; however, it was found to
be exceptionally bright at long wavelengths. From these observations the Planck
temperature proved to be 2000 , 2250 and 2500 K. This is analogous to the fact
that the maximum of the radiation energy of the star Wolf 359 corresponds to 2500 K.
In the visual and photographic regions the spectrum of Wolf 359 resembles that of
EV Lac in normal conditions.

Several attempts were made to measure the value of the Balmer jump,

$D = \log(J_{3646-}/J_{3646+})$ at the time of the flare. On account of the complexity of the spectrum in that region the quantity D cannot be determined with sufficient accuracy. The available data indicate that the value of D at the moment of the flare is not very large. For two flares of AD Leo and one flare of EV Lac, for example, it was found that $D \gtrsim 0.60$ [9]. However, other estimates for flares of EV Lac gave $D \sim 0.15$ to 0.43 at maximum brightness and at the beginning of the descending branch of the lightcurve, and $D \sim 0.15$ for one rather strong flare of AD Leo [5].

Usually the value of the Balmer jump is a good indicator for the determination of the electron temperature of the medium in those cases where it is certain that the radiation is of purely recombination origin (model of nebulae). But for the flare stars such an interpretation of the Balmer jump cannot be used. The spectrum of the radiation in the region of the Balmer jump for flare stars is composed of at least three components: continuous emission of non-thermal nature, ordinary thermal radiation of the star's photosphere, and recombination radiation of the chromosphere (of rather complicated composition) or of the surrounding gaseous medium. For these reasons the value of the Balmer jump found from direct measurements will be smaller than what we would have in the case of pure recombination. At the same time, the smaller the Balmer jump, the lower the temperature of the gas. Hence the uncorrected Balmer jumps will give us at best an upper limit of the temperature of the gas. For instance, for $D < 0.2$ the electron temperature of the medium must be higher than 40 000 °K.

It is of particular interest to discuss profiles of emission lines with central absorption at very high spectral resolution (0.1-0.2 Å). Recordings of a number of such lines for EV Lac, Wolf 359, and Ross 614 were obtained by Worden and Peterson [19]. Figure 9.10 gives the profile of one of these lines--Hα. The saddle-shaped depression at the top of this line is interesting; it gives a clear indication that this line has been formed in an optically thick medium and was scattered repeatedly before leaving the star. This explanation of the saddle-shaped profile appears more probable (as compared with possible Zeeman splitting in strong magnetic fields) considering that such profiles do not differ significantly from one star to another. According to the theory of formation and transfer of emission lines in an optically thick medium, the central depression must be deeper with greater optical depth in the line t_i [16]. In this case the cavity is not very deep but it corresponds to a value $t_i \sim 10$, in any case it is smaller than 100. Theoretically it is clear that the chromosphere of a flare star at a given moment is not transparent in one of the lines belonging to a subordinate series. In other words, there is a reason (high density of the radiation in Lyman alpha, or great efficiency of excitation by electron collisions?) leading to overpopulation of the second energy level of hydrogen.

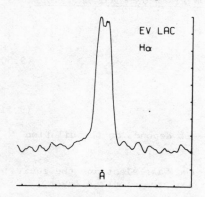

EV LAC

Hα

Å

Fig. 9.10. Profile of the emission line Hα in the spectrum of EV Lac, obtained with high spectral resolution. The deepening at the top of the line is seen.

Indications of depressions at the peaks of emission lines can also appear in cases where the spectrograms were obtained with moderate dispersion. As an example a microphotometer tracing is shown in Fig. 9.11 of the spectrum of the strong flare of UV Cet mentioned above [3], in which the <u>asymmetry</u> of the tops of the various emission lines is clearly seen. Moreover it has been established that this asymmetry appears and disappears at certain moments during the development of the flare (Fig. 6.6), and therefore the medium is not always opaque in a given line, but only at times around the maximum of the flare. In any case the length of time during which the asymmetry of the lines is observed is several times shorter than the total duration of the flare itself.

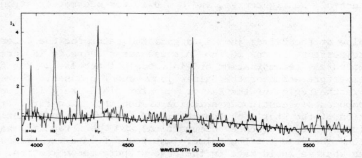

Fig. 9.11. Microphotometric tracing on the intensity scale of one spectrogram obtained at the time of maximum of a strong flare of UV Cet. The asymmetry at the tops of some emission lines is seen.

12. DEPENDENCE OF INTENSITY OF EMISSION LINES ON FLARE AMPLITUDE

We can write (9.17) in the following form:

$$\frac{W_\lambda}{\lambda} = K_\lambda \frac{F_2(\tau)}{C_\lambda(\tau)} \tag{9.50}$$

where

$$K_\lambda = W_{\lambda_i} \frac{3\gamma^4}{4} \frac{e^x - 1}{x^4} \quad ; \tag{9.51}$$

here K_λ is dimensionless and does not depend on τ—it depends on the dilution coefficient W and the star's temperature T_*.

For a given stellar temperature and the energy of the fast electrons the equivalent width depends only on the effective optical depth τ of the electron layer, i.e. on the flare amplitude ΔV:

$$\frac{W_\lambda}{\lambda} \sim f(\tau) \sim \psi(\Delta V) . \tag{9.52}$$

Unfortunately W_λ/λ cannot be expressed in ΔV in an explicit form. But implicitly this dependence may be derived quantitatively if use is made of what is known about the form of the dependence of the functions F_2 and C on τ, on the one hand, and of ΔV on τ on the other hand. Figure 9.12 is based on the results of computations of this kind; it gives the family of curves of the dependence of W_λ/λ on τ for the case $T_* = 2800$ K and $\lambda^2 = 10$. The numerical values of the functions $F_2(\tau)$, $\Delta V(\tau)$ and $C_\lambda(\tau)$ have been taken from Tables 4.1, 6.3 and 4.2 respectively.

In Fig. 9.12 for a given value of K_λ we can find the equivalent width W_λ of a given line if ΔV is known from the observations or vice versa. K_λ can vary from one flare to the next, or from one star to another; in both cases it is a question of variations either of the dilution coefficient W or of the effective temperature or both at the same time (which is less likely).

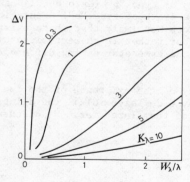

Fig. 9.12. Theoretical relation between equivalent width of an emission line W_λ and flare amplitude ΔV for $T_* = 2800$ K and $\gamma^2 = 10$. The numbers next to the curves give the values of K_λ in arbitrary units.

The general character of Fig. 9.12 shows that for large values of K_λ significant variations of the equivalent width can occur practically without fluctuations in the total brightness of the star. On the contrary, in other stars with smaller values of K_λ, significant brightness fluctuations (by more than one stellar magnitude) can occur practically without variations of the equivalent widths of the emission lines.

Considerations from Fig. 9.12 may be directly related to stars of T Tau type.

With regard to the stars of UV Cet type, there exist data [5] confirming the conclusion drawn above, namely: considerable fluctuations are observed in the intensities of the emission lines (large values of K_λ in Fig. 9.12). In other words, in those cases the behavior of the emission lines does not depend on the fluctuations in brightness nor on the appearance of a flare in the continuous spectrum of the star.

13. HARO'S TWO TYPES OF FLARE STARS

During his study of the spectrophotometric character of flare stars in stellar aggregates and associations Haro found that there exist two extreme types of flare stars [6].

Type "a": During the flare the continuous spectrum becomes stronger only in the blue and ultraviolet parts of the spectrum; changes in the red region of the spectrum, from 6100 Å up to Hα, are practically absent. The emission lines, in particular Hα, are very strong during the flare.

Type "b": During the flare not only the blue and ultraviolet spectral regions are strengthened, but the visual region as well. However, the emission lines are very faint.

Haro distinguished these two extreme types by taking photographs of flare stars through an objective prism with small dispersion, so that at the same time the behavior of the star in the Hα line emission and the neighboring red continuum (up to 6100 Å) could be followed.

The direct observations show that in the case of type "a" the increase of intensity in blue and ultraviolet light before the maximum is exceedingly quick, and the fall, although slower, on the whole is relatively fast. As a rule, at the end of the flare the Hα emission disappears. Most of the flare stars in Orion belong to this group.

In type "b" the growth of intensity before maximum is considerably slower than in the case of type "a"; it lasts from 40 to 50 minutes. The return of the star from maximum to the normal state lasts 5 or 6 hours, but the Hα emission is visible even after that, for one or two days longer (although it is much weaker). Very few flare stars in Orion belong to this type--Haro 66, 99, 149, 153, and also 8 and 177.

It is very curious that three of these "slow" stars of type "b" (Haro 66, 149 and 153) at the same time were found to be "fast" flare stars of type "a"! Hence it follows that the division into types "a" and "b" does not imply that there exist two different physical categories of flare stars, but rather it characterizes different forms of flares, which may occur in one and the same star during different flares.

The fast electron hypothesis allows the possibility of the existence of the two types of flare stars. It seems that two facts play a decisive role: the dilution of the ionizing radiation, and the spectral class of the flare star. We discuss these in more detail.

14.   EFFECT OF DILUTION OF RADIATION

The equivalent width of an emission line also depends on the dilution coefficient W, i.e. on the mean distance of the cloud of fast electrons from the stellar photosphere. Therefore we may write (9.17) as follows:

$$\frac{W_\lambda}{\lambda} = Wf(\tau) , \qquad\qquad (9.53)$$

where the function $f(\tau)$ depends, for given r_i, γ, and T, only on τ. For different τ we have different flare amplitudes. Evidently it is always possible to find combinations of W with Δm, or with τ, for which the equivalent width will be the same. This means that in principle we can have a given equivalent width for very small as well as very large values of the amplitude Δm, depending on how much larger or smaller the dilution W of the radiation is.

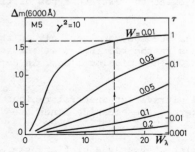

As an illustration Fig. 9.13 gives the theoretical curves of the amplitude of the brightness fluctuations at $\lambda \sim 6000$ Å versus equivalent width for a number of values of the dilution W of the radiation. As seen from the figure, in one and the same flare star Haro's two types of flares can occur. The same intensity of an emission line can be found for almost no fluctuations in brightness at $\lambda \sim 6000$ Å ($\Delta m \sim 0$) as well as for considerable fluctuations in brightness ($\Delta m \gtrsim 1$; broken lines in Fig. 9.13). The first case corresponds to large values of W, the second to small values. Thus flares of type "a" differ from flares of type "b" in that in the first case the primary material from the star's core is ejected to a relatively small distance from the surface of the star, where the fast electrons are formed; in the second case these processes occur much farther out from the star. It is probable that the composition and proper-

Fig. 9.13. "Effect of dilution of the radiation": one and the same equivalent width of the emission line W_α (in arbitrary units) corresponds with large flare amplitudes Δm, as well as with amplitudes which are nearly zero at $\lambda \sim 6000$ Å.

ties of the matter are the same in the two cases "a" and "b". We may conclude that the intervals of time from the moment the primary material leaves the star's surface until the fast electrons are generated from this material must be the same for both types of flares. Consequently, the only thing which will distinguish a flare of type "a" from a flare of type "b" is the value of the original velocity of ejection of the matter from the stellar interior. In the case of a flare of type "a" this velocity will be smaller; in the case of a flare of type "b" it will be larger.

15. EFFECT OF SPECTRAL CLASS

The theoretical brightness amplitude of the fluctuations, that is the value of the coefficient C_λ for $\lambda \sim 6000$ Å, is found to depend on the star's temperature. For instance, for low temperatures we have $C(6000$ Å$) > 1$, and for sufficiently high temperatures $C(6000$ Å$) < 1$. Hence there must exist a certain temperature for which $C(6000$ Å$) = 1$; in such a case there will be no significant fluctuations of brightness in the continuous spectrum of the star at $\lambda \sim 6000$ Å. However, the intensity of an emission line depends, as before, strongly on τ. Therefore for a certain star we shall observe a rather strong emission line appearing and disappearing whereas the brightness in the continuous spectrum around 6000 Å will hardly change. This corresponds with flare stars of Haro's type "a".

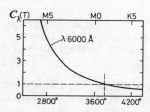

Fig. 9.14. "Effect of spectral class": determination of the spectral class of a star with zero flare amplitude at $\sim$6000 Å from the condition $C_\lambda(T_*) = 1$.

It is not hard to find the required temperature or spectral class of a star with zero amplitude of the brightness fluctuations. It is sufficient to construct the curve giving the dependence of C_λ on T_* for $\lambda = 6000$ Å and to determine the temperature for which $C_\lambda = 1$. This has been done in Fig. 9.14, constructed from the data in Table 4.2; the curve $C_{6000}(T_*)$ intersects the horizontal line corresponding to the value $C_{6000} = 1$ at $T_* \sim 3750$ K, which corresponds approximately with spectral class M0. The computations show that for stars of class M0 large values of the equivalent width of an emission line are possible, but the amplitude of the star's brightness fluctuations around 6000 Å will normally not exceed 0$\overset{m}{.}$3.

Although under certain conditions flares of both types, "a" and "b" may occur in the same star (effect of the dilution of radiation), it is possible that there exists a homogeneous group of stars for which only flares of type "a" are possible. These may be stars of spectral class $\sim$ M0.

When the appearance and disappearance of emission lines are accompanied by considerable fluctuations of brightness in the continuous radiation of the star at 6000 Å; then $C(6000$ Å$) > 1$, corresponding with a star of class M6. Such cases correspond with flare stars of Haro's type "b".

16. ON THE NATURE OF SLOW FLARES

The term "slow" flares came up for the first time in connection with observations of flare stars in associations and aggregates. Flares of Haro's type "b", for example, are at the same time "slow", when the growth of the flare from the start till maximum lasts rather long, from ten minutes up to an hour. But in such cases the flare amplitude is also very large, up to 8^m or more (in U wavelengths). According to the interpretation given by V. A. Ambartsumian slow flares may appear when liberation of the energy from the stellar interior takes place in the sub-photospheric layers. This means that in such cases the ionizing radiation cannot get out of the photosphere and therefore we must not expect the emission lines to be enhanced. Unfortunately the lack of reliable observational data concerning the behavior of emission lines during slow flares makes any conclusions uncertain.

Another type of "slow" flare was introduced in connection with stars of UV Cet type. The chief properties of such flares are as follows:

a. Extremely small values of flare amplitude (fractions of a stellar magnitude).

b. A nearly Gaussian form of the lightcurve, that is, practically the same rate of increase and decrease of the star's brightness during the flare, with a rather flat maximum.

c. The emission lines dominate (recombination radiation), and the continuous emission is extremely small; moreover, it appears during a short time only around flare maximum. This indicates that slow flares are accompanied by strengthening of the emission lines. At the same time this fact shows that, however small the power (amplitude) of the slow flare may be, the ionizing photons (or high-energy particles) are produced directly in the atmosphere of the star, and not in the sub-photospheric layers. An example of a lightcurve of a "slow" flare is given in Fig. 6.4 (flare 53 AP).

The fast electron hypothesis gives a natural explanation of "slow" flares with the three properties mentioned above. First of all, "slow" flares are weak flares. These are the flares which occur at the backside or at the star's hemisphere which is not accessible to the observer.

Normally a weak flare occurring at the backside of the star will not be detected; it will be screened by the star itself. However the back half of the chromosphere obtains from this flare a corresponding portion of ionizing radiation, which leads into a general strengthening of the emission lines. The wave of disturbance, spreading in all directions, soon appears in the half of the chromosphere accessible to the observer. The observer records the processes taking place in a ring-shaped part of the chromosphere. The chromospheric radiation is due to recombinations, and therefore the emission lines will predominate--a conclusion which is in agreement with the observations. A certain part of the initial radiation of such a flare in the form of continuous emission may diffuse through the ring-shaped "window" of the chromosphere towards the observer. All this must occur through the diffusion of the fast electrons from the backside of the star to the side of the observer under the influence of the general magnetic field of the star.

Here we have only described the "slow" flares in a qualitative way. It is possible to state the problem in a quantitative form; then we must state and solve the non-stationary problem of the behavior of the ionization front during the development of the flare, taking into account all initial and boundary conditions. This problem is a general one and it may be posed as follows: what are the observational aspects of a weak flare taking place at the reverse side (backside) of the star? We especially stress weak flares because the consequences of a strong flare at the backside of the star are clear; they will hardly differ from the consequences of flares of the same strength occurring on the observable hemisphere of the star.

One of the results of the solution of this problem is, in particular, that we find the mathematical form of the lightcurve of the "slow" flare. This curve will differ significantly from the ordinary lightcurves of flares with a very sharp maximum, and will rather resemble a Gaussian curve.

17. THE FORBIDDEN LINES

The well known forbidden lines, N_1, N_2 [OIII], are not seen in the spectra of stellar flares; the line 4068 [SII] is an exception; it appears relatively often, and rarely faint lines of [Fe II] are also observed. How can we explain these circumstances?

For each forbidden line there exists a certain critical electron density n_e^o, for which this line can be excited. For values of n_e larger than n_e^o the forbidden line will be weaker, and for still larger values it will not be excited at all (the predominant role is played by collisions of the second kind). The values of n_e^o for some forbidden lines are given in Table 9.8 (cf [1]).

TABLE 9.8   Maximum Permissible Electron Densities n_e^o for Excitation of Some Forbidden Lines

Line	n_e^o (cm^{-3})
N_1, N_2 [O III]	10^6
6548 [N II], 6584 [N II]	0.7×10^6
3727 [O II]	10^4
4068 [S II]	10^8

The critical density n_e^o for the line 4068 [S II], is rather high--of the order 10^8 cm^{-3}. Therefore, in those regions of the stellar atmosphere where the electron density during the flare reaches values of 10^8-10^9 cm^{-3}, the lines N_1, N_2, and 3727 [O II] cannot be excited, whereas for the line 4068 [S II] these conditions are admissible. It appears that something analogous occurs for the lines of [Fe II] as well.

These considerations make it possible to estimate a lower limit of the electron density in the outer layers of the chromosphere at the time of the flare: for most of the cases it seems that it is larger than 10^8-10^9 cm^{-3}.

18.   POSSIBILITY OF OBSERVING THE FORBIDDEN LINE 4363 [OIII]

According to the computations the critical electron density for the forbidden line --4363 [O III]--must be of the order of 10^8 cm^{-3}, the same as for the line 4068 [S II]. If the conclusions about the absence of some forbidden lines and the presence of others made above are valid, we should observe in the flare spectra besides the line 4068 [S II] also the line 4363 [O III]. However, up till now the line 4363 [O III] has not yet been detected in flare spectra.

Probably the reason for this is the faintness of the line itself and the difficulty of distinguishing it against the general background of the continuous emission of the star. We note that on account of the relation [17]:

$$\frac{E_{(4363)}}{E(H_\gamma)} \sim \frac{n(O^{++})}{n_e} T_e^{-\frac{1}{2}} \exp\left(-\frac{\chi}{kT_e}\right) \qquad (9.54)$$

the relative intensity of the line 4363 [O III] depends on the behavior of the ratio $n(O^{++})/n_e$ during the flare, where $n(O^{++})$ is the concentration of the twice-ionized oxygen ions. This behavior is not well known. This question requires more detailed analysis.

19.   EXCITATION OF HELIUM EMISSION LINES

Joy and Humason were the first to discover helium emission lines in the flare spectra
of flare stars; on a spectrogram obtained by them during the flare of UV Cet on 25
September 1948 they noticed the faint lines $\lambda\lambda$ 4471 and 4026 of neutral helium and
the line λ 4686 of ionized helium [14].

More detailed descriptions of the behavior of helium lines in the spectra of flares
of UV Cet and AD Leo appeared at a much later time [4].   Ordinarily, outside the
flare, there are no helium lines in the spectrum of AD Leo. But during one of the
strong and long lasting flares of this star (18.V.1965) seven spectrograms clearly
showed the line λ 4471 of He I.   The equivalent width of this line W(4471) is 4.5 Å.
On the same spectrograms the line λ 4026 of He I was seen as well. Both these lines
reach their limiting intensity immediately after the maximum of the flare. After
that the line λ 4471 of He I gets weaker, but the line 4026 of He I disappears rather
quickly. With regard to the line λ 4686 He II, this was not detected on the spectrograms mentioned above.

In the case of UV Cet, again at the time of one of the strongest flares of this
star (24. IX.65), the line 4686 He II was observed with certainty practically during
the whole duration of the flare; it had an equivalent width of 3 to 7 Å.   Also the
line λ 4471 He I with an equivalent width 6–12 Å was observed.

It is important to note that in both cases the emission lines of helium appeared
later than the lines of hydrogen and calcium, and disappeared considerably earlier
than the lines of hydrogen and calcium.   This may indicate the possibility that
there exists a correlation between the length of time during which the lines are
seen in the spectrum and the ionization potential of the atom causing these lines.
The behavior of the line 4921 He I during one of the flares is remarkable: it was
found to appear twice during the same flare, in the form of two short lived bursts
in the spectra.   In other words, the appearance and disappearance of this line had
a sudden origin.

At the time of the other flare mentioned above, apart from the usual emission lines
of hydrogen and ionized calcium, the lines 4026, 4471 and 5876 He I were observed,
but no lines of ionized helium were seen.

It should be stressed that the helium lines are not always visible; their excitation
appears to be correlated with the absolute power of the flare. During one flare,
for example, of YZ CMi it was found possible to obtain a good spectrogram of the
blue region of the spectrum, on which emission lines of hydrogen and ionized calcium were clearly seen, but no lines at all of neutral or ionized helium.

The observations confirm that helium lines, in particular lines of ionized helium,
appear in strong flares only; in faint flares the lines of He II at least are not
observed at all.   Thus, for instance, during the flare of AD Leo (18.V.65) for which
$\Delta B = 1\overset{m}{.}8$, the line 4868 He II was not seen at all. But during another, considerably
stronger, flare of UV Cet (24.IX.1965, $\Delta V = 1\overset{m}{.}9$), where the flare amplitude in B
light was twice as large, $\Delta B = 3\overset{m}{.}8$, the line 4686 He II was quite clearly seen.
Considering that the line 4686 He II was absent even during the flare of UV Cet
(14.X.72), for which $\Delta U = 5\overset{m}{.}2$, and therefore $\Delta B \approx 2\overset{m}{.}5$ (Fig. 8.2), the conditions
for excitation of the line 4686 He II must be rather strict.   It appears that
$\Delta B \gtrsim 3^m$ must be taken as a condition for the appearance of the line 4686 He II in
the spectrum of a stellar flare.

It is possible that the emission lines of helium are more sensitive to the physical
conditions in which they arise than the lines of hydrogen. If this is the case, it
will be possible to use the helium lines as sensitive indicators for decoding the
processes taking place in the outer regions of the star during the flare.

Below we shall discuss that using the fast electron hypothesis and maintaining the energy characteristics of the electrons themselves, a natural explanation can be given for the appearance of emission lines of neutral and ionized helium during stellar flares.

The emission lines of neutral and ionized helium are generated at the same place as the hydrogen lines, that is in the star's chromosphere. Furthermore, the required ionizing radiation (He^+ and He^{++}), comes from the non-thermal bremsstrahlung of the fast electrons and falls on the chromosphere from outside, from the direction of the cloud or shell of fast electrons. The general expression for finding the intensity of the ionizing radiation in such a case is given by Eq. (9.11). In particular, if we substitute $f_c(He^{++}) = 0.08$, we have for the intensity of the radiation ionizing helium twice:

$$E_{br}(\tau) = 1.30 \times 10^{25} \, r_* \tau^2 \quad \text{ergs/s} \tag{9.55}$$

Part of this energy--$\Gamma(4686)$--is transformed into the emission line 4686 He II. Therefore we can write, for the intensity of this line, taking into account the dilution W of the radiation:

$$E(4686) = \Gamma(4686) E_{br}(\tau) W \frac{n(He)}{n(H)} , \tag{9.56}$$

where the last term at the right takes into account the fact that in the absorption processes in the frequency region of the ionization of helium the hydrogen atoms take part as well.

The value of the transformation coefficient $\Gamma(4686)$ is not known. However, considering that the line 4686 He II corresponds with the first line of the Paschen series of hydrogen (λ 18751 P_α), for which Γ_i is of the order of 0.01 or less, we may find from (9.56), taking $n(He)/n(H) = 0.1$:

$$E_{(4686)} = 1.30 \times 10^{22} \, r_* \tau^2 W \quad \text{ergs/s} \tag{9.57}$$

This is the total energy radiated by the whole star in the line 4686 He II during a flare of intensity τ. For finding the equivalent width of this line the level of the continuous spectrum $E_*(\tau)$ of the star during the flare at the same wavelength must also be known. This level can be determined in the usual way, that is, taking into account the Compton component of the radiation of the flare. Therefore we have:

$$E_*(\lambda) = 4\pi r_*^2 C_\lambda(\tau) B_\lambda(T_*) , \tag{9.58}$$

where λ = 4686 $\overset{\circ}{A}$. From (9.57) and (9.58) we have for the equivalent width of the line 4686 He II:

$$W(4686) = 0.37 \times 10^{10} \frac{W}{r_*} \frac{\tau^2}{C_\lambda(\tau)} \quad \text{cm} \tag{9.59}$$

where $C_\lambda(\tau)$ is given by (4.28) or (4.46) (Tables 4.2 and 4.4). Let us apply this equation to UV Cet ($r_* = 0.56 \times 10^{10}$ cm, $T_* = 2700$ K). We have:

$$W(4686) = 7 \times 10^8 \frac{\tau^2}{C_\lambda(\tau)} W \overset{\circ}{A} . \tag{9.60}$$

Table 9.9 gives the values of W(4686) found from this formula for a number of values of τ and two values of W. From these data the following conclusions can be drawn:

a) For faint flares ($\tau < 0.001$) the equivalent width of the line 4686 He II will be less than 1 $\overset{\circ}{A}$ and hence the line can practically not be observed in flare spectra.

TABLE 9.9 Equivalent Width (Theoretical) of the Emission Line
4686 He II in the Spectra of Flare Stars

τ	W(4686) (Å)	
	W = 0.1	W = 0.01
0.0001	0.06	0.01
0.002	1.2	0.1
0.001	5	0.5
0.005	70	7
0.01	170	17

On the other hand, the great majority of flares of UV Cet correspond to the case
$τ \gtrsim 0.002$. Therefore the presence of the line 4686 He II in the flare spectrum must
be considered to be a rare phenomenon; it can occur only in strong flares, that is
when $τ > 0.001$. This conclusion agrees well with what we know about the behavior
of this line in spectra of flares of various strengths.

b) For a given star the value of the equivalent width of the line 4686 He II depends
only on W--the dilution coefficient, since the strength of the flare, $τ$, can be
found for each individual case by a completely independent method--from the observed
value of the amplitude of the flare in some wavelength region. This circumstance
gives the interesting possibility of finding the numerical value of W, that means
the distance to which the fast electrons extend on the average, during a given flare.
Thus, for instance, at the time of the flare of UV Cet (24.IX.65) we had $\Delta U = 5^m2$,
which corresponds to $τ \approx 0.006$, and W(4686) = 3 to 7 Å. With these data we find
from Table 9.9 that $W \approx 0.01$, i.e. the cloud or shell consisting of fast electrons
extended to a distance five times the radius of the star--a quite reasonable value,
considering the strength of the flare.

If one talks about the excitation of emission lines of neutral and ionized helium
one important circumstance should be kept in mind: in both cases the line intensity
is proportional to $τ^2$. At the same time, as we have seen above, the intensity of
the hydrogen lines is proportional to the first power of $τ$. This implies that after
flare maximum the intensity of the helium lines must decrease much faster than that
of the hydrogen lines. For the same reason the helium lines must appear in the flare
spectrum later than the hydrogen lines. As a result the helium lines will be visible
in the flare spectrum for a considerably shorter time than the hydrogen lines.

20. LYMAN-ALPHA EMISSION IN FLARE STARS

The emission line of Lyman-alpha (L_α) of hydrogen is always present in each flare
of a star. Contrary to helium lines, which are excited only during strong flares,
the L_α line of hydrogen can appear even in very faint flares.

The fast electron hypothesis enables us to describe the main properties of L_α emission
in flare stars. That is the possibility of computing the expected strength of the
L_α emission, to find the equivalent width and to determine the form and dimensions
of the profile of the line itself. All three of these parameters can be found from
direct observations (outside the earth's atmosphere). Therefore we shall consider
them somewhat in detail.

The strength of the L_α-radiation. According to the theory of the hot gaseous nebulae,
if the medium is completely opaque in the Lyman-alpha line, about 75% of all L_c

energy finally is converted into radiation in the L_α line, as a result of fluorescence processes.

The chromospheres of flare stars are after all completely opaque in the L_α line; this follows from the fact that the L_c radiation, which ionizes hydrogen, falling from outside upon the chromosphere, penetrates up to the depth $\tau_c \sim 1$, while t_α, the optical depth in the L_α line, is of the order 10^4. Hence the problem of finding the strength of the L_α emission at the time of the flare is simplified.

The total energy radiated by the star in the Lyman-alpha line per second is noted as $E_\alpha(\tau)$. By definition we have:

$$E_\alpha(\tau) = 0.75\ E_c(\tau)\ W \quad \text{ergs s}^{-1} \qquad (9.61)$$

where $E_c(\tau)$ is the total amount of energy radiated by the star at the frequencies of hydrogen ionization; it is composed of two components--Compton and bremsstrahlung processes, that is

$$E_c(\tau) = E_{co}(\tau) + E_{br}(\tau)\ , \qquad (9.62)$$

where $E_{co}(\tau)$ and $E_{br}(\tau)$ are given by (9.7) and (9.12) (cf. Table 9.1). For the flux of L_α emission on the Earth we have:

$$N_\alpha(\tau) = \frac{E_\alpha(\tau)}{4\pi r_*^2 h\nu_\alpha}\ \text{photons cm}^{-2}\ \text{s}^{-1}. \qquad (9.63)$$

Table 9.10 gives the computed values of $E_\alpha(\tau)$ and $N_\alpha(\tau)$ for UV Cet as a function of the strength τ of the flare.

TABLE 9.10 Expected Power of the L_α Emission $E_\alpha(\tau)$ and the Flux $N_\alpha(\tau)$ of L_α Radiation Reaching the Earth from UV Cet at the Time of Flares of Varying Strength (τ)

τ	$E_\alpha(\tau)$ ergs s^{-1}		$N_\alpha(\tau)$ photons cm^{-2} s^{-1}	
	W = 0.1	W = 0.01	W = 0.1	W = 0.01
0.0001	4.0×10^{25}	4.0×10^{24}	0.003	0.0003
0.001	4.5×10^{26}	4.5×10^{25}	0.03	0.003
0.01	1.0×10^{28}	1.0×10^{27}	1.0	0.1

The absolute values of the L_α flux at the Earth are small--less than 1 photon cm^{-2} s^{-1} --but they are predicted to be larger than the radiation in the emission line H_α by a factor of about 15. Therefore it appears that it will not be difficult to observe the L_α line photographically. We note that the strength of the L_α emission during strong flares ($\tau \sim 0.01$) is comparable with the radiative power of UV Cet in U radiation in its quiescent state ($E_U = 2.3 \times 10^{27}$ ergs s^{-1}).

The expected L_α flux for the flare star--AD Leo--is several times larger than that found for UV Cet.

The equivalent width of the L_α line. Although the strength of the L_α emission in the two cases--inverse Compton effect and non-thermal bremsstrahlung--is nearly equal, the levels of the continuous spectrum at $\lambda 1200$ Å are completely different.

Hence we have quite different values for the equivalent width of this line, $W(L_\alpha)$, in the case of weak flares (inverse Compton effect) and of strong flares (bremsstrahlung). In the case of weak flares we have for $W(L_\alpha)$ from (9.17):

$$\frac{W(L_\alpha)}{\lambda_\alpha} = W\Gamma_\alpha \; \frac{3\gamma^4}{2} \; \frac{e^{x_\alpha} - 1}{x_\alpha^4} \; \frac{\tau}{C_\alpha(\tau)} \tag{9.64}$$

where $x_\alpha = h\nu_\alpha/kT_*$. Substituting here the value of the function $C_\alpha(\tau)$ for short wavelengths from (4.31) we find:

$$\frac{W(L_\alpha)}{\lambda_\alpha} = W\Gamma_\alpha \; \gamma^8 \; \frac{e^{x_\alpha/\gamma^2} - 1}{x_\alpha^4} \; , \tag{9.65}$$

i.e. for weak flares the equivalent width of the L_α line does not depend on τ. Substituting $\Gamma_\alpha = 0.75$, $\gamma^2 = 10$ and $T_* = 2800$ K, we find $W(L_\alpha) \approx 200\ W$ Å.

In the case of strong flares we have from (9.18):

$$\frac{W(L_\alpha)}{\lambda_\alpha} = W\Gamma_\alpha \; \frac{f_c(H^+)}{[\omega f(\omega,\gamma)]_\alpha} , \tag{9.66}$$

which means that also in this case $W(L_\alpha)$ does not depend on τ. Substituting $f_c(H^+) = 7\times10^{-4}$ and $[\omega f(\omega,\gamma)]_\alpha = 1.19\times10^{-4}$ for the L_α line, we find: $W(L_\alpha) = 5.4\times10^3\ W$ Å.

Thus in both cases—for weak flares ($\tau < 0.001$) and for strong flares ($\tau > 0.001$) —the equivalent width of the L_α line depends on one parameter only, the dilution coefficient W. Table 9.11 gives the expected values of $W(L_\alpha)$ as a function of W.

TABLE 9.11 Expected Values of the Equivalent Width of the L_α Line for Strong and Weak Stellar Flares

W	$W(L_\alpha)$	
	Strong flare ($\tau > 0.001$)	Weak flare ($\tau < 0.001$)
0.1	540	20
0.01	54	2

Thus, observations and measurements of the L_α line during stellar flares will be interesting because they may allow us to find the quantity W—the dimensions of the zone where the fast electrons are distributed around the star, in each individual case.

Profile of the L_α line. The profile of the L_α line of a flare star should look more or less as shown in Fig. 9.15. This Doppler profile has been constructed on the basis of the essential features of the energy transfer in lines through a medium of large optical depth ($\sim 10^4$) by means of many scattering acts [16]. The dimensions of the separate parts of the profile were found for an electron temperature of the chromosphere of $\sim 20\ 000$ K. The central narrow wedge-shaped absorption line is due to absorption in the interstellar medium with temperature ~ 100 K, a hydrogen density of $\sim 0.1\ cm^{-3}$, and a distance to the star ~ 3 pc. The emission line

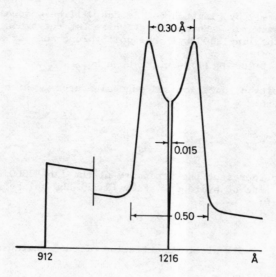

Fig. 9.15. Expected profile of the Lyman alpha line in
flare stars. The profile was computed for a value of the
optical depth of the chromosphere in the line ∿10⁴, an
electron temperature ∿20 000 K and a temperature of the
interstellar hydrogen ∿100 K.

itself is superposed on the continuous flare spectrum, the level of which depends
on the strength of the flare and rises towards the short wavelengths, up to 912 Å,
where it suddenly drops.

In this analysis of strength, equivalent width, and profile, of the L_α line in flare
stars no account was taken of the influence of the L_α background, due mainly to the
scattering of photons from the Sun by the hydrogen geocorona. This background should
certainly be taken into account in the preparation and execution of experiments outside the Earth's atmosphere with the purpose of recording the L_α line of flare stars.

21.   THE EMISSION LINE 2800 Mg II IN THE SPECTRA OF FLARE STARS

Theory predicts the possibility that the emission line of the well known ultraviolet
doublet of ionized magnesium 2800 Mg II (λλ 2793 + 2803 Å) be present in the spectra
of flare stars. Moreover, as we shall see below, this doublet should be the strongest emission line except the L_α line, in the entire spectrum of the flare star, from
1000 Å to 10 000 Å. The conditions in the chromospheres of flare stars suggest
that the main mechanism of excitation of the emission line 2800 Mg II is electron
collisions of the second kind.

The high efficiency of electron collisions of the second kind in this case is not
only due to the fact that the line itself (the doublet) 2800 Mg II is a resonance
line, but also to the fact that the excitation potential of this resonance level is
quite small, 4.4 eV.

Thus the transition of once ionized magnesium atoms from the ground state 1 to the
resonance level 2 is caused by inelastic electron collisions. But the period during
which these ions stay in the resonance level is very short ($A_{21} = 5.2 \times 10^8$ s⁻¹). As a

result the transitions $1 \to 2$ by electron collisions will be compensated by spontaneous transitions $2 \to 1$ downwards. Hence we may write for the amount of energy radiated by the chromosphere in the line 2800 Mg II per unit volume and per second:

$$E(2800 \text{ Mg II}) = n_1^+(\text{Mg}) b_{12}^o n_e h\nu_{12} , \tag{9.67}$$

where $n_1^+(\text{Mg})$ is the density of once ionized magnesium atoms, which are in the ground state, and b_{12}^o is equal to:

$$b_{12}^o = 8.54 \times 10^{-6} \frac{\Omega(1,2)}{\omega_2} T_e^{-\frac{1}{2}} e^{-\frac{\chi_{12}}{kT}} . \tag{9.68}$$

It is more convenient to represent the intensity of the line 2800 Mg II relative to the intensity of the Hβ line of hydrogen. From (9.31) the volume coefficient of radiation in this line is:

$$E_\beta = n_e^2 \varepsilon_\beta^o, \tag{9.69}$$

where ε_β^o is given by Eq. (9.32) after substitution of the constants for the line Hβ.

From (9.67) and (9.69) we find:

$$\frac{E(2800)}{E_\beta} = \frac{n_1^+(\text{Mg})}{n_e} \frac{b_{12}^o h\nu_{12}}{\varepsilon_\beta^o} . \tag{9.70}$$

Under the conditions of the chromospheres of flare stars the following condition holds: $n_e \approx n(\text{H})$, where $n(\text{H})$ is the density of the hydrogen atoms. With regard to magnesium, it is easily seen that this will be practically ionized once, and therefore $n_1^+(\text{Mg}) \approx n(\text{Mg})$. Then from (9.70) the upper limit of the relative intensity of the line 2800 Mg II is:

$$\frac{E(2800)}{E_\beta} = \frac{n(\text{Mg})}{n(\text{H})} \frac{b_{12}^o h\nu_{12}}{\varepsilon_\beta^o} . \tag{9.71}$$

We note that in this form the ratio $E(2800)/E_\beta$ no longer depends on the electron density of the chromosphere n_e.

For the Mg^+ ion we have: $\Omega(1,2) = 10.5$ [18], $\omega_2 = 4$, $\chi_{12} = 4.4$ eV. Taking $T_e = 20\,000$ K for the electron temperature of the chromosphere, we find from (9.71):

$$\frac{E(2800)}{E_\beta} = 1.5 \times 10^6 \frac{n(\text{Mg})}{n(\text{H})} , \tag{9.72}$$

where $n(\text{Mg})/n(\text{H})$ is the relative abundance of magnesium. Assuming that this is the same in flare stars as in ordinary stars, that is of the order 10^{-5}, we finally find the ratio of the intensities of 2800 Mg II and Hβ:

$$\frac{E(2800 \text{ Mg II})}{E_\beta} = 15 . \tag{9.73}$$

The value found for the intensity of the doublet 2800 Mg II may seem unusually large; however, such large intensities are ordinarily proper for _forbidden_ lines in gaseous nebulae, that is lines which are excited by electron collisions. In our case this effective mechanism for the excitation of emission lines--electron collisions--also works in relation to a _permitted_ line.

The resonance lines (doublet) of ionized calcium H and K are also present in the spectra of flare stars. The two doublets, 2800 Mg II and H and K Ca II, arise as a result of the same types of atomic transitions. But the cosmic abundance of magnesium is about 10 to 20 times that of calcium. Therefore under the same conditions the intensity of the doublet 2800 Mg II must be many times larger than the intensity of the doublet H + K of Ca II. On the other hand, the observed value of the ratio E(H+K Ca II)/Hβ in flare stars is of the order unity. Hence the intensity of the doublet 2800 Mg II must be 10 to 20 times larger than the intensity of the Hβ line. Of course, here we are assuming that the H + K Ca II lines are excited by electron collisions too.

From the relative intensities (9.72) we can find the ratio of the equivalent widths as follows:

$$\frac{W(2800)}{W_\beta} = \frac{E(2800)}{E_\beta} \frac{[C_\lambda B_\lambda]_{\lambda=4686}}{[C_\lambda B_\lambda]_{\lambda=2800}} \; , \tag{9.74}$$

where $C_\lambda = C_\lambda(\tau,\gamma,T_*)$ is given by formula (4.28), and B_λ is the Planck function for a given effective stellar temperature T_*. Performing the corresponding computations we find from (9.74) for $T_* = 2800$ K:

$$\frac{W(2800)}{W_\beta} = 0.56 \frac{E(2800)}{E_\beta} \approx 8 \; ; \tag{9.75}$$

this means that the equivalent width of the doublet 2800 Mg II exceeds the equivalent width of the line Hβ.

The strength of the doublet 2800 Mg II proves to be rather large even if $n(Mg)/n(H) = 10^{-6}$, namely: $E(2800)/E_\beta \approx 1.5$ and $W(2800)/W_\beta \approx 0.8$.

As we have seen, the H_α line of hydrogen is somewhat stronger than H_β in flare stars (cf. Table 9.7). Hence we come to the conclusion that the doublet 2800 Mg II must be the strongest emission line (except L_α) in the spectra of flare stars over the entire spectral range which is accessible to observation (at least in principle).

Due to this strength the doublet 2800 Mg II will also be visible during the quiescent state of the star, outside flares. The equivalent width of 2800 Mg II in the quiescent state of the star will be considerably larger than the value expected at the time of a flare. As a flare disappears we see the progressive weakening of all emission lines, the doublet 2800 Mg II may prove to be the only line which will still be visible after all lines of the Balmer series of hydrogen, among which H_α and H_β, have disappeared completely.

In certain cases (for instance, in stars with a smaller flare frequency) the presence of another interesting emission line is expected in the spectra of flare stars, namely 2852 Mg I of neutral magnesium. This will be correlated with 2800 Mg II: as the latter weakens, 2852 Mg I must become stronger.

That the doublet 2800 Mg II--in emission, as well as in absorption--plays a very important role in the physics of stellar atmospheres and in the interstellar medium was well known earlier. Therefore the preparation and execution of special experiments with the aim of discovering and measuring the line 2800 Mg II in many non-stationary objects should be given a high priority.

22.   MACROSTRUCTURE OF THE SPECTRUM OF FLARE STARS IN THE ULTRAVIOLET

From the results of the analysis given of the continuous spectrum (Chapter 5) and of the emission lines (the present chapter) a general picture can be given for the macrostructure of the spectrum of flare stars in the ultraviolet--from 912 Å to the limit of the visible region, i.e. to 3000 Å.

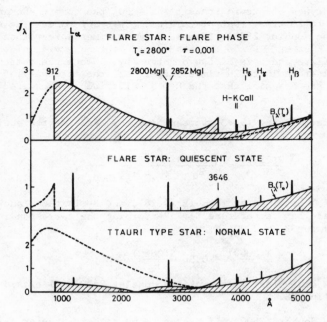

FIG. 9.16 General picture of expected spectra in the ultraviolet (912–3000 Å) for flare stars at the moment of the flare
(top), in the quiescent state (center), and also for stars of
T Tau type in their normal state (bottom). The strongest
emission lines and the Balmer continuum are shown as well. The
spectra were constructed according to the hypothesis of fast
electrons for a flare of strength $\tau = 0.001$.

Such a theoretical spectrum, based on the hypothesis of fast electrons, is shown
in Fig. 9.16. In the upper part the spectrum of a flare with strength corresponding
to the value $\tau = 0.001$ is given. The full-drawn line gives the continuous spectrum
which appears only at the time of the flare and is superposed on the normal (Planck)
spectrum of the star B_λ (T_o), corresponding to its effective temperature T_o (broken
line). The cutoff of the continuous spectrum at 912 Å is due to absorption by interstellar hydrogen, the influence of which is very important even in the case of flare
stars of UV Cet type near the Sun. The absorption by interstellar dust is small
and was not taken into account in this case. This diagram shows that the expected
level of the continuous spectrum in the region of wavelength ∿1000 Å must be considerably higher than its level in the optical spectral range, even for average
flares.

The strongest emission lines are superposed on the continuous flare spectrum. In
the ultraviolet these are: the L_α line of hydrogen, the resonance doublet 2800 Mg II
of ionized magnesium, and possibly, the resonance line 2852 Mg I of neutral magnesium.
The magnesium lines must be considerably stronger than the H and K lines of ionized
calcium and even stronger than H_β.

The diagram in the middle of Fig. 9.16 gives the expected shape of the spectrum of a
flare star in its quiescent state, outside flares, in the same wavelength range. In
this case there must be practically no continuous component of the spectrum in the
ultraviolet, if the weak continua of various series, among which those of hydrogen,
are not considered. However emission lines are present; although they become weaker

in the course of time, they do not completely disappear before the next flare.
Over a considerable part of the spectrum, from 912 Å to 3000 Å, the lines L_α and
2800 Mg II, and also 2852 Mg I, will stand out clearly against a dark background.   In
the quiescent state the Lyman continuum is generated, but it will be completely
absorbed by interstellar hydrogen and therefore will not be detectable.

In the lowest part of Fig. 9.16 the expected spectrum in the ultraviolet is shown
for another class of unstable stars--of T Tau type--in their normal state.   These
objects are, as will be shown in the following chapter--flare stars which are
permanently or extremely often at the flare stage.   Therefore the main spectrum of
T Tau type stars in the normal state in the region 912-3000 Å (broken line in the
lowest diagram) is essentially similar to the spectrum of an ordinary flare star
at the time of the outburst.   However since the T Tau stars are at relatively large
distances from us, the absorption by interstellar dust with its characteristic
maximum at 2200 Å will be large, as a result of which the observed level of the
continuous spectrum in the ultraviolet will be strongly suppressed.

We conclude that it is extremely important and useful to perform spectrophotometric
and wide-band photometric observations of flare stars and T Tau type stars in the
ultraviolet, i.e. outside the Earth's atmosphere with the aid of powerful orbiting
observatories specially constructed for this purpose.

REFERENCES

1. G. A. Gurzadyan, Planetary Nebulae, Gordon and Breach, N.Y., 1969.
2. P. F. Chugainov, Izv. Crimean Obs. 38, 200, 1967.
3. B. W. Bopp and T. J. Moffett, Ap.J. 185, 239, 1973.
4. R. E. Gershberg and P. F. Chugainov, Astron. J. USSR 43, 1168, 1966; 44, 260,
       1967.
5. R. E. Gershberg and N. I. Shachovskaja, Astron. J. USSR 48, 934, 1971.
6. G. Haro and E. Chavira, ONR Symposium, Flagstaff, Arizona, 1964.
7. G. A. Gurzadyan, Doclady Academy Nauk SSSR 172, 1046, 1967.
8. G. A. Gurzadyan, Doclady Academy Nauk SSSR 130, 287, 1960.
9. W. E. Kunkel, Flare Stars, Thesis, Texas, 1967.
10. B. J. Bray and R. E. Loughhead, The Solar Chromosphere, Chapman and Hall,
        London, 1974.
11. J. W. Chamberlain, Ap.J. 117, 387, 1953.
12. A. H. Joy, Stellar Atmospheres, Ed. J. L. Greenstein, 1963.
13. V. P. Vyazanitsin, Izv. GAO 147, 19, 1951.
14. A. H. Joy and M. L. Humason, PASP 61, 133, 1949.
15. J. L. Greenstein and H. Arp, Ap.J. Lett. 3, 149, 1969.
16. V. V. Ivanov, Radiative Transfer and the Spectra of Celestial Bodies (in
        Russian), Moscow, 1969.
17. G. A. Gurzadyan, Astron. J. USSR 35, 520, 1958.
18. V. V. Sobolev, Radiative Transfer in the Atmospheres of the Stars and Planets,
        Moscow.
19. J. K. Van Blerkom, J. Phys. B. Atom. Mol. Phys. 3, 932, 1970.

CHAPTER 10
T Tauri Type Stars

1. FLARE OF T TAU TYPE STARS

In December 1953 Haro observed a flare with amplitude $1^m.5$ of a faint star of unknown spectral class (in the Orion stellar association); later this star was called VY Ori (Haro 18 = Brun 112) [1]. After about two years he recorded a second flare of this star with an amplitude of more than 2^m. About the same time Herbig obtained a spectrogram of that star made with the slit spectrograph of the Lick Observatory; the object turned out to be a star of T Tauri type, with very strong ultraviolet excess and intense emission lines [2]. Thus it was shown for the first time that flares can occur in stars of T Tau type--objects considerably younger than the representative flare stars near the Sun.

Subsequent investigations showed that many flare stars in Orion are objects with $H\alpha$ emission and, vice versa, many of the stars known to have $H\alpha$ emission began to appear in lists of flare stars. According to Haro [3] 34 of 176 flare stars in Orion, i.e. about 20%, are also objects with $H\alpha$ emission. This should be taken as a lower limit since the decision if there is $H\alpha$ emission or not is based on spectro-grams obtained with an objective prism. Since the stars with $H\alpha$ emission are faint, slit spectrograms can be made of them very rarely; therefore it is hard to establish that they are all stars of T Tau type.

The importance of this discovery--flares of stars of T Tau type--lies in the fact that this greatly enhances the significance of the flare phenomenon itself. Up to that time it was known that flares occur in a group of stars scattered in the neighborhood of the Sun, with ages of the order of 10^8 years. Now stars which are in their very earliest stage of formation are found to show the same phenomenon.

It was soon discovered that the characteristics of the flare itself in the case of T Tau type stars do not differ from those usually observed in other stars which occur in associations. Hence we can ask the question: Is the same non-thermal mechanism of energy generation responsible for flares in ordinary flare stars and T Tauri stars?

From an analysis of the observational data V. A. Ambartsumian suggests that the processes resulting into flares in T Tau and UV Cet stars are identical [4]. How-ever, whereas in the UV Cet stars the production of the additional energy has an impulsive character, in the T Tau stars it has a more or less stable character and is accompanied by a certain change in the temperature of the star. According to V. A. Ambartsumian, in T Tau stars the energy from the stellar nucleus may be liberated in different layers of the photosphere. If this process takes place in deep layers of the photosphere, then the variations of the brightness will be less sudden, and if the energy is liberated above the photospheric layers, the variation of the brightness will be more sudden and will be accompanied by strengthening of the continuous emission. Here a difficulty arises, namely: if the transfer method of the energy from the stellar interior is the same in the UV Cet and T Tau stars, then it is difficult to understand why this energy cannot be produced in the photospheric layers of UV Cet stars as well. In UV Cet stars the energy is liberated in the outer parts of the star only, above the photosphere. It is not very probable that the age difference of these two groups of stars has any importance on this question.

If the fast electron hypothesis is adopted, however, it is not necessary that in the T Tau stars the additional energy be produced in different layers of the photosphere. In that case the photosphere can also be heated from the outside, under the influence of the radiation of Compton origin falling upon it. The general picture then is the following: We assume that above the photosphere of a T Tau star there is a medium in the form of an envelope consisting of fast electrons and fast protons. The envelope is regenerated by new fast electrons and protons from the matter which is ejected from the star's interior. However, as opposed to the ordinary flare stars in which the ejection of this material has an impulsive character, in the T Tau stars this process takes place at a more or less constant rate, during the entire time the star is in the T Tau state. In those cases where the rate at which the matter from the stellar interior is ejected into the outer regions of the star varies, the strength of the additional radiation of non-thermal nature and the degree of heating of the photosphere may vary as well.

Hence, according to this model T Tauri type stars are stars flaring permanently with very high frequency [5]. Such flares occur with a more or less constant strength, characterized, as before, by the effective optical thickness τ of the surrounding electron shell for Thomson scattering processes. Fluctuations in the value of τ give corresponding fluctuations in the available total amount of energy of Compton origin. Part of the radiation, directed towards the observer, will be recorded as continuous emission of non-thermal nature and the remainder, directed towards the star and absorbed by its chromosphere and photosphere, which later is re-radiated in the form of thermal energy.

The possibility of a "permanent flare" is related to convection activity, which in the youngest stars, such as the T Tau stars are, is considerably higher than in the ordinary flare stars of UV Cet type. The requirement that the energy from the stellar interior be liberated in the regions above the photosphere in the form of a "permanent flare" is also a consequence of the fact that there must be an active source of radiation ionizing hydrogen to explain the presence of emission lines (which are even stronger than in UV Cet stars) in the spectra of T Tau type stars. In other words, if the additional energy is produced in the deep photospheric layers it will be hard to understand how the ionizing radiation can reach the chromosphere.

The validity of the fast electron hypothesis for T Tau type stars will be discussed in the following sections of this chapter.

2. MAIN PROPERTIES OF T TAU TYPE STARS

A detailed description of the properties of T Tauri type stars is given by Herbig [6]. Here we mention the most important properties of these objects. The question whether a certain star belongs to the T Tau type is after all the subject of spectroscopic investigation. Therefore in the following special attention will be given to spectroscopic criteria.

1. Stars of T Tau type are irregular variables. The fluctuations of the brightness, rapid as well as slow ones, are entirely irregular. In individual cases there are indications that the fluctuations of brightness may show a certain periodicity with a period up to several days (RW Aur, S CrA, RU Lup). But the available photometric data are not sufficiently numerous and homogeneous for definite conclusions to be drawn from them. However, certain regularities of a different character have been

found. Thus, for instance, in one case the variable star is more often bright than faint (class I according to Parenago [7]), in another case, on the contrary, the variable star is more often faint than bright (class III). Sometimes a variable star has usually its average brightness (class II), and finally there exist varia- bles (class IV) which do not show any signs of belonging to any of these classes.

2. Stars of T Tau type belong to spectral types F8 - M2. A subclass later than M2 probably occurs, but it is very hard to obtain slit spectrograms of such stars because they are very faint.

3. The absolute magnitudes of T Tau type stars indicate that they are normal stars. However, the classification of their luminosities requires further study.

4. The hydrogen lines and the H and K lines of ionized calcium are seen in emission. The emission spectrum is very similar to a flare spectrum.

5. Fluorescence emission lines λ 4063 HeI and λ 4132 HeI are present (they were discovered only in the spectra of T Tau type stars), as well as lines of Fe II, Ti II; the latter are the faintest emission lines in the spectra of these stars.

6. Usually, but not always, the forbidden lines λ 4068 [S II], and λ 4076 [S II] are seen; sometimes also the lines λ 6717 [S II], 6731 [S II], more rarely the doublet λ 6300 [OI] and λ 6363 [OI]. The lines of [Fe II] are either very faint or altogether absent.

7. Sometimes T Tau type stars of late classes show emission lines of He I, and rarely lines of He II.

8. The case that the strong line λ 6707 Li I occurs in a star's spectrum in absorption may be an additional criterion that the star is of the T Tau type.

9. The T Tau type stars have unusually large infrared excesses, up $6 - 8^m$.

The emission lines are ordinarily superposed on the continuous spectrum which can either have no absorption lines at all (a pure continuum with one or two depress- ions characteristic for stars of late classes) or be normal, with absorption lines, corresponding to the late subclasses of F, classes G and K, or to the early sub- classes of M. That there are no T Tau stars of subclasses earlier than F8 seems real, but it is not yet definitively established that there exist no stars with subclasses later than M2-M3.

Stars with the spectral characteristics of T Tauri very often lie close to or within small gas-dust nebulae. Then, the nebulae themselves also vary in brightness. However, the association with nebulae is rather a secondary phenomenon. The fact that there is a relation of a star with a nebula does not mean that the star is of T Tau type.

The most interesting and perhaps the most characteristic property of T Tau type stars is the unusually large infrared excess in their radiation. This property was discovered by Mendoza [8] from UBVRILKM photometry of stars of T Tau type. In this investigation it was found that [8-10]:

(a) An ultraviolet excess occurs in all stars, but it is very large in late type stars. For instance, for DF Tau it is equal to $E (U - B) = 2^m2$.

(b) A blue excess is characteristic for stars with a faint ultraviolet excess. For instance, for the same star DF Tau $E (B - V) = 0^m6$.

(c) A red excess is seen in all stars.

(d) An infrared excess of 1 to 5^m occurs in all stars. For some of them this excess is very large and reaches 8^m, e.g.

$$\text{for T Tau} \quad E (V - M) = 6\overset{m}{.}2$$

$$\text{for R Mon} \quad E (V - M) = 8\overset{m}{.}2 .$$

Hence T Tau stars are "normal" in the visual region of the spectrum only. In the spectral regions of very short or very long wavelengths the radiation becomes "abnormal"--the observed radiation fluxes exceed considerably the levels customary for dwarf stars of the same temperature [11,12].

3. THE PERMANENT FLARE OF T TAURI STARS

Some years after the idea was suggested of the possibility of a permanent flare in stars of T Tau type [5], Rodono succeeded in obtaining observational proof of its validity [13]. This was done for a star well known for its flare activity, a member of the Hyades aggregate, H II 2411. During 551 hours of homogeneous photographic observations of this star (in the region of the Pleiades) about 50 flares of this star were recorded, which gives for the average flare frequency 0.09 fl/hr [14]. This star holds the record for the number of observed flares (104 flares from data up to 1976 [15]) among the flare stars in stellar aggregates. The star H II 2411 is of class dM4e (V = $14\overset{m}{.}18$), and all of its parameters--absolute luminosity (M_V = $11\overset{m}{.}18$), color characteristics (B – V = $+1\overset{m}{.}59$, U – B = $+1\overset{m}{.}04$ [16]), the presence of emission lines, its position in the U – B vs. B – V diagram, etc.--indicate that it belongs to T Tau type. Following this star (for 15 hours) with an electrophotometer with high time resolution (1 second), sensitive in B rays and mounted on the 91 cm and 208 cm telescopes of the McDonald Observatory, Rodono observed three flares of this star. In one of these flares (No. 2, 28.XI.72) the lightcurve--shown in Fig. 10.1--shows a strange microstructure, not at all similar to the ordinary light-curves of flares of UV Cet type stars. The strangeness lies in the appearance of an obvious high-frequency oscillation of the brightness which seems to be superposed on some average level of the lightcurve. This strange lightcurve prompted the observer to change the method of measuring and recording the flare, in particular the size of the entrance diaphragm was changed (from 0.5 mm to 2 mm), and the time of integration (from 1 to 2 seconds). The results remained the same; as before, a high-frequency oscillation was observed.

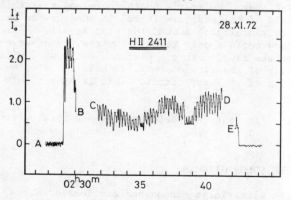

Fig. 10.1. Lightcurve of flare No. 2 of the star H II 2411 (Hyades cluster), obtained with the 208-cm telescope of the McDonald Observatory in B light. The integration time was 1 second from A to B and 2 seconds from C to D. The high-frequency oscillations, with amplitudes much larger than the measuring errors, are clearly visible.

Analysis of all possible factors which might lead to such an artificial effect--statistical modulation of the detection of the photons, causing a pulsation of the stellar picture at the edges of the diaphragm, atmospheric turbulence, etc.--do not provide an explanation. The observed oscillation was found to be rather periodic, and the amplitudes of the brightness fluctuations proved to be much larger (40-70%) than could be accounted for by atmospheric scintillation.

The average period of the observed oscillations was found to be very constant, and equal to 13.08 ± 0.6 s. If these oscillations are real acts of flaring, this gives for the flare frequency about 280 fl/hr or around 6700 fl/day--two orders higher than we found for the flare frequency of stars of UV Cet type.

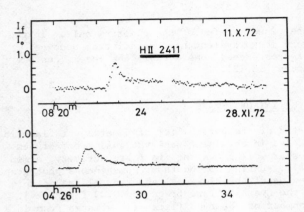

The lightcurves of the other two flares (No. 1, 11.X.72 and No. 3, 28.XI.72) with amplitudes 0^m6 and 0^m5 (in B light) do not show such oscillations (Fig. 10.2). However, due to limited data it is hard to say to which degree the observations were influenced by differences in observing conditions. With regard to the interpretation of the light-curve for flare No. 2 (Fig. 10.1), the question remains in how far the observed amplitudes of the oscilla-tions were limited by instrumental factors, in particular the time constant. For the moment we can only establish the fact that the discovery made by Rodono does not contradict the hypothesis of a permanent flare in stars of T Tau type. Further , the constancy of the oscillations leads us to believe that in the case discussed there is a certain real periodicity of

Fig. 10.2. Lightcurves of flares No. 1 and 3 of the star H II 2411, obtained with the 91-cm and 208-cm telescopes. The integra-tion times are 3 seconds and 2 seconds, respectively. There are no high-frequency oscillations.

the phenomenon itself, due to certain other factors. Let us assume that there has appeared a cloud of fast electrons high above the star's photosphere and that this cloud, drawn along by the magnetic field of the star, is forced in some way to per-form a circular motion (with the velocity of light) around the star itself. Making one revolution in ~13 seconds, this cloud could appear to emit towards the observer at every revolution. During that time the cloud covers a distance of about 4 million kilometers, which gives for the diameter of the circle about 1.2 million km--con-siderably more than the assumed diameter of the star itself. This is as yet only a pure hypothesis.

The example of H II 2411 convinces us of the extreme usefulness of new observations of T Tau type stars.

4. T TAU TYPE STARS IN THE COLOR DIAGRAM

Most of the stars of T Tau type have an ultraviolet excess and are located above and to the right of the main sequence in the U - B vs. B - V diagram. In some cases their "distance" from the main sequence reaches up to $1^m5 - 2^m$ (along the U - B axis). A comparison of the observed colors of T Tau stars with a theoretical U - B vs. B - V diagram, constructed with the assumption of the fast electron hypothesis, may certainly be interesting.

The most reliable and sufficiently homogeneous data about colors of T Tau stars and similar objects have been collected by Walker [17] for the very young cluster

NGC 2264 and by Smak [18] for the Taurus cluster. The results of their observations are shown in the theoretical U – B vs. B – V diagram (Fig. 10.3) corresponding with the fast electron hypothesis. We note that the curve for subclass M0 coincides with the limiting curve of the U – B vs. B – V relation, derived for the case of a hot gas. The overwhelming majority of the stars shown in Fig. 10.3 proved to be outside that limit. At the same time they are freely distributed in the region of the diagram within the boundaries determined by the fast electron hypothesis.

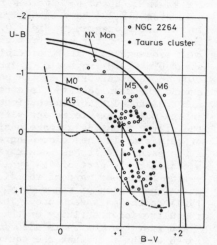

Fig. 10.3. Stars of T Tau type (from the clusters NGC 2264 and Taurus) in the theoretical U – B vs. B – V diagram.

Walker also made colorimetric measurements of members of another young cluster--NGC 6530 [19]. However, the fraction of stars in that cluster which are relatively active in the ultraviolet is considerably smaller.

In all cases mentioned above the T Tau stars are located in the color diagram where usually flare stars of UV Cet type are found at the time of a flare. Hence it may be concluded that the additional radiation in these two categories of objects, T Tau stars and flare stars, may be of the same nature. In the quiescent state the flare stars usually lie around the lower branch of the main sequence. Only at the time of the flare are they lifted above this branch. But the T Tau stars are "living" permanently in the upper part of the diagram. In other words, the state of activity which is incidental for flare stars is constant for T Tau stars. Therefore it would be better to formulate this state of affairs somewhat differently: the T Tau stars are stars in a <u>permanent</u> flare state.

The T Tau stars shown in Fig. 10.3 are scattered all over the diagram, from its lowest part corresponding with flare activity close to zero ($\tau \approx 0$), up to $\tau \sim 0.1$ and even somewhat higher. In fact the diagram gives the degree of activity, i.e. the value of τ, for each individual star. Stars are encountered with widely differing activities--τ ranges from 0.01 to 0.0001 or less. At the same time a rather large concentration of stars is found in the lower part of the diagram. The impression is given that each T Tau star at the time of its birth appears in the upper part of the diagram, corresponding with very high ultraviolet activity, of somewhat permanent character. However, judging from the low concentration of stars in the upper part of the diagram, these stars must move downward, to a region of lower activity, later still further downwards, and so on until they reach the main sequence.

Thus for T Tau stars the passage from the upper part of the diagram to the lower part indicates an evolution in time.

5.   THE T TAU AND RW AUR STARS

These stars are the most distinct representatives of the stars of T Tau type.

<u>T Tau</u>. A classical star of T Tauri type, of spectral class dG5e. Its brightness

varies between $9^{m}_{.}5$ and $13^{m}_{.}0$ (in photographic light).

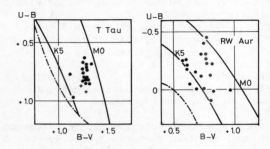

Fig. 10.4. Observed color indices of T Tau and RW Aur (dots) in the theoretical U – B vs. B – V diagram.

The known U – B and B – V values for this star [18,20] have been given in the color diagram in Fig. 10.4; the distance of the star from the main sequence is quite noticeable, about $0^{m}_{.}5$ – $0^{m}_{.}8$ (in the U – B direction). The scattering of the points around the mean position is after all real and due to fluctuations in the main ultraviolet activity of the star. From the average position of this star in the color diagram it is seen that it has gone through a considerable part of its evolution --from the upper part of the diagram (the region of stars of NX Mon type) to the main sequence. That the ultraviolet activity of the T Tau star fluctuates is also indicated by the fact that the equivalent widths of the emission lines in its spectrum suffer relatively fast variations over a considerable range. As an illustration Table 10.1 gives sufficiently reliable

TABLE 10.1 Equivalent Widths (Å) of Some Emission Lines in the Spectrum of T Tau at Various Periods of Observation

Date	Hα	Hβ	Hγ	Hδ	K (Ca II)
4.XII.70	43.2	9.3	1.20	--	8.9
17.XII.70	100.6	2.5	0.35	0.48	3.9
19. I.71	--	7.3	0.94	0.35	20.0
20. XI.71	95.5	5.2	1.20	0.37	6.9
11. I.72	--	5.0	1.10	0.22	8.6

measurements of equivalent widths of some emission lines in the spectrum of T Tau, made at various times [22]. As we see, the equivalent widths of individual emission lines fluctuate by a factor 2-3, or even more. However it is hard to judge from these data to which degree these fluctuations are due to variations of the level of the continuous spectrum, and to what extent they are due to variations of the intensity of the lines themselves. In the spectrum of T Tau the following forbidden lines had been detected earlier: λ 6717 [S II], λ 6731 [S II], λ 6300 [OI], λ 6363 [OI], and recently the rather faint lines λ 6548 [N II] and λ 6584 [N II] [22]. The presence of these forbidden lines indicates that there exists a region around the star, apparently part of a larger envelope, where the electron density is of the order 10^{6} cm^{-3}.

According to all of the data the star T Tau is a potential flare star. However, for a long time no reliable flares were detected from it. Its first flare was observed on 15.II.74 [21] by the photoelectric method by simultaneous observations in four spectral regions--U, B, V and R (λ_{eff} = 7000 Å). The amplitudes at flare maximum were found to be: $\Delta U = 1^{m}_{.}30$, $\Delta B = 0^{m}_{.}31$, $\Delta V = 0^{m}_{.}28$ and $\Delta R = 0^{m}_{.}17$. The mean duration of the flare was about 20 minutes. The amount of energy liberated

during that flare was estimated to be of the order 10^{35}–10^{36} ergs (for a distance
of the star equal to 170 pc) or, averaged over the flare period, around 10^{32}–10^{33}
ergs/s. This is a much more powerful process than that which is observed in
ordinary flares of stars of UV Cet type.

RW Aur. Spectral type dG5e; the brightness fluctuates in photographic light
between 9.0 and 12.0.

This star also lies above (by about 1^m) the main sequence in the U – B vs. B – V
diagram (Fig. 10.4). However, the scatter of the observational points [6] is
larger than in the case of T Tau, which implies stronger fluctuations of the ultraviolet activity of the star around the mean value.

I. Salmanov [22] made synchronous spectrophotometric and photometric observations
of this star during several years. This data gives the fluctuation in brightness
ΔV, and the equivalent width W_λ of some emission lines during various periods of
activity of the star. About 120 emission lines were identified and measured in the
spectrum of RW Aur. The results for Hγ (12 observations in three years) are given
in Fig. 10.5 in the form W(Hγ) vs. ΔV (dots); the full lines are the theoretical
W(Hγ) vs. ΔV relations, computed in the same way as in the case of Fig. 9.12
(Chapter 9), only here T_* was taken equal to 4200 K (the effective temperature of

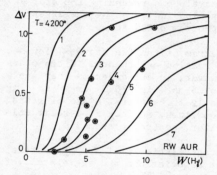

Fig. 10.5. Observed values of W(Hγ) vs. ΔV (dots in circles)
for RW Aur in the theoretical diagram of W(Hγ) vs. ΔV (full-
drawn lines), constructed on the assumption of the fast
electron hypothesis. The numbers next to the curves indicate
the following combinations of K_λ and W:

Curve	1	2	3	4	5	6	7
K_λ	0.03	0.064	0.08	0.13	0.16	0.23	0.32
W	0.005	0.01	0.0125	0.02	0.025	0.035	0.050

a star of class G5). We see that the observational points are found scattered
within reasonable limits of the initial parameters. And nearly all points lie between curves 3 and 5, i.e. between values of K_λ from 0.08 to 0.16 and values of
the dilution coefficient W from 0.0125 to 0.0250, i.e. within the limits of small
fluctuations of the effective temperature and a small scatter of the effective
distances around the star up to which the ensemble of fast electrons extends.

Regardless of the large number of emission lines in the spectrum of RW Aur their
contribution to the total brightness of the star is relatively small: in stellar

magnitudes it is from 0^m10 to 0^m40, depending on the activity of the star (in the spectral range λ 4000–4800 Å).

General analysis of the spectrograms of RW Aur shows that three types of spectra can be distinguished, which may appear at various periods of stellar activity:

a) The normal emission spectrum with moderate or strong emission lines. Strong continuous emission, hardly any absorption lines in the visible region of the spectrum.

b) Intermediate spectrum with weak emission lines, but with moderately strong absorption lines. Faint or completely absent continuous emission.

c) Extreme spectrum, when both emission lines and absorption lines are weak. There is some continuous emission, but not strong.

In the spectrum of RW Aur the following forbidden lines had been observed before: λ 4068 [S II], λ 4076 [S II], λ 4359 [Fe II], λ 4452 [Fe II]. Recently the line λ 4363 [O III] [22] was discovered in the spectrum of this star—this is the first time that this line is seen in spectra of stars of T Tau type. It is interesting to note that this line is observed only during periods of great activity. All forbidden lines can arise in a medium where the electron density is of the order 10^8 cm^{-3}. Hence the gaseous envelope surrounding the star RW Aur is considerably more dense than T Tau. This is also indicated by the absence of the forbidden lines of [O I] and [N II] in the spectrum of RW Aur.

Returning to the color diagram of the two stars—T Tau and RW Aur—we must point out a discrepancy between observation and theory. Often stars of T Tau type belong to spectral classes K–G. However, in our diagrams they are within the limits corresponding to the subclasses M6–M5 (Fig. 10.3) or M0–M5 (Fig. 10.4). However, there is no contradiction. Usually the spectral class of stars of T Tau type is determined from their absorption lines, and not from their colors. The colors of these stars correspond with earlier spectral classes.

Analysis of the color characteristics of T Tau type stars on the one hand and flare stars belonging to stellar associations and aggregates (cf. Chapter 11) on the other hand leads us to the important conclusion that there exists without any doubt a similarity in the general structure of the U–B vs. B–V diagrams of the following three categories of objects: (a) stars of T Tau type; (b) flare stars in Orion, Pleiades and other aggregates; (c) flare stars of UV Cet type at flare maximum.

6. PARTICULARLY ACTIVE STARS OF T TAU TYPE

There exist stars of T Tau type with unusually strong ultraviolet radiation. A typical representative of this category of objects is NX Mon, a well-known variable star in the cluster NGC 2264 in Monoceros. In order to give an idea of the power of the radiation of this star in the ultraviolet it is sufficient to say that according to its U light it is comparable with O-type stars, whereas in visual light its color corresponds to stellar class F–K. This follows from the results of four measurements by Walker [17] given below:

V	15^m63	16^m10	16^m10	15^m87
U – B	-0^m76	-1^m10	-1^m02	-1^m21
B – V	$+0^m32$	$+0^m43$	$+0^m71$	$+0^m58$

At another active time of this star, it was found that U – B = -1^m35, B – V = $+0^m57$

(photoelectric measurements [23]).

Another object resembling NX Mon is a variable star in Orion – BC Ori (= Haro 119); for this star it was found that U – B = –0.93, B – V = +1.02 [23]. Strong ultraviolet radiation was also observed in HS Ori, CE Ori, AU Ori, YY Ori; these are all members of the Orion association. How large is the fraction of the stars which are particularly active in the ultraviolet among the objects of T Tau type? In order to answer this question Haro and Herbig [2] made special observations by the multiple-image method, making three images of the star on one plate, consecutively in blue, yellow and ultraviolet light. As a result they came to the following conclusions:

1. Among 175 stars with Hα emission in Orion, 28 (16%) have unusually high activity in the ultraviolet as compared with normal dwarf stars. In the cluster NGC 2264, 14 (19%) of 73 stars with Hα emission are uncommonly bright in the ultraviolet. From these data it follows that the high percentage of objects which are bright in the ultraviolet is the same for the two groups of T Tau stars.

2. Strong activity in the ultraviolet is only found in stars with emission lines. Slit spectrograms of these stars show that they belong to T Tau type. An ultraviolet excess is discovered in stars of which the intensity of the Hα emission is estimated to be from "average" to "very strong". Strong ultraviolet excess does not occur at the same time as weak Hα emission.

3. The distribution of the energy in the spectral region $\lambda > 3800$ Å for stars which are active in the ultraviolet (NX Mon, AU Aur, BC Ori, etc.) shows that they belong approximately to class M or late subclasses of K. The spectra of these stars show no structure; there are no absorption lines at all. In the region $\lambda < 3800$ Å the energy distribution differs from that which is usual for stars of late type classes. The growth of intensity in the continuous spectrum starts at 3750 Å and, reaching a maximum around 3700 Å, falls towards the short wavelengths. In general the presence of the emission line Hα in the spectrum may be considered as the criterion for the presence of ultraviolet emission.

4. Some differences have also been found among stars which are very active in the ultraviolet. For instance, the spectrum of HS Ori (= Haro 46) differs from the spectra of NX Mon, AU Ori and BC Ori. The continuous spectrum of HS Ori resembles in the long-wavelength region the continuous spectrum of a star of class F; the intensity increase around 3750 Å is less noticeable than in the other stars. In RW Aur there is no such difference. At the same time, both in the energy distribution and in the character of the bright lines the spectrum of HS Ori is similar to that of RW Aur (HS Ori is the first star with an emission spectrum of the type of RW Aur, found in Orion). It is remarkable that at maximum brightness HS Ori is nearly three magnitudes fainter than the brightest T Tau stars in Orion, whereas at maximum RW Aur is comparable with the brightest objects in the Taurus clouds.

5. Some stars with ultraviolet excess show significant fluctuations in Hα emission. At the same time the intensity of the continuous spectrum may vary as well, which follows from Joy's observations [24] of the star YZ Tau. As pointed out above (§14 and §15 of Chapter 9), theoretically different combinations are possible here. In particular, there can be noticeable fluctuations in the intensity of Hα, practically without significant variations of the star's brightness in the continuous spectrum.

An unusually strong continuous emission in the region of short wavelengths is proper to the following three types of objects:

a) stars with more or less permanent continuous emission;

b) some types of flare stars in which the emission undergoes frequent, and perhaps continuous, variations;

c) flare stars which show emission at the moment of the flare only.

Stars of T Tau type which are particularly active in the ultraviolet occupy an extreme place in the sequence of this type of objects. Therefore it is useful to distinguish among the T Tau stars this particular group showing an exceptionally high emission at short-wavelengths. Let us call these objects "stars of NX Mon type".

Stars of the NX Mon type are concentrated in the U – B vs. B – V diagram farthest from the main sequence. The positions of some of them--NX Mon, BC Ori, LH$_\alpha$ 67--are shown in Fig. 10.6. The positions of two other stars--UX UMa and SS Cyg--which do not have a direct relation with T Tau type stars, yet are interesting on account of their color characteristics are also shown. UX UMa is an eclipsing system of Algol type, with a period of revolution of 4 hours 43 minutes. One of the components of this system belongs to class sdB, but the class of the second component is not known. In normal conditions the color of the system is U – B = -0.81, B – V = +0.11. However in one case the values U – B = -1.58 and B – V = -0.10 were recorded [25].

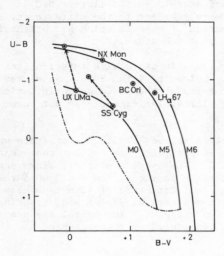

Fig. 10.6. Stars with abnormal color indices in the theoretical U – B vs. B – V diagram.

SS Cyg is a double system too--a variable of the type of U Geminorum. The spectral classes of the components are sdB and G5 and the period of revolution is 6 hours 38 minutes. In SS Cyg considerable fluctuations of brightness accompanied by color variations have been observed.

The two stars, UX UMa and SS Cyg, differ from each other in many respects. But there is also a similarity between them. This appears in their position in the color diagram and, in particular, in the character and limits of the color variations. Are these variations due to the appearance of fast electrons in the atmosphere of one of the components of these binary systems? In normal conditions UX UMa and SS Cyg are located in the color diagram in a position close to the region corresponding to the "hot gas" model. Only in periods of enhanced activity they leave this region and reach the zone where the stars of NX Mon type lie.

Spectrograms of SS Cyg have been obtained of the continuous radiation of the flare in the region 4100-3550 Å [26]. Here the distribution of the energy in the continuous spectrum differs considerably from what we have in the case of Planck radiation with infinite temperature.

7. THE VARIATIONS OF U – B AND B – V WITH TIME

It is interesting to study the character of the color variations of a given unstable star in the U – B vs. B – V diagram. As an example the results of photoelectric measurements by Varsavsky [27] for a group of stars of T Tau type in Taurus will be analyzed. The color indices U – B and B – V found by him from the first series of observations for 25 stars of this type are given in the color diagram (Fig. 10.7).

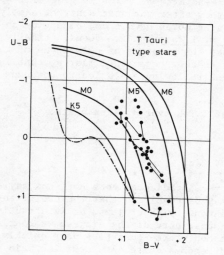

Two months later a second series of measurements was made for eight stars of this group; these results are also shown in Fig. 10.7. The dots of the two series of observations have been connected by lines. These lines show the displacement of the star in the diagram during the fluctuations of its ultraviolet activity. There is the general tendency of the star to be displaced along the theoretical tracks giving the color variations during changes of flare activity. Sometimes the observed lines are nearly parallel to the theoretical tracks. In such cases the color variations are only caused by fluctuations of the non-thermal activity of the star, not accompanied by variations of its temperature. But there are also cases when the observed tracks are not parallel to the computed curves. Then they may be decomposed into two components--parallel to the main curve and perpendicular to it. The parallel component evidently gives the fraction of the non-thermal radiation in the total color variations, and the perpendicular component gives the fraction of the thermal radiation, i.e. the variations of the star's temperature. Judging from the data in Fig. 10.7 the contribution of the thermal component is small. However, the observational material is not sufficient for making definitive conclusions.

Fig. 10.7. Stars of T Tauri type in the theoretical U - B vs. B - V diagram.

The same conclusions follow from an analysis of the observations by Smak for another group of T Tau type stars [18]. From this group of stars fourteen were measured several times. For eight of them--among which the star T Tauri--the variations of U - B vs. B - V were found to be parallel to the theoretical tracks; for these stars the color variations are due to fluctuations of the non-thermal component of the radiation. In the case of the remaining six stars, thermal and non-thermal radiation play equally important roles.

8. THE EVOLUTION OF STARS OF T TAURI TYPE

We have seen that in the theoretical U - B vs. B - V diagram the stars of T Tau type occupy a wide space in the diagram, from the main sequence to the uppermost boundary of the diagram. At the same time the position of each star corresponds to a certain degree on its activity, i.e. with a certain value of τ (Fig. 10.3). In the uppermost part we have a small group of stars which are particularly active in the generation of non-thermal emission, which we have given the name "objects of NX Mon type"; for them $\tau \sim 0.1$ to 0.01. Somewhat lower we have the stars with average activity, with $\tau \sim 0.01$ to 0.001. Still lower and close to the main sequence the stars with moderate activity are scattered, for which $\tau \lesssim 0.0001$. Finally, on the main sequence (or close to it) lie the ordinary flare stars of UV Cet type.

Any star of UV Cet type is first of all an unstable object. This means that it cannot maintain the degree of its activity--the given value of τ--for an infinite time. During a time comparable with the lifetime of the star it must gradually change its place in the diagram, performing a kind of drift from the upper boundary of the diagram to the lower one.

Thus it seems to be most reasonable to explain the typical distribution of the T Tau stars over the color diagram as due to evolution--an attempt which is generally speaking not new [28]. The initial state is immediately after the star's birth, namely the state of a star of NX Mon type. Later the star moves downwards, occupying an intermediate position, where the ordinary stars of T Tau type are found. Finally, when the star reaches the main sequence, it loses the property of permanence of the flare state; however, it maintains the property to show occasional flares, that is, it becomes a typical flare star. As a result the following evolutionary sequence is described [5]:

Stars of NX Mon type → ordinary stars of T Tau type → flare stars,

or, in symbols:

NX Mon → T Tau → UV Cet.

The distribution of the T Tau stars in the color-color diagram does not show any signs of greater and lower density. Hence it may be concluded that the evolution along the path NX Mon → T Tau → UV Cet is continuous, without jumps or interruptions. Considering that there are fewer points in the upper part of the diagram we conclude that the initial phase of the evolution--the NX Mon state--must pass very quickly. Afterwards the rate of evolution slows down and becomes quite slow near the main sequence.

The time during which the star is in the NX Mon phase is thus the shortest. According to Haro, the stars of NX Mon type constitute about 15% of all T Tau stars in associations. This means that the time during which a star lives in the NX Mon state must be nearly one order smaller than the lifetime of normal T Tau stars. The latter is of the order of 2×10^5 years. Therefore the age of stars of NX Mon type will be of the order of 2 to 3×10^4 years.

For T Tau type stars in the cluster NGC 2264 Walker [17] constructed a more detailed color diagram. Careful study of this diagram brings to light the following interesting properties in the distribution of the T Tau stars as a function of the degree of their activity:

a) When a T Tau type star lies far from the main sequence, it is as a rule variable and vice versa.

b) When a star of T Tau type is far from the main sequence, there are almost always hydrogen emission lines in its spectrum, and vice versa.

As an example of such a star NX Mon itself may be used. In its spectrum strong emission lines Hα, Hβ, Hγ, Hδ are seen; moreover, its total brightness is variable; the amplitude of the brightness fluctuations during Walker's period of observation was $0^m.57$ in visual light.

The evolutionary sequence NX Mon → T Tau → UV Cet was derived from analysis of the color characteristics of the stars. There are, however, other facts confirming the reality of this sequence; we shall discuss them in the following sections.

9. BRIGHTNESS FLUCTUATIONS

The T Tau type stars have been classified by us as objects in a state of "stable instability," so called because of the permanency of their flare activity. Therefore one should not expect strong fluctuations in their brightness during a short time. For making a definite picture of the character of the brightness variations one must have observational data in two or more spectral regions during a period covering at least several years.

From the two-color observations performed for stars of the T Tau group the following were established [29]: The photovisual amplitudes of the brightness fluctuations of the stars T, UX, RY, XZ, DN, GK Tau were found to be small, of the order $0^{m}.4$ to $0^{m}.6$; at the same time the amplitudes in photographic light proved to be three to four times larger. For thermal processes the ratio of the amplitudes in photographic and photovisual light is only 1.25. All these stars, with the exception of XZ Tau, do not have very strong emission lines. Therefore in this case it is a question of strengthening of just the photographic region of the continuous spectrum. For the remaining stars of this group the character of the variations in ΔB and ΔV indicates possible variations of the temperature of the star.

In another case [18] for 8 out of 14 stars of T Tau type, observed two and three times, the ratio $\Delta B/\Delta V$ was larger than 1.25 and in some cases it reached values of 2 to 3.

A number of photometric (sometimes multi-channel) observations of T Tau type stars were made synchronously with a determination of the intensity of the emission line $H\beta$ [22,30-32]. Analysis of these data leads to the following conclusions:

First, all observers agree that the class of T Tau type stars is inhomogeneous with regard to the character of the variability. Further, the character of the relation between the brightness variations in a certain part of the continuous spectrum and the intensity of some emission line can be completely different for different stars. Thus, for instance, for the least active stars of T Tau type the amplitudes of the brightness variations are somewhat larger in the red part of the spectrum than in the blue part, and in this case the strength of the $H\alpha$ line increases with increasing brightness.

In the case of some more active stars, when the increase of brightness is larger in the blue part of the spectrum than in visual light, the $H\alpha$ emission increases at minimum brightness (RW Aur).

Finally, for stars which are most active in the ultraviolet part of the spectrum the $H\beta$ emission is considerably enhanced at minimum brightness (DF Tau). In addition, for a whole series of cases there is a positive correlation between the $H\alpha$ emission and the star's brightness in the ultraviolet. As we have seen in Chapter 9, all these correlations between intensity of an emission line and stellar brightness in the continuous spectrum can be understood if the fast electron hypothesis is adopted, without additional assumptions.

The results of multi-channel photometric measurements of T Tau type stars also give evidence of an unusual law of absorption of light in the dust clouds surrounding these stars. It is suspected that possibly the absorption of light in such envelopes is neutral (non-selective), i.e. there may occur in them dust particles of a large range of dimensions. From other data the absorption in the medium surrounding the star seems selective, and for instance in the case of V 380 Ori, the absorption law even has the form λ^{-2} [22]. Multi-channel photometry with very high time resolution shows an interesting property of T Tau type stars--the so-called rapid fluctuations of brightness of the star in the ultraviolet [31]. Such fluctuations can occur in a time shorter than 30 minutes. During a time shorter than half an hour the values of $U-B$ and $B-V$ can vary rather considerably (for instance, from $-0^{m}.70$ to $-0^{m}.49$). These facts suggest evidence of active processes taking place in the outer regions of the star, including ejection of matter from the stellar interior, scattering, decay of unstable nuclei, etc. This seems to indicate that in the T Tau type stars the internal convection encompasses the entire star as well as its surface layers.

10. ENERGY LOSSES BY T TAURI TYPE STARS

In the case of a permanent state of flare, which is characteristic for the T Tau type stars, the total amount of energy P_t leaving the star during a time t by means of fast electrons will be

$$P_t = P t , \qquad (10.1)$$

where P is given by the relation (6.6), but here it represents the amount of energy lost by the star in one second as a result of the "exodus" of fast electrons. More correctly, one should start from the idea that the fast electron, before leaving the star (if it leaves it at all) may be bound for some time by the stellar magnetic field in the outer regions of its atmosphere. As a result the number of electrons flowing away from the star per second (the "flow" of electrons) will be considerably less than P/ε, where ε is the energy of one electron. Assuming, however, that the outflow of fast electrons occurs with an intensity P ergs/s, we can determine an upper limit to the total amount of energy lost by a star of T Tau type during its existence.

For a stellar radius $r_* \sim 1 R_0$ and energy of the fast electrons $\gamma = 3$, we have from (10.1):

$$P_t = 10^{40} \tau t \text{ ergs } . \qquad (10.2)$$

The "flare activity" of normal T Tau type stars is characterized by a value of $\tau \sim 0.001$ (cf. §3) and this activity lasts for $t \sim 10^5$ years. Hence we find

$$P_t \sim 10^{49} \text{ ergs } .$$

For very active T Tau stars which are objects of NX Mon type, the "flare activity" is $\tau \sim 0.01$ and it lasts 10^4 years. This gives once more

$$P_t \sim 10^{49} \text{ ergs.}$$

In §15 of Ch. 6 we found for the total amount of energy lost by a flare star of UV Cet type, for which $\tau \sim 0.001$ and $t \sim 10^8$ years

$$P_t \sim 10^{46} \text{ ergs.}$$

From these results, however approximate they may be, an important conclusion can be drawn. First of all, the star loses as much energy in the first ten thousand years as in the following hundred thousand years. Furthermore, during its whole period of normal "flare" activity, i.e. 10^8 years, it loses a thousand times less energy than in the first ten thousand years.

If we consider, instead of the total amount of the energy loss, the specific loss, i.e. the energy lost by a star of a given type per year, the distribution of the "specific loss" is characterized by the following table, in which the energy loss of a star of NX Mon type is taken as unity:

Star of NX Mon type	1
Normal stars of T Tau type	0.1 - 0.01
Flare stars	$10^{-5} - 10^{-6}$

From these data it follows that the property of "flares", either permanent or ordinary, is present foremost in stars of T Tau type. In the typical flare stars this property has in a certain sense the character of a remnant since it plays a negligible role in its exchange of energy. It is not very probable that these conclusions will be changed when more accurate data become available.

We do not know well the absolute luminosities of the T Tau stars themselves. As a guess let $L(M5)/L_\odot \approx 10^{-2}$. Then the energy emitted per year by a T Tau type star by means of radiation $\sim 10^{39}$ ergs/year. Starting from this estimate and also considering the results derived above, we may compile a list of the distribution of the energy losses depending on the form of this loss--light radiation or particle radiation--in the early stages of development of the star; this is given in Table 10.2.

TABLE 10.2   Total Energy Losses of Various Types of Stars by Light Emission and by Ejection of Fast Electrons

Star	Age (years)	Total loss of energy by light emission E_t (ergs)	Total loss of energy by ejection of fast electrons P_t (ergs)	P_t/E_t
Type NX Mon	10^4	10^{43}	10^{48}	10^5
Type T Tauri	10^5–10^6	10^{44}–10^{45}	10^{48}	10^3–10^4
Flare	10^8	10^{47}	10^{46}–10^{45}	0.1–0.01

The data in the last column of Table 10.2 are remarkable--the ratio of the total energy emitted by the star in the form of ejection of fast electrons (P_t) and the total energy liberated by light emission (E_t). This ratio is very large--of the order 10^5--for stars of NX Mon type, and considerably smaller than one for ordinary flare stars. For the ordinary flare stars the emission of light is the predominant form of liberation of stellar energy. On the contrary, for stars of T Tau type, and in particular of NX Mon type, the main form of liberation of the energy from the stellar interior is corpuscular radiation--the ejection of fast electrons; the light emission plays an insignificant role for them.

From different arguments Poveda [33] also concludes that the ejection of particles with high energy--corpuscular radiation--plays an important role and has predominant significance in the total balance of energy exchange in T Tau stars.

Poveda puts forward the idea of corpuscular radiation of T Tau type stars with the aim of explaining their infrared excesses. According to this idea the infrared excesses are caused by heating of the dust envelope surrounding the star by a corpuscular stream emitted by the star. A similar interpretation of the infrared excess was also suggested by Huang [34]. It is important to remark that in this case the infrared excess has a non-thermal origin.

In connection with the non-thermal nature of the infrared excess of T Tau type stars, the measurements of total luminosity of some of these stars by Mendoza [10] should be recalled. By total luminosity L_* we mean the integrated radiation of the star in the range from 0.3 to 50 µm. These luminosities were found to be very high, of the order of 10 $L_\odot$. Thus, $L_* \approx 40\ L_\odot$ for T Tau, about 5 $L_\odot$ for RW Aur, $\sim 270\ L_\odot$ for V 380 Ori, $\sim 800\ L_\odot$ for R Mon, etc. These numbers exceed the luminosity of a T Tau type star assumed above in our calculations by three to five orders of magnitude. But it should not be forgotten that the estimate $L(M5) \sim 10^{-2}\ L_\odot$ refers

to radiant energy of purely _thermal_ origin, whereas L_* is completely or nearly
completely of non-thermal origin.

The data in Table 10.2 are only estimates and can be made more accurate in the
future. However, it should be recognized that these conclusions are certainly
not compatible with the customary ideas about ways and forms of liberation of
stellar energy. In these conclusions the flare phenomenon itself gets a new
significance. It appears that every star has, in the initial phase of its development, not only an excess of mass, which it must get rid of in some way or another,
but also an excess of energy. And especially the flare--permanent or episodic--
is the main form in which the star gets rid of the extra energy in its
interior, in the same way the ejection of gaseous material is the main form of the
excess mass loss.

11. EMISSION LINES

Emission lines in T Tau type stars are excited in principle in the same way as in
ordinary flare stars. The emission lines themselves are formed in the chromosphere
of the star. Short wave radiation--either of Compton origin or due to radiative
processes in the gas-dust cloud--falling from outside on the chromosphere causes
the ionization in it. In both cases, however, the ionizing radiation is non-
thermal. The shell or cloud of fast electrons is formed in the beginning above the
chromosphere, but probably below the gas-dust cloud surrounding the star. Then
the fast electrons reach the gas-dust cloud, inducing transition radiation, and at
the same time heating the dust particles. The character of the emission lines indicates that gaseous matter does not flow uniformly from the surface of a star of T Tau
type. For instance it has been established that in T Tau stars considerable spectral
variations occur during 24 hours and smaller variations in the course of only a few
hours [35]. However, the velocity of the outflow is relatively small--of the order
of 100 km/s. The amount of mass lost by the star as a result of such an outflow
is about 10^{-5} $M_\odot$ per year, which appears to be somewhat too high [6].

If the L_c radiation penetrates sufficiently deep into the chromosphere, the excitation of the hydrogen emission lines may also take place in denser layers. In this
case self-absorption will be unavoidable, as a result of which the decrement of the
emission lines leaving the chromosphere will differ from the decrement for an
optically thin medium (planetary nebulae). In particular, the decrement will be
less steep. However, the radiation in the emission lines in the case of T Tau stars
must pass through the gas-dust cloud surrounding the star before it reaches the
observer. This distorts somewhat the relative line intensities. Hence, with
regard to its steepness, the Balmer decrement of T Tau stars will be intermediate between the decrements of flare stars and of planetary nebulae. This is confirmed
by the data shown in Table 10.3--showing averaged values of the Balmer decrement
for four stars of T Tau type--V 380 Ori, T Tau, RW Aur [22] and NX Mon [44]. The
scatter is rather large, but even for the same star the Balmer decrement varies,
depending on the activity of the star, within rather wide limits. It seems firmly
established that in a very short time ($\sim$ one hour) the spectral properties of some
T Tau stars (in particular RW Aur) may vary radically--even to the degree that an
absorption line disappears and an emission line appears, and vice versa. Such rapid
variations are related to the ejection of gaseous material. To a certain degree
this activity is also indicated by the varying emission lines.

As an example Fig. 10.8 shows the profiles of the Hα line in four T Tau type stars
[22]. For the stars RW Aur and RY Tau, for instance, the line profile has the shape
of a saddle, shifted to the violet. In the case of SU Aur the profile is split.
But in the case of FU Ori we have a line profile characteristic for stars of P Cyg
type, i.e. a faint emission line with a strong deep absorption component. In T Tau

TABLE 10.3 Average Balmer Decrement (Emission Lines) for Some Stars of
T Tau Type

	V 380 Ori	T Tau	RW Aur	NX Mon
H_β	1.00	1.00	1.00	1.00
H_γ	0.30	0.15	0.32	0.50
H_δ	0.10	0.04	0.10	0.32
H_ϵ	0.07	--	--	--
H_8	0.04	--	--	0.27

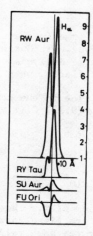

Fig. 10.8. Profiles of the
emission line Hα in the
spectra of four T Tau type
stars. The vertical scale
is given in units of the
level of the continuous
spectrum.

the emission in Hα very often has a simple structure
[45,22], although sometimes this profile shows an
intricate structure with a faint violet emission
component [46]. Thus the profiles of the emission
lines of T Tau type stars may be of three types:
single profiles (T Tau, V 380 Ori), two component
profiles with the absorption being shifted towards
the violet (RW Aur, RY Tau, SU Aur), and of P Cyg
type, with faint emission, but with a strong absorption component (FU Ori).

There is still another type of variations of the profiles of the emission lines; this has been found for
the star V 380 Ori [22]. Namely, in one observational
period the hydrogen emission lines show positive displacements of +80 km/s on the average, but during
another period they show negative shifts of the same
size, -80 km/s. These variations are accompanied
also by some variations in central intensity and
equivalent width of the line.

The most probable interpretation of this fact is the
following: gaseous matter has been ejected with a
velocity (∿100 km/s), significantly smaller than the
escape velocity from the star. Later this matter
falls back on the photosphere of the star. Walker
[36], when studying the structure of spectral lines
for a number of T Tau type stars (YY Ori, SY Ori, XX Ori), also found a strong
absorption component at the red side of the hydrogen emission lines. However,
Walker considers, quite erroneously, this fact as proof that there is accretion--
namely that interstellar matter falls from the outside upon the star. Simply he
"caught" the star just at the moment when gaseous matter, ejected earlier, was
falling back on to the star. There are data indicating that the change of sign of
the displacement of the spectral line (from positive to negative) may take place
in the course of one day [47]. This reminds us very much of what we observe for
the Sun when gaseous matter is ejected and then falls back on the photosphere (only
the scales of the ejection are in the case of V 380 Ori greater).

The idea of the "permanent" flare of T Tau type stars as the principal form of their
activity was based exclusively on analysis of colorimetric data of these stars.
Now the behavior of the emission lines confirms this idea. The emission lines give

evidence of stormy processes taking place continuously in the atmospheres of these stars.

The model atmosphere of T Tau type stars which has been described also provides the possibility of explaining some strange features concerning the distribution of the continuous energy in their spectra. In the ultraviolet and in the infrared regions of the spectrum these stars radiate considerably more energy than would follow from Planck's law for a given effective temperature (cf. Ch. 1, Fig. 1.1). In our case the ensemble of fast electrons around the star leads to the generation of additional radiation in the ultraviolet. At the same time the gas-dust cloud surrounding the star, which is heated by the corpuscular stream and by non-thermal radiation of Compton origin, is itself a source of additional radiation with maximum energy in the far infrared region. The sum of these two kinds of radiation, ultraviolet and infrared, gives the observed spectrum. The development of a quantitative theory for such a combined atmosphere certainly deserves a great deal of interest.

12. EMISSION LINES IN WEAK FLARES

In most unstable stars, among which are the T Tau type stars, sometimes considerable fluctuations of the intensity of the emission lines are observed, without significant fluctuations of the star's brightness in integrated light. Analysis of equation (9.17) shows that such a phenomenon is possible in faint flares, when $\tau \ll 0.01$.

As an example we give the results of computations for the $H\alpha$ line in the case of a star of class M0 ($T_* = 3600\,°K$). For $\tau \leq 0.01$ we can take in (9.17) $C_i(\tau,\gamma,T) \approx 1$. Then we have

$$\frac{W_\alpha}{\lambda_\alpha} = W\Gamma_\alpha \; \frac{3\gamma^4}{4} \; \frac{e^{x_\alpha} - 1}{x_\alpha^4} \; J_3\left(\frac{x_o}{\gamma}\right) \frac{x_o}{2} \tau \; . \tag{10.3}$$

Taking $W = 0.2$ and $\Gamma_\alpha = 0.14$, we find, with $\gamma^2 = 10$, $W_\alpha = 5.5 \times 10^4$ Å.

In Table 10.4 the values of W_α are given for a number of values of τ. The amplitudes of the fluctuations of the brightness in U, B and V radiation are given as well; these have been computed by the usual methods (Ch. 6) and for the same values of τ.

TABLE 10.4 Equivalent Width of the H_α Line and Amplitudes of the Fluctuations in Brightness of the Star in U, B and V Light for Weak Flares

τ	W_α(Å)	ΔU	ΔB	ΔV
0.01	55	1^m5	0^m36	0^m08
0.005	27	1.0	0.20	0.03
0.002	11	0.5	0.12	0.01
0.001	5	0.3	0.04	0.007

It follows from this table that clearly visible fluctuations in the intensity of the emission line are accompanied by observable fluctuations of the brightness in V and B light, but the absolute values of the amplitudes are quite small--of the order 0^m01 - 0^m3. On spectrograms obtained with an objective prism such large fluctuations in the intensity of the emission lines will be visible even with the naked eye, whereas the fluctuations of brightness in B and V light can only be discovered with great effort by means of photometric measurements. The fluctuations

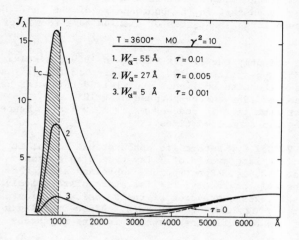

Fig. 10.9. Distribution of the continuous
radiation of Compton origin in the visible
region, and the region ionizing hydrogen,
for weak flares of MO type stars.

of brightness are large only in U
light. The structure of the continuous spectrum of the star in
faint flares is shown in Fig. 10.9;
this gives curves of the energy
distribution (on a wavelength
scale) in the continuous spectrum
of a star of class MO for three
values of τ: 0.01, 0.005 and
0.001. The broken vertical line
indicates the limit of hydrogen
ionization; the intensity of the
Hα line is proportional to the
energy to the left of this limit.

We have seen that for weak flares
the intensities of the emission
lines are particularly sensitive
to the strength of the flare.
Hence these lines may be used as
indicators of the fluctuations
of the non-thermal activity of
the star.

13. FUORs

In Orion, around λ Ori--a star of class 07--there is a large region of ionized
hydrogen with a radius of about 3°. On the eastern edge of this region lies a
small dark nebula called B 35. This nebula is interesting because fifteen objects
have been discovered in it with H II emission and with other indications of variability [37]. One of these is the star FU Ori.

Until 1936 FU Ori was known as a variable star with weak brightness fluctuations
from m_{pg} = 15^{m}3 to 16^{m}3. But at the end of 1936 the brightness of this star began
to increase, reaching 10th magnitude in photographic light. At the end of 1937 it
was even brighter than magnitude 10; later it reached m_{pg} = 10^{m}5 and since that
time its brightness has hardly varied (or rather, an exceedingly slow weakening
has been noticed).

The moment of the jump in brightness, or of the "flare" of FU Ori, has not been
determined with sufficient accuracy. Nevertheless, it may be taken as an established
fact that the star's brightness increased by about a hundred times in a few months.
It is even more striking that this sharp and strong increase in brightness has been
maintained in this star for more than thirty years!

The spectrum of FU Ori before its sudden flare is not known. It was studied in
detail after the "flare" [38,22], and it was found that two systems of spectral
lines exist. One of these systems is characterized by unusually strong Balmer
lines in absorption, indicating that the star belongs to the subgiants of class G.
The second system consists of emission lines characteristic of a gaseous envelope;
all these lines are shifted towards the shorter wavelengths by 80 km/s, indicating
that gaseous material is flowing away from the star. The profile of the Hα line
itself is of P Cyg type (Fig. 10.8).

In the spectrum of FU Ori a very strong line λ 6706 of Li I was discovered, from which it may be concluded that the relative abundance of lithium in the photosphere of FU Ori is about eighty times larger than that in the photosphere of the Sun. We note that only in T Tau type stars such large abundances of lithium were found (cf. Ch. 16).

The FU Ori event was for a long time the only one of its kind. In 1969 a similar event occurred in another star--V 1057 Cyg = Lick Hα 190, which lies in the diffuse nebula NGC 7000, in a group of stars with emission lines. Until 1969 this star had a magnitude $m_{pg} = 16^m.0$. About the end of 1969 the brightness of V 1057 Cyg suddenly started increasing and after some time it reached $m_{pg} \approx 10^m.0$ [39]. Hence the star's brightness increased more than a hundred times.

Contrary to FU Ori, the spectrum of V 1057 Cyg before its outburst was known; it indicated that this star belonged to the late dwarfs of T Tau type with emission lines and hardly any absorption lines [40,41]. Spectrograms obtained after the outburst show properties which are usually characteristic for stars of relatively high luminosity. In particular, a shift was discovered of the Hα emission line relative to the absorption component, corresponding to 420 km/s. Hence it follows that in this star too gaseous matter has been flowing outwards forming a gaseous envelope around the star.

In the two cases of FU Ori and V 1057 Cyg the brightness increased more than a hundred times. Contrary to ordinary flare stars where, after the flare outburst, the star returns relatively quickly to its initial state, no such return has taken place in these two stars. These stars have passed suddenly and relatively quickly from one state to another. In order to distinguish them from ordinary flare stars V. A. Ambartsumian proposed to call them FUORs [42].

A hundredfold increase in brightness is not very rare for flare stars. In this respect FUORs are objects resembling flare stars. But they differ from typical flare stars by the following two characteristics:

a) FUORs have a slower flare; their rate of increase in brightness is about $0^m.0001$ min^{-1}, which is at least two to three orders of magnitude slower than the inter-mediate flare stars.

b) FUORs maintain the "flare" state for very long, for tens, or maybe even for hundreds of years.

Although FUORs have slower outbursts, an increase in brightness by a factor of more than a hundred during a few months is quite rapid since during this time some reconstruction of the internal state of the star must take place, leading to an important increase of the integrated strength of the sources of radiation [42]. One can assume that there are intensive and constantly active sources of energy in the outer regions of the star. This means that in FUORs the same mechanism for the production of continuous emission, and excitation of emission lines, may exist as in T Tau type stars--a permanent or perpetual flare. The requirement that the sources of energy must be situated in the outer regions of the star is also very important in the case of FUORs; this even predicts an observational fact--the presence of faint emission lines in the spectra of FUORs. It is hard to imagine that the excitation of the emission lines takes place in the chromosphere or in the gaseous envelope of the star since the sources producing the additional energy, including the ionizing radiation, are distributed in sub-photospheric layers.

It should be stressed that in the case that the sources producing the energy are situated outside the photosphere the fast electron hypothesis can explain the observed growth of the brightness of FUORs during their slow flare. Thus, for instance, if we take approximately $\Delta m_{pg} = (\Delta U + \Delta B)/2$, we find from Fig. 6.3:

$\Delta m_{pg} \approx 9^m$ for $\tau = 0.1$ and $\Delta m_{pg} \approx 4\overset{m}{.}5$ for $\tau = 0.01$. The observations give, as we have seen above, $\Delta m_{pg} \approx 5\overset{m}{.}5$ to 6^m.

FUORs imply the idea that a star passes from a state of low luminosity to a state of high luminosity. Let us assume that the observer cannot observe the state of low luminosity of the star, but records the state of high luminosity with all the signs that indicate instability. Then unavoidably we may conclude that among the T Tau type stars there must be FUORs. One should look for such FUORs among the very active stars of NX Mon type, about which we have spoken above.

That the FUORs actually are T Tau type stars, or become such stars after their slow outburst, is also confirmed by the results of infrared observations. Such observations of FU Ori and V 1057 Cyg were made by Mendoza [43] by means of a nine-channel electrophotometer, covering the range from U (0.35 μm) to L (3.4 μm). Fig. 10.10

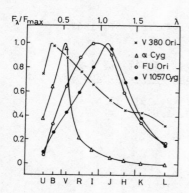

gives the energy distribution in the spectra of these stars, and also of V 380 Ori, and for comparison of α Cyg (Deneb)-- a star of class A2. The nearly complete agreement of the spectral curves of FU Ori and V 1057 Cyg is striking, in particular in the region of wavelengths longer than 1 μm. Both of these curves resemble the ones for T Tau (Fig. 1.2) with strong infrared excess characteristics (this holds also for V 380 Ori). In these stars a significant excess is also observed in the ultraviolet part of the spectrum. The quantities U – B, for instance, are equal to $-0\overset{m}{.}23$, $+1\overset{m}{.}00$, and $+0\overset{m}{.}58$ for V 380 Ori, FU Ori and V 1057 Cyg, respectively.

Fig. 10.10. Distribution of the energy in the continuous spectrum, from 0.34 μm to 3.4 μm, for two FUORs (FU Ori and V 1057 Cyg), one star of T Tau type (V 380 Ori), and one hot star of class A2 (α Cyg).

We shall return to discuss FUORs in more detail in Chapter 12.

REFERENCES

1. G. Haro, E. Chavira, Vistas in Astronomy, VIII, 89, 1965.
2. G. H. Herbig, G. Haro, Bol. Obs. Tonant. y Tacub. 12, 33, 1955.
3. G. Haro, Flare Stars. Stars and Stellar Systems. Vol. VII, Nebulae and Interstellar Matter, Ed. B. M. Middlehurst and L. H. Aller, 1968, p. 141.
4. V. A. Ambartsumian, Non-Stable Stars, University Press, Cambridge, 1957, p. 177.
5. G. A. Gurzadyan, Bol. Obs. Tonant. y Tacub. 35, 255, 1970.
6. G. H. Herbig, Advances in Astr. and Astrophys., Ed. Z. Kopal, Academic Press, Inc., New York, 1962, p. 47.
7. P. P. Parenago, Proceedings GAIS 25, 1954.
8. E. E. Mendoza, Ap. J. 143, 1010, 1966.
9. E. E. Mendoza, Ap. J. 72, 311, 1967.
10. E. E. Mendoza, Ap. J. 158, 977, 1968.
11. G. H. Herbig, IAU 13th General Assembly, Prague. Agenda and Draft Report, 1967, p. 509.
12. L. V. Kuhi, Interstellar Ionized Hydrogen, Ed. Y. Terzian, Benjamin Press, New York, 1968, p. 13.

13. M. Rodono, Astron. Astrophys. 32, 337, 1974.
14. G. Haro, E. S. Parsamian, Bol. Obs. Tonant. y Tacub. 5, 41, 1969.
15. G. Haro, Bol. Inst. Tonant. 2, 3, 1976.
16. B. Iriarte, Bol. Obs. Tonant. y Tacub. 4, 79, 1967.
17. B. Walker, Ap. J. Suppl. 2, 365, 1956.
18. J. Smak, Ap. J. 139, 1095, 1964.
19. M. F. Walker, Ap. J. 125, 656, 1957.
20. F. Lanouvel, C. Flogere, J. des Obs. 40, 37, 1957.
21. N. N. Kilyachkov, V. S. Shevchenko, USSR A. J. Lett. 2, 494, 1976.
22. I. R. Salmanov, Thesis, 1975. Shemacha Astr. Obs.
23. H. Johnson, B. Iriarte, Ap. J. 127, 510, 1958.
24. A. H. Joy, Ap. J. 102, 168, 1945.
25. M. C. Zuckermann, Ann. d'Ap. 24, 431, 1961.
26. M. F. Walker, G. Chincarini, Contr. Lick Obs., No. 262, 1968.
27. C. Varsavsky, Ap. J. 132, 354, 1960.
28. G. Haro, Bol. Obs. Tonant. y Tacub. 14, 1, 1956.
29. H. S. Badalian, Commun. Byurakan Obs. 25, 49, 1958; 31, 57, 1962.
30. P. P. Petrov, Izv. Crimean Obs. 54, 69, 1976.
31. P. Kuan, Ap. J. 210, 129, 1976.
32. L. V. Kuhi, PASP 78, 430, 1966; Astr. Astrophys. 15, 47, 1974.
33. A. Poveda, A. J. 72, 824, 1967; Bol. Obs. Tonant. y Tacub. 26, 15, 1965.
34. S. S. Huang, A. J. 72, 804, 1967.
35. E. B. Weston, L. H. Aller, Memoires Liege Obs. XV, 251, 1955.
36. M. F. Walker, Stellar Evolution, New York, 405, 1965.
37. G. Manova, Astr. J. USSR 36, 187, 1959.
38. G. H. Herbig, Vistas in Astronomy VIII, 109, 1966.
39. G. Welin, Astr. Astrophys. 12, 312, 1971.
40. G. H. Herbig, Ap. J. 128, 259, 1958.
41. G. H. Herbig, E. A. Harlen, IBVS No. 543, 1971.
42. V. A. Ambartsumian, Fuors. Preprint BAO No. 3, 1971.
43. E. E. Mendoza, Bol. Obs. Tonant. y Tacub. 37, 135, 1971.
44. K. Bohm, Memoires Liege Obs. XX, 271, 1958.
45. L. V. Kuhi, Ap. J. 140, 1409, 1964; 143, 991, 1966.
46. E. A. Dibay, Astrofizika 5, 249, 1969.
47. L. V. Kuhi, Mercury 4, No. 5, 10, 1975.

CHAPTER 11
Flare Stars in Stellar Associations

1. FIRST SURVEYS

The first attempts to look for flare stars in regions of the Galactic System far
from the neighborhood of the Sun were undertaken at the end of the 1940s by Joy
[1], who drew attention to rapid flares of variables of T Tau type, which are
members of the Taurus cluster. Already, in the early stages of observation of flare
stars in the neighborhood of the Sun, the idea began to develop that the flare
phenomenon occurs in relatively young stars which are still at the stage of formation. Therefore a large number of flare stars might be expected in places where
the process of star formation is taking place, i.e. in stellar associations and
young star clusters. However, at that time not a single flare star was known
which was a member of a stellar association. This could be explained by their
relatively large distance; the nearest of them is at a distance of the order of 100
parsecs, and one can hardly expect to discover by chance a dwarf star at the moment
of the flare. It was therefore necessary to perform special observations in order
to prove or reject the possibility of flare stars occurring in stellar associations.

Haro, who was convinced of the evolutionary significance of the flare phenomenon,
was one of the first to look, with much perseverance and for many years, for flare
stars in associations. Already at the time when he studied stars with emission
lines in Orion, Haro [2] drew attention to the fact that in some of them the spectral
properties of ordinary flare stars were rather strongly expressed. Later Haro with
his collaborators discovered many flare stars in Orion, using 26-31" Schmidt telescope of the Tonantzintla Observatory.

The success obtained in Orion prompted astronomers to widen the field of search.
As a result, very soon appeared a small list of stellar associations and young star
clusters or "aggregates," according to Haro's terminology, in which flare stars had
been discovered.

The flare stars in aggregates are normally discovered photographically, by the
method of multiple exposures, where a chain is used of several (usually 4-6) images
of a star on the same plate with exposure times of 10-15 minutes for each image
and intervals between exposures of the order of seconds (cf. Fig. 6.29). Notwithstanding its evident simplicity, this method suffers from a number of drawbacks.
First of all, the influence of numerous photographic effects and defects works so
strongly that, in order to prove reliably a flare in some star, the increased
density of the blackening should occur in at least two successive images of the
star. This means that all flares lasting less than 10 minutes are lost. For the
same reason weak flares as regards their amplitude cannot be recorded either.
Essentially, when we speak about flare stars in aggregates, objects are considered
for which flares have been recorded with an amplitude of more than 1^m, seldom
$0.6-0^m7$, and which last for more than 15-20 minutes. Finally, on account of the
long exposure times (10-15 minutes) the true amplitude of the flare, reached at the
moment of its maximum, cannot be determined.

As a result of these and other drawbacks there is an unavoidable strong selection in the observational material, with respect to the statistics and the physical characteristics of flare stars in aggregates. And this explains the difficulties which arise every time one tries to compare physical characteristics and statistical indicators of flare stars in aggregates on the one hand and of flare stars scattered in the neighborhood of the Sun on the other hand, since the latter are studied practically only by photoelectric methods.

According to data up to 1976 flare stars have been found in ten aggregates; a list of them is given in Table 11.1. It also gives their total number in the aggregate considered, and the brightness of the brightest flare star in the aggregate. Further, column 5 gives the earliest and the latest spectral class of flare stars of which the spectral classes are known, and the 6th column shows the ages of the aggregates. Most of the data in this table have been taken from the survey by Haro [3], and the remainder from the sources indicated in the last column of the table.

TABLE 11.1   Flare Stars in Stellar Aggregates (Associations, Open Clusters, Nebulae)

Stellar aggregate	Total number of fl. stars (up to 1976)	Distance* (pc)	Brightest star (V)	Limiting spectral classes	Age of aggregate* (years)	References
Orion	325	470	12.4	K0–M3	3×10^5–10^6	[3,4,5]
NGC 2264	13	740	15.4	K0–M	10^6	[3,6]
Taurus Dark Clouds	11	...	12.5	K6–M5	10^6:	[3,37]
NGC 7000	51	700	15.50	...	3×10^6	[12–14]
Pleiades	469	125	12.04	K2–M	5×10^7	[3–9]
Coma Berenices	4	80	14.9	M3–M	5×10^8	[3]
Praesepe	33	160	14.21	M	$4\times10^{8**}$	[6,11]
Hyades	3	42	14.50	M3–M5	$6\times10^{8**}$	[3]
NGC 7023	6	290	...	...	...	[10]
Ophiuchus Dark Cloud	4	120(?)	...	K–M	...	[43]

 * Distances and ages of aggregates have been taken mainly from [27], pp. 260, 278.
 ** 9×10^8 years according to [31].

From the data in Table 11.1 an interesting property of flare stars in aggregates can already be established; the younger the stellar aggregate, the earlier the average spectral class of the star is found to be. In the young aggregates (Orion) there are very many objects of class K among the flare stars, whereas these do not occur in old aggregates (Hyades).

The largest numbers of flare stars have been discovered in Orion, in the Pleiades, and also in Cygnus (NGC 7000). At the same time these aggregates have been studied better than the others; therefore it is useful to consider them in detail.

2. THE FLARE STARS IN ORION

The first three flare stars in Orion were discovered by Haro and Morgan in 1953 [15]. The observers drew the attention to the great similarity of the lightcurves of flares of these stars with lightcurves of flare stars of UV Cet type. This was also the start of systematic searching for flare stars and of their study in the region of Orion, with the Large Orion Nebula (NGC 1976) as center. According to the data up to 1976, 325 flare stars have been discovered in Orion. The chief merit for this work lies with Haro and his collaborators from the Tonantzintla Observatory, and also with Rosino and his collaborators from the Asiago Observatory (Italy). Lists of these stars have been given in [3-6]; there the number of the star, the coordinates, the brightness of the star, the flare amplitude and spectral class (when known) are shown.

The most complete analysis of the collected observational material about the flare stars in Orion has been made by Haro in the papers [3-5]. In short the main properties of the flare stars in Orion are the following:

1. The earliest spectral class of a flare star in Orion is K0, and even late G type; the latest is M2-M3. For instance, the star T45 has been classified as G-Ke, the star T208 as Ge, but the star T176 as M3e. But these data cannot be thought to be complete, considering that spectral classes are known for less than 10% of the flare stars. Especially interesting are the spectral classes of the extremely faint (19-20^m) flare stars in Orion.

As regards the subclasses M5-M6, it appears that they do not occur in Orion. As a matter of fact it is usually difficult to classify stars of class M0 and the subclasses close to this. But stars of subclasses M5-M6 can easily be recognized even on spectrograms of moderate dispersion due to the characteristic bands of titanium oxide in their spectra.

2. Since most, if not all, of the flare stars in Orion are members of the association (distance from the Sun ∿470 parsecs), their absolute magnitudes, without correction for interstellar absorption, are found to lie within the limits M_V = +4.5 to +13, i.e. from subgiant to dwarf. Examples of flare stars in Orion with different apparent magnitudes and different flare amplitudes in U rays are given in Table 11.2.

TABLE 11.2 Examples of Flare Stars in Orion with Various Brightnesses in V and U Rays and Flare Amplitude ΔU

Flare star Tonanzintla No.	V minimum	U minimum	Flare amplitude ΔU
T146	12.9	14.6	1.4
T 86	13.9	15.8	2.4
T209	14.2	17.2	1.0
T102	14.4	17.0	1.5
T200	15.3	17.8	2.4
T186	16.4	18.5	5.3
T195	16.6	18.8	5.8
T236	16.7	18.7	5.6
T177	17.5	∿19.7	∿8.1
T242	17.6	19.0	4.6
T 7	19.5	>21.0	>6.2

3. About 30% of the flare stars in Orion belong to the irregular "normal" variables of T Tau or RW Aur type with Hα emission. These stars show significant fluctuations of brightness in their "normal" state, on which the lightcurve of the flare is superposed. Among them also are found stars with strong emission in the ultraviolet (XX Ori, NS Ori, SU Ori, etc.).

4. About 25% of the flare stars in Orion show repeated flares. The distribution of the numbers of stars with respect to the number of repeated flares discovered is given in Table 11.3.

TABLE 11.3 Distribution of the Numbers of Flare Stars Over the Number of Flares k in Orion and the Pleiades

Number of flares k	Number of flare stars n_k	
	Orion	Pleiades
1	236	250
2	60	78
3	21	49
4	4	24
5	1	13
6	1	11
7	1	4
8	0	4
9	0	6
10	0	3
11	1	1
...	...	...
$\sum n$	325	447

5. In the quiescent state, outside flares, the Hα emission line can be discovered on plates taken with an objective prism (for 30% of the flare stars in Orion). However, it should be noted that emission lines cannot be discovered by means of an objective prism for UV Cet type stars.

6. From the 325 flare stars discovered in Orion seven cases have been recorded of "slow" flares, where the duration of the development of the flare from the moment of its first appearance till maximum is 45 minutes or more. A bright representative of this type of objects is the star T177 (cf. Ch. 6). But there are stars (T66, 149, 153) which have "slow" as well as "fast" flares.

7. The flare amplitudes in U rays are on the average one-third larger than the flare amplitudes in photographic light, and the latter in their turn are considerably larger than the amplitudes in photovisual light. In the infrared region (∿8400 Å) no noticeable fluctuations of the brightness whatever were discovered on photographic plates.

Since, as mentioned above, the flare stars in Orion and generally in aggregates are discovered photographically and with exposure times of the order of 10 minutes, the maxima of the lightcurves will be strongly flattened, so that the observed amplitudes will be considerably smaller than the true amplitudes. This means that in those cases (T177) where flares are recorded with amplitude 8 - 8^m5, the true value may reach 10^m, if a flare is recorded by a photometric method with a time constant of the order of a few seconds. The time during which such a flare grows is usually of the order of 40-60 minutes. This means a 10000-fold increase of brightness per hour, i.e. about a three-fold increase per second. It must be realized that this is a very high rate of increase of brightness of a star (cf. §8, Ch. 1), especially if we consider that this rate is maintained for a whole hour.

The largest flare amplitudes--of the order 6-8^m in U rays--in Orion are recorded for the faintest stars--fainter than 19-20^m. The faintest flare stars in Orion may be fainter than 21^m (in photographic light). Searching for flare stars with brightness of the order 21^m and fainter is exceedingly difficult.

Although the Great Orion Nebula is at the center of the system of flare stars (cf. §7 of this chapter) and there is every reason to suspect a large increase of their number in the nebula itself, it is extremely difficult to discover them, on account of the strong background of the nebula itself. It would be useful to perform special observations for discovering these flare stars through color filters, transparent for radiation between the strongest emission lines of the nebula.

The flare stars in Orion, particularly as a system of objects of one type, have not been studied completely as yet. But there are many reasons to turn Orion into an object requiring special attention. First of all, Orion is the youngest stellar association; there are very many stars with Hα emission and stars of T Tau type in it, i.e. objects even younger than the flare stars. In Orion the objects which are most active in the ultraviolet are found (the stars of NX Mon type), which we cannot find in the Pleiades. There are also the Herbig-Haro objects, i.e. objects younger than the stars with Hα emission and T Tau stars, and which are undoubtedly in the very earliest stages of stellar formation that we know and observe at all. In Orion there are very many peculiar stars of unknown nature and nebulae of tiny dimensions. Many unexpected results may be obtained by observations of the very faint stars in Orion--fainter than 20^m. Finally, one of the richest O-association (around λ Ori) occurs in Orion, with a large number of hot stars with emission lines (O, Be, Wolf-Rayet), with a very large diffuse nebula (NGC 1976), the complex around ε Ori, the reflection nebula M78, the dark nebula Lunds 1630, Barnard's Loop, an enormous amount of interstellar gas-dust material ($\sim$60.000 $M_\odot$ [30]), etc. Evidently the beginning of a new and interesting stage in the study of the Orion aggregate itself as well as of flare stars in particular must be made possible by the application of electrophotometric methods of recording the radiation of the very faint stars, and also of performing observations in the ultraviolet part of the spectrum--up to 1000 Å, with the help of space observatories.

3.   FLARE STARS IN THE PLEIADES

The first flare star in the Pleiades was discovered photoelectrically by Johnson and Mitchell in 1957 [16]; it was the star H II 1306 (or T17 according to the later numeration of the Tonantzintla Observatory), which has been mentioned in the earlier chapters. Subsequently Herbig [17] and Haro [18] undertook a search for T Tau type stars and stars with Hα emission up to 16-17^m (visually). Their results were negative--in the Pleiades region they could not detect stars with Hα emission (within the limits of sensitivity of the objective prism method). On the basis of these results they concluded, in accordance with [16], that probably many of the faint variable stars in the Pleiades belong to the type of the flare stars.

This assumption was soon confirmed by observations made at the Tonantzintla [18,19] and Asiago [6] Observatories. In the period 1963-1964 61 flare stars were discovered in the Pleiades, in an area of 4 x 4° around Alcyone. Later other observatories joined in the search for flare stars in the Pleiades, including the Byurakan Observatory, that of Budapest (Konkoly), Alma Ata (USSR), Sonneberg (DDR) and others. According to the data up to 1976 the total number of flare stars discovered in the

Pleiades was 469; lists of these stars are given in [3-7].

In the case of the Pleiades the question whether a certain star belongs to the cluster is quite troublesome. Here careful analysis of the proper motions of the stars on the one hand and search for new spectroscopic criteria on the other hand are required. As a result of this it appeared for instance that the star H II 2411, which lies in the sky not far from Alcyone and for which a record number of flares --104 (up to 1976) [5]--were observed, actually is one of the three known flare stars in the Hyades cluster.

Kraft and Greenstein [20] followed the second way--looking for spectroscopic criteria; they succeeded in showing that if an intensive emission line K of Ca II is present in the spectrum of a faint star which lies in the Pleiades field, this is quite sufficient to believe this star to be a member of this cluster.

Working with a high-dispersion spectrograph, Kraft and Greenstein established the presence of the emission line K Ca II in the spectra of at least 39 members of this cluster, and in 14 of them hydrogen emission was observed as well. Nineteen known flare stars were found among this sample. To these should be added 13 stars in the Pleiades in the spectra of which McCarthy [21] discovered (with a slit spectrograph) the Hα line in emission. As a result the total number of stars with Ca II and Hα emission lines reaches 52, among which 27 with Hα emission.

As we see, there do occur stars with Hα emission in the Pleiades, among which are flare stars, contrary to the first conclusion [17,18]. However, the total number of such stars is here three times less than in Orion (compared with the number of flare stars).

Wilson [22], who has much studied the behavior of the H and K lines of Ca II in the spectra of stars of different classes, came to the interesting conclusion that in general the intensities of these lines can be used as a good criterion for chromospheric activity in a certain star. In its turn, this activity depends inversely on the star's age; the stronger the line, the younger the star. Haro [4] confirms the validity of this rule in flare stars: in Orion and NGC 2264, the youngest aggregates, there are flare stars of spectral class K0 and later, in which the intensity of the Ca II emission line is much greater than in the stars of the corresponding classes in the Pleiades. It also follows from the measurements by Kraft and Greenstein that the emission lines K2 of Ca II are nearly twice as strong in the spectra of flare stars in the Pleiades as in the spectra of stars of the Hyades cluster, which is considerably older than the Pleiades cluster.

Developing these considerations Haro comes to the interesting possibility that the degree of chromospheric activity may be used as a criterion for determining whether a certain star belongs to the group of flare stars. Any star with spectral class between K0 and M0, which has strong chromospheric activity, must also be a flare star. With regard to the practical applicability of this criterion, unfortunately it cannot be widely used; it is necessary to have the possibility of obtaining spectrograms of the star with sufficiently high dispersion. As Wilson remarks, in some stars the very faint K lines of Ca II can be discovered with a dispersion of 10 Å/mm, but cannot be detected if one works with a dispersion of 38 Å/mm. It is clear that when one works with dispersions of 100 or 200 Å/mm (objective prisms) only very strong Hα or K Ca II lines can be discovered.

The total number of flare stars known in the Pleiades (up to 1976) is very large, about one and a half times larger than in Orion. However, this is the result of observational selection; the effective time of observation was in the case of the Pleiades simply three times longer than in the case of Orion. On the other hand, there are many arguments favoring the idea that the total number of flare stars in

Orion must be much larger than in the Pleiades. It is also characteristic that the flare stars in the Pleiades region with the largest number of repeated flares—several tens or even a hundred (H II 2144, T55, H II 2411)—are not true members of the Pleiades aggregate itself. They appear to be typical stars of UV Cet type, but not connected with the aggregate and not lying in the immediate neighborhood of the Sun. This is shown in particular by the great similarity of the lightcurves of the flares of H II 2411, obtained by the photoelectric method [38], with the lightcurves of flares of stars of UV Cet type.

For each aggregate a star may be determined with a certain limiting brightness, brighter than which the stars show no flares. From the <u>absolute luminosity</u> one can determine the <u>earliest</u> spectral class of the star earlier than which no flares occur. For the Pleiades the brightest flare star is T59b; it has V = 12.04, which corresponds to an absolute magnitude $M_V \approx 6.5$; it is of spectral class K3 Ve [5]. Altogether there are no less than ten flare stars brighter than V = 13 in the Pleiades; all of them belong to class K, or rather K2-K3 (in one case only K5 has been estimated).

The faintest flare star in the Pleiades appears to be fainter than 21^m in photographic light, or fainter than 20^m in V rays. Thus the absolute magnitudes of flare stars in the Pleiades vary from $M_V = +6.5$ to $M_V = +15.0$.

In Orion as well as in the Pleiades "slow" flares were recorded for several stars, where the increase of the brightness before flare maximum is not so quick as in ordinary flare stars but lasts quite long, from ten to 45 minutes [39]. In such cases the decrease of the brightness after the maximum also lasts a long time, up to 8-10 hours. As regards the nature of the "slow" flare, this question has been discussed in the preceding chapter, where it was shown in particular that the assumption that the energy from the stellar interior becomes free in the sub-photospheric layers of the star cannot be valid. "Slow" flares are rather a <u>permanent</u> ejection (instead of an impulsive ejection) of the carriers of the energy from the stellar interior, i.e. an ejection which continues without stopping for a long time —up to 45 minutes or one hour. In other words, the instantaneous structure of such a permanent ejection does not differ in any way from the impulsive ejection in the case of ordinary flares of UV Cet type stars, only at a certain time its frequency becomes much higher, so that one gets the impression of a continuous process.

That the picture here described is not far from probable also follows from the fact that often "slow" as well as "fast" flares are recorded for <u>one and the same star</u>, and with the same flare amplitude. This means that the nature of the two types of flares is the same; the difference lies only in their dynamics.

In the Pleiades region at least four cases were observed of exceedingly strong flares of very faint stars during which the amplitudes of the flares were more than 8^m in U rays. These cases are the following:

Star	U	ΔU
T18	18.5	8.5
T26	20.8	8.5
T153	>22.0	8.5
T53 b	>22.0	>8.5

The question whether these stars are physical members of the Pleiades cluster itself remains open. It is possible that they are simply stars of the galactic background, lying at even greater distances than the Pleiades, but this has no relation to the phenomenon of star flares itself.

4. THE AGGREGATE IN CYGNUS (NGC 7000)

The first flare star in the region of the diffuse nebula in Cygnus NGC 7000 ("North America") was discovered on November 14, 1971 by Haro and Chaviro [32] astonishingly quickly, during five hours of total observing time, the flare amplitude was rather large--4$\overset{m}{.}$5 in U rays. This star was found not far from the FUOR V 1057 Cyg, discovered by Velin [33] the year before, in 1970.

Considering that this region of the sky is characterized not only by an abundance of dark clouds and luminous nebulae, but also by a great number of stars with Hα emission, and stars of T Tau type, by the presence of a T association (Cyg TI) etc. [34-36], the assumption that this region may be a potential focus of flare stars is more than probable. Actually, systematic observations started at the Byurakan Observatory in 1972 were crowned with success: during a few years several tens of flare stars were discovered in the region of NGC 7000 and the neighboring nebulae IC 5068-70 (51 stars up to 1975, of which 7 were discovered at Tonantzintla, 4 in Asiago, and the remaining at Byurakan). The number of stars with Hα emission (late types) was brought to nearly one hundred and fifty [12,13]. The aggregate in Cygnus became the richest one in number of flare stars after Orion and the Pleiades. But it lies nearly one and a half times as far away as Orion ($\sim$700 parsecs). Possibly this is the reason that the average flare frequency is the lowest (0.00020 fl/hr), less than for Orion (0.00051 fl/hr), the Pleiades (0.00035 fl/hr) and in particular Praesepe (0.00087 fl/hr).

In the region of the Cygnus aggregate no stars of NX Mon type, which are particularl active in the ultraviolet, are found. This means that this aggregate is older than Orion (10^6 years). But in the aggregate NGC 7000 there are very many stars with Hα emission, which makes it younger than the Pleiades (5×10^7 years). Apparently the most probable value for the age of the aggregate in Cygnus can be taken to be 3×10^6 years, considering that it resembles Orion in flare activity. As we shall see later, this estimate is also confirmed by the data of colorimetric measurements of flare stars and stars with Hα emission.

5. FLARE STARS IN OTHER AGGREGATES

The young cluster NGC 2264 (age $\sim 6 \times 10^6$ years) was suspected to contain many flare stars because of the fact that there occurs a group of T Tau type stars in it. Actually, two of the 13 flare stars discovered in this cluster proved to be stars of T Tau type. There are no qualitative differences whatever between the flare stars of this cluster and those of Orion. In Orion as well as in NGC 2264 there are T Tau stars which are at the same time flare stars, and inversely, there are flare stars which show no emission lines on low-dispersion spectrograms. In NGC 2264 there are some very faint T Tau type stars with very strong ultraviolet emission (of NX Mon type). However in this aggregate no Herbig-Haro objects [5] have been discovered.

A good representative of the older aggregates is the Praesepe cluster--its age is estimated at 4×10^8 to 9×10^8 years [31]. In the period 1965-66 Rosino [6] discovere 13 flare stars in this cluster. Later I. Jankovich [11] looked for flare stars in Praesepe; he found 20 more flare stars, so that the total number (up to 1975) becam 33. The average flare frequency in this aggregate turned out to be the highest known (0.00087 fl/hr). The mean value of the color indices was found to be $U - B = 0.85$, $B - V = 1.39$, which does not differ much from the values in the Pleiade ($U - B = 0.99$, $B - V = 1.34$). The expected total number of flare stars in Praesepe, found by the method of repeated flares ($n_2 = 5$) is about 100 (cf. §13 of this chapter).

At the end of the 1940's relatively many--23--stars with Hα emission were discovered in the region of Camelopardalis-Scorpio (α = 16^{h}23^m, δ = -24°20') [41,42]. This led Haro and Chavira [43] to undertake special searches for flare stars in this region of the sky. In 1974, after a relatively short time of observations (43 hours), four flare stars were discovered in the direction of the Ophiuchus Dark Cloud, with flare amplitudes from 2^m to 4^{m}5 in U rays (Table 11.1). Further study of this region promises to give interesting results.

6. FLARE STARS OF THE GALACTIC BACKGROUND

Within a sphere with a radius about 10 parsecs around the Sun about 50 flare stars of UV Cet type have been counted. The average number projected on an arbitrary part of the sky with area Ω square degrees will be 50 Ω/41253. Let us assume that up to considerable distances from the Sun the spatial concentration of the flare stars is constant. Then, on any aggregate occupying a section of the sky of Ω square degrees N$_\phi$ flare stars of the galactic background will be projected, with

$$N_\phi = 50 \ \frac{\Omega}{41253} \ \left(\frac{r_*}{10}\right)^3 \approx \frac{\Omega}{800} \ \left(\frac{r_*}{10}\right)^3 , \qquad (11.1)$$

where r$_*$ is the distance of the aggregate from us. For instance in the case of the Pleiades we have: Ω ≈ 20 sq. degr., r$_*$ = 125 pc. With these data we find from (11.1) N$_\phi$ ≈ 50. According to Haro's estimate [5] about 20% of the total number of the flare stars in the Pleiades, i.e. about 86 stars (with N = 431) are background stars. But it should be remembered that here stars are counted which may lie farther away than the Pleiades, whereas formula (11.1) gives the number of field stars at distances up to the aggregate considered. Therefore the true number of background stars in the case of the Pleiades will be less than 86. This is already of the same order as we found above by extrapolating the conditions in the neighborhood of the Sun to the distance of the Pleiades.

The fact that the number of background flare stars for the Pleiades turned out to be the same as would be found on the assumption that the field of flare stars surrounding the Sun is homogeneous and constant over the whole space up to distances of 100-150 pc from the Sun, is a convincing argument that there exists no local system or "association" of flare stars around the Sun. Even if we do not consider the rather varied composition (as regards age, evolution, kinematics, etc.) of the flare stars scattered in the solar neighborhood, it suffices to say that any flare star will have escaped from the place of its birth by not less than 100 pc in 10^7 years, or 1000 pc in 10^8 years, and therefore it may appear in the neighborhood of the Sun from the neighboring aggregates.

Application of formula (11.1) to the Orion aggregate (r$_*$ = 470 pc, Ω ≈ 20 sq. degr.) seems to lead a large number of field stars, namely N$_\phi$ ≈ 2500 (!), nearly eight times the total number of flare stars discovered in Orion. However, the number of field flare stars in Orion is probably 50, i.e. about 15% (this number is found from comparison of Figs. 11.1 and 11.2). The reason for this large discrepancy is understandable: with the limiting magnitude 17^{m}5 of the telescopes used for discovering flare stars in Orion, only stars can be discovered with absolute magnitude less than +9^m or +10^m; flare stars with M$_V$ = +13 are quite rarely detected and only accidentally. However, in our compilation of flare stars (Table 1.1) the fraction of stars with M$_V$ < +10^m is 3/50, and with M$_V$ < +9 even 1/50. Moreover, in the case of Orion it is especially important to take into account the interstellar absorption. Evidently, more detailed quantitative analysis of this problem (to find the true number of field flare stars), may lead to interesting results.

7. THE SPATIAL CONCENTRATION OF FLARE STARS

The last column of Table 11.4 gives some impression of the spatial concentration of
flare stars in the neighborhood of the Sun, within a sphere with a radius of 10
parsecs, as well as in the aggregates of Orion and the Pleiades. In the case of
Orion the computations have been made both for the entire area, of diameter about
5°, and for its densest central part, within a sphere with a diameter of about 2°,
in both cases without taking into account the influence of the field stars. As we
see, the difference in spatial concentration is found to be one order of magnitude.
In the case of the Pleiades the influence of the field stars, about 20% of the total
number of flare stars, was taken into account.

TABLE 11.4 Space Concentration of Flare Stars in the Neighborhood of the Sun and
in the Orion and Pleiades Aggregates

| Flare stars | Radius of sphere | | Total number of flare stars | Space concentr. of flare stars stars/pc^3 |
	Ang.	Linear pc		
In the neighbor-hood of the Sun	--	10	50	0.01
Orion (a)	2°.5	20	321	0.01
Orion (b)	1°.0	8	183	0.09
Pleiades	2°.5	5.5	469	0.56

From the data in this table the space concentration of flare stars in Orion is found
to be 5 to 6 times smaller than in the Pleiades (case b). However this difference
does not seem real. On account of the large distance of Orion from us, as compared
with the Pleiades, and also on account of the fact that the Great Orion Nebula lies
in the center of the region, not all flare stars in it can be discovered (this remark
refers as well to case (a)). At the same time there is no reason to expect equal
space concentration for flare stars in all aggregates, in particular when they are
at different age phases.

As regards the value found for the concentration of the flare stars scattered in the
neighborhood of the Sun (0.01 star/pc^3) this must be considered to be quite reliable;
the process of discovering the nearest flare stars, at distances not more than 10
parsecs, appears to be practically perfect, and more accurate determination of the
distances of individual stars cannot give fundamental corrections to this estimate.

If it is assumed that everywhere in the Galactic System the concentration of flare
stars is equal to that found for the solar neighborhood, the total number of flare
stars in our Galactic System may be found, which is of the order 10^9 (for a volume
of the Galactic System of 1.5×10^{11} pc^3). In order words, the flare stars constitute
about 1% of the total number of stars of all types in the Galaxy. This must be considered to be a large quantity, if one thinks of the quite specific properties of
flare stars.

8. THE APPARENT DISTRIBUTION OF FLARE STARS

In the preceding chapter the conclusion was drawn that in any aggregate the stars
of T Tau type and in general those with Hα emission must be younger than the flare
stars. If this conclusion is valid, we must in particular observe a considerable

difference in the apparent distribution or the dimensions of the systems of the two
categories of objects in a given aggregate. At the same time this difference can be
discovered with certainty only in cases where the number of stars of the two types
in an aggregate is sufficiently large. In this respect the most suitable objects are
Orion and the Pleiades; analysis of the apparent distributions of flare stars and
stars with Hα emission in them lead to interesting results [23].

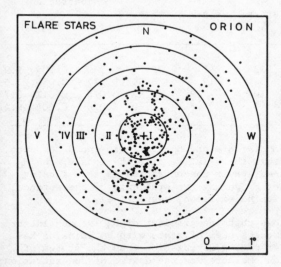
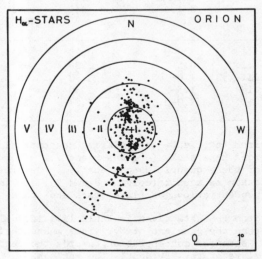

Fig. 11.1. Apparent distribution of
flare stars in the Orion association.
The cross indicates the center of the
Great Orion Nebula (Trapezium). The
distance between the concentric zones
is 0°.5. The number of flare stars in
zones I-V are shown in Table 11.5.

Fig. 11.2. Apparent distribution of
stars with Hα emission in Orion (cf. Fig.
11.1).

Let us first consider the Orion association. Figures 11.1 and 11.2 give charts of
the apparent distributions of flare stars and stars with Hα emission. In the center
of these charts lies the Large Orion Nebula, or rather the stars of the Trapezium,
indicated by a cross.

In both charts the influence of the light absorbing matter on the character of the
apparent distributions of the stars is clearly seen. This is recognized in particular
from the asymmetry of the stellar distributions, and also from the increase of the
total number of stars in directions which are relatively transparent. However,
this absorption, which distorts the apparent distribution of the types of stars we
are interested in, will have practically no influence on the character of their
<u>relative</u> distribution. Therefore the results of the analysis of this relative dis-
tribution, and especially the variation of the ratios of their surface concentrations with increasing distance from the center of the Orion association, will have
a real physical sense.

With this in mind the whole Orion region, of dimensions 5x5°, was divided in five
concentric zones, from I to V, in the way shown in the diagrams. The dimensions of
the zones in angular measure, their areas in square degrees, and the total number
of flare stars N_f and of stars with Hα emission $N_α$ in each of these zones are given
in the first five columns of Table 11.5.

TABLE 11.5. Apparent Distribution of Flare Stars and Stars with Hα Emission in Orion

Zone	Boundaries from center degrees	Area of zone sq. degr.	Total no. flare stars N_f	Hα stars N_α	Concentration stars/sq. degr. flare stars	Hα stars	Relative concentration N_f	N_α	$\dfrac{N_f}{N_\alpha}$
I	0–0.5	0.78	84	134	108	171	1	1	1
II	0.5–1.0	2.36	99	82	42	35	0.39	0.205	1.9
III	1.0–1.5	3.93	75	34	19	8.65	0.177	0.050	3.5
IV	1.5–2.0	5.50	43	10	7.8	1.82	0.072	0.011	6.7
V	2.0–2.5	7.07	20	5	2.8	0.70	0.026	0.004	6.4

From these data the relative concentration of flare and Hα stars was found for each zone separately (columns 6 and 7). From this the relative distributions N_f and N_α of these quantities can easily be determined, their concentration in zone I being taken as a unit (columns 8 and 9). In fact N_f and N_α are gradients of the surface concentrations of flare stars and stars with Hα emission, respectively.

From the data given in Table 11.5 it follows that the gradient of N_α is much larger than the gradient of N_f. This means that the system of stars with Hα emission has a stronger concentration towards the center of the association than the system of flare stars. Evidently the ratio N_f/N_α can characterize the degree of concentration of the flare stars with respect to the stars with Hα emission; the numerical value of this ratio is given in the last column of the table.

It is seen that the ratio N_f/N_α increases rather quickly from the inner to the outer zones. This means that the further we go from the center of the association, the larger the relative number of flare stars, or the smaller the relative number of stars with Hα emission. For instance, in Zone IV the relative number of flare stars is more than four times the number of Hα stars.

Considering all flare stars and stars with Hα emission in Orion together as one system, we can state that the flare activity of the stars of the Orion association increases with increasing distance from its center.

Let us assume that the stars are born by condensation of diffuse gas–dust matter. Then it is to be expected that the process of star formation will be more intensive in the central parts of that cloud than at its periphery, if only because the density of the matter is considerably larger in the central regions. Immediately after its birth the star appears in the state of a star of T Tau type or of a star with Hα emission. The latter are rather quickly transformed into stars of flare type. Judging from the data in Table 11.5, we see that at the periphery of the Orion association already about 80% of the initial number of Hα stars have been transformed into flare stars.

An analogous result is obtained in the case of the Pleiades. Comparing in the same way as in Figs. 11.1 and 11.2 the charts of the apparent distribution of flare stars and Hα stars for the Pleiades and dividing the whole area into five concentric zones we come to an interesting result: in zone V and further there is not a single star with Hα or Ca II emission, while in this zone nearly 20% of all flare stars are found (Fig. 11.3).

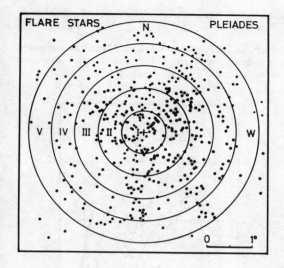

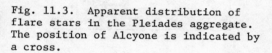

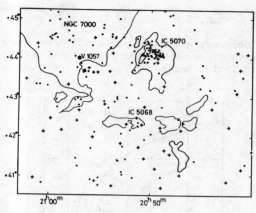

Fig. 11.3. Apparent distribution of
flare stars in the Pleiades aggregate.
The position of Alcyone is indicated by
a cross.

Fig. 11.4. Apparent distribution of 47
flare stars (crosses) and about 150 late-
type stars with Hα emission (dots) in the
Cygnus aggregate (NGC 7000 and IC 5068 - 70).
Four of the flare stars proved to be outside the region of the sky considered.

Hence, the relatively small number (52) of stars with emission lines and their
total absence in the periphery of the aggregate (i.e. $N_f/N_\alpha \to \infty$) is one of the
characteristics of this aggregate, distinguishing it from Orion. But the Pleiades
aggregate is older than the Orion association by more than a factor of ten (Table
11.1). From this it follows that at the present time the process of star formation
in the Pleiades is mainly concluded, and that by far most of the stars with emission
lines have passed to the state of flare stars. This conclusion--the rapid decrease
of the fraction of Hα stars towards the periphery of an aggregate--leads us to the
problem of star formation itself and, in particular, to the state and spatial distribution of the primary matter out of which the stars have been formed. This
question will be discussed in §12 of the present chapter.

A somewhat different picture is observed in the case of the aggregate in Cygnus.
In the region of the nebulae NGC 7000 and IC 5068-70 stars with Hα emission of late
types are encountered mainly in groups, whereas the flare stars are more or less
homogeneously scattered (Fig. 11.4). Also noticeable is a connection of the stars
with Hα emission with the densest parts of the nebulae. These contain no compact
components, as reflected in the character of the apparent distribution of flare
and Hα stars. With all these differences from Orion and the Pleiades there is,
however, no contradiction in the conclusion that also in the case of the Cygnus
aggregate the stars with Hα emission finally become flare stars.

9.   FLARE STARS IN THE COLOR-COLOR DIAGRAM

Of special interest are the results of colorimetric observations of flare and Hα
stars, and also the comparison with the theoretical U - B vs. B - V diagram, as deduced
from the fast electron hypothesis. In the case of Orion such observations were made
photographically by Andrews [24] and Gasparyan [25] for a large number (∿250) of
flare stars. Moreover this work also gives colorimetric characteristics for 200 stars

with Hα emission in Orion. All measurements have been made in the normal state of the star, i.e. outside flares.

The observational results [25] for stars with Hα emission and for flare stars have been drawn in the theoretical color-color diagram in Figs. 11.5 and 11.6. From an analysis of these diagrams the following conclusions can be drawn:

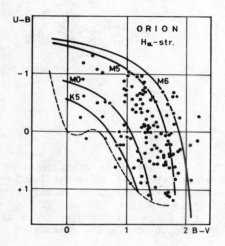
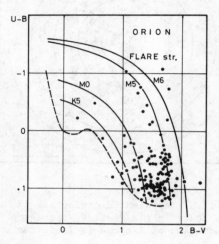

Fig. 11.5. Stars with Hα emission in Orion in the theoretical color-color diagram U – B vs. B – V.

Fig. 11.6. Flare stars in Orion in the theoretical color-color diagram U – B vs. B – V.

a) First the distribution of the Orion stars with Hα emission in the color-color diagram does not differ from what we had above for the T Tau type stars (Fig. 10.3). This confirms once more the idea which has often been brought forward that stars of T Tau type and stars with Hα emission are nearly identical. In particular, practically all stars with Hα emission do not lie on the main sequence, they are scattered remarkably homogeneously all over the color-color diagram, occupying also the region of the NX Mon type stars.

b) The flare stars in Orion also appear above the main sequence. However, the mean deviation of the system of flare stars from the main sequence is considerably smaller than that of the system of Hα stars.

Both conclusions, (a) and (b), form a new and convincing proof that stars with Hα emission constitute a preceding stage and that they turn into flare stars in the course of their evolution. The flare stars approach more and more the main sequence during their development. The color-color diagram for the flare stars in the Pleiades is given in Fig. 11.7, constructed from the data in [26]. Comparing this diagram with Fig. 11.6, we see that in the case of the Pleiades, whose stars are considerably "older" than those in Orion, the average distance of the system of flare stars from the main sequence is significantly smaller than in the case of Orion.

In Fig. 11.6 it is remarkable that there occur practically no flare stars in the region of the diagram corresponding to spectral classes M5-M6. According to this diagram the flare stars in Orion must belong to subclasses earlier than M2-M3. This conclusion is in agreement with the one made by Haro from the spectral characteristic

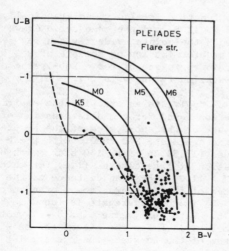

Fig. 11.7. Flare stars of the Pleiades in the theoretical color-color diagram $U - B$ vs. $B - V$.

of the stars, according to which there exist in Orion no flare stars of subclasses later than M2-M3. In our case this conclusion is made as a result of the comparison of the colorimetric data with the theoretical color-color diagram constructed on the basis of the fast electron hypothesis.

Stars which are in the upper part of the diagram (Fig. 11.6), i.e. for which $U - B$ is negative, also show flares, but they cannot have flares with a large amplitude compared to the stars which are closer to the main sequence. A star of class M5, which lies on the main sequence, and has normal color indices, may flare theoretically with an amplitude up to $9-10^m$ in U rays. In the case of stars with abnormal color indices, the maximum flare amplitude cannot be so large, only of the order of $3-4^m$. This conclusion holds for all stars of NX Mon type (cf. Chapter 10, §6). The results given in the preceding chapter allow us to find the exact theoretical relation between color indices and flare amplitude and thus, to predict the limiting amplitude for a given flare activity of the star characterized by the color indices.

10. A METHOD FOR DETERMINING THE AGE OF AN AGGREGATE

We reached the interesting conclusion that the average deviation of a system of a certain type of stars from the main sequence depends on the age of the system itself, in such a way that, the larger this deviation, the younger the system, and vice versa. This leads to the idea: might it not be possible to construct an empirical relation between the mean deviation measured along the $U - B$ axis, calling this $d(\overline{U - B})$, and the age of the system? Evidently, once this relation has been found in quantitative form, it will be possible to use it for the determination of the age of any aggregate or of a system with non-stationary stars in it, if from the observations the quantity $d(\overline{U - B})$ is known.

Apart from Orion and the Pleiades there are sufficiently reliable colorimetric data about two more aggregates--in Cygnus (NGC 7000) [13] and Praesepe [11]. The ages of these aggregates are also more or less accurately known. With these data the required empirical relation between mean deviation from the main sequence $d(\overline{U - B})$ for a given system of non-stationary stars and its age may be represented in the following form:

System of stars	$d(\overline{U - B})$	Age (years)
Orion, Hα stars	$1\overset{m}{.}0$	3×10^5
Orion, flare stars	0.6	10^6
NGC 7000, flare stars	0.6	3×10^6
Pleiades, flare stars	0.16	5×10^7
Praesepe, flare stars	0.0	4×10^8

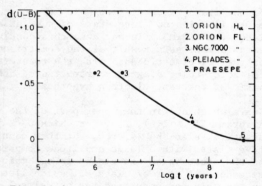

1. ORION Hα
2. ORION FL.
3. NGC 7000   "
4. PLEIADES   "
5. PRAESEPE

Fig. 11.8. Empirical relation between average distance of the color index $d(\overline{U - B})$ for a given category of stars (flare stars or stars with Hα emission) from the main sequence and age t of the aggregate.

This function is also given graphically in Fig. 11.8, and the mean curve drawn through the points evidently can be used for a determination of the age of a certain aggregate or for a sub-system of non-stationary stars in it.

Let us apply, as an example, the empirical relation found for the determination of the age of the sub-system of the Hα stars in the aggregate NGC 7000. With the U – B values [13,33,36] for ∿150 stars with Hα emission of late classes (i.e. for which B – V > +1.0), the mean distance of the system of the Hα stars from the sequence $d(\overline{U - B})$ in this aggregate is found to be ∿0.76. With this value of $d(\overline{U - B})$ we find from Fig. 11.8 for the age of the sub-system of the Hα stars the age of ∿10^6 years.

The method proposed for the determination of the age of an aggregate or groups of well-defined types of non-stationary stars is undoubtedly superior because it is not based on the characteristics or peculiarities of one or two stars alone, but those of a whole system of completely homogeneous types of stars, thus excluding the influence of accidental errors. The individual properties of the separate exotic types of stars, e.g. the presence or absence of stars of NX Mon type or of Herbig-Haro objects in a given aggregate may be used as an additional criterion only for checking the reliability of the age of the aggregate found by means of the curve in Fig. 11.8.

11.   THE RELATION BETWEEN FLARE STARS AND STARS WITH EMISSION LINES

One can hardly doubt that flare stars; rapid irregular variables, which are stars of class dMe; and stars of T Tau type, are groups of similar objects and that there is a genetic connection between them [3]. These types of stars form an evolutionary series, a fact which we have shown above, from the apparent distributions of flare stars and stars with Hα emission in Orion, and also from the character of the U – B vs. B – V diagrams for these groups of stars.

Which are some other indicators connecting the objects together? Haro [3] distinguishes the following:

1) These stars have a tendency to be accumulated in groups, in particular in young star clusters, in which interstellar matter occurs in considerable quantities.

2) For irregular variables and flare stars it is characteristic that they occur only in regions rich in interstellar matter. Some physical properties of the stars lying far from the clouds of interstellar matter may differ from the properties of stars immersed in these clouds. For instance, there is not a single star known of T Tau type outside nebular regions, and hardly a flare star of class earlier than K. This may rather be a consequence of evolution than interaction with the surrounding medium

3) Rapid and irregular variables of all types, in particular those which are connected with aggregates, may be found above the main sequence. However, some irregular variables and flare stars do occur around or even below the main sequence.

4) Typical T Tauri and RW Aurigae stars can show flares, and vice versa, many flare

stars can at the same time be "normal" irregular variables.

5) The spectroscopic properties at the time of the flare are for all flare stars the same as for stars of T Tau type in ordinary conditions. Just as for many irregular variables, the spectra of some of the flare stars show emission lines of hydrogen and ionized calcium. However, there are also stars for which no emission lines can be detected, even at the time of maximum brightness.

6) The kinematic properties of the irregular rapid variables, including also stars of class dMe and flare stars, are, according to preliminary data, similar.

7) Flare stars may be expected to occur in each isolated group of stars which contains rapid irregular variables.

8) The later the spectral class of the brightest flare star in a given group of stars, the fewer "normal" irregular variables it contains, and the smaller the amplitude of their brightness fluctuation.

To summarize, the flare stars in aggregates--in associations and young clusters-- and also in the neighborhood of the Sun, belong to one and the same physical family and the differences which are found between them should be taken to be the effects of stellar evolution.

12. THE SPATIAL DISTRIBUTION OF FLARE STARS IN AGGREGATES

Even a quick glance at the charts of the apparent distributions of flare stars and stars with $H\alpha$ emission (Figs. 11.1, 11.2, 11.3) shows that there may exist certain laws in the space distribution of these stars within a given aggregate. Evidently, if such a law is found, this will have a direct relation with the origin and evolution of these stars.

We shall start from the simplest assumption, namely that the system of the type of non-stationary stars considered has a spherical-concentric symmetry and that the space concentration of the stars in it varies according to the law:

$$n(x) \sim x^{-n} \tag{11.2}$$

where x is the distance from the center of the system, and n characterizes the rate of decrease of the number of stars per unit volume with increasing distance from the center. If we call N(r) the total number of stars projected on the sky within a ring of unit width and radius r (in projection on the sky) from the center of the system, then we have for the ratio $N(r)/N_o$

$$\frac{N(r)}{N_o} = \int_r^{r_o} \frac{x^{1-n} dx}{\sqrt{x^2-r^2}} , \tag{11.3}$$

where r_o is the radius of the sphere (the circle) outside which there are no stars, and N_o is the total number of stars in the central zone (all linear quantities in (11.3) are expressed in r_o as a unit).

The ratio $N(r)/N_o$ is therefore the apparent distribution of the stars within the aggregate, which can easily be found from the observations. In the case of Orion, for instance, we have: for the flare stars $N(r)/N_o = N_f$, and for the stars with $H\alpha$ emission $N(r)/N_o = N_\alpha$, where the numerical values of N_f and N_α are given in the 8th and 9th columns of Table 11.5.

Furthermore if a number of discrete values of the parameter n are put in (11.3), from n = 0   (case of constant density of stars) to n = 3, curves can be constructed repre-

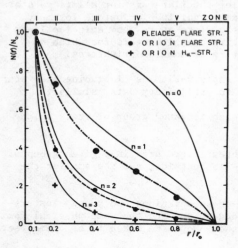

senting the variation of the function $N(r)/N_O$ with r for each value of n separate. these curves are shown by full-drawn lines in Fig. 11.9. Now we need only put in this figure the observed values of $N(r)/N_O$ for a given type of star, which has been done for flare stars in Orion and in the Pleiades, and also for stars with Hα emission in Orion

Finally the assumption of a concentric-spherical structure for the aggregate, as well as the assumed law $n(x) \sim x^{-n}$, are idealizations of the problem. The influ-ence of the interstellar absorption within the aggregate itself has not been taken into account. Nevertheless, the distri-butions of the two categories of stars in Orion and the Pleiades follow the computed curves remarkably well. The best agree-ment of the observations with theory is found for the following values of n (cf. Fig. 11.9):

Fig. 11.9. The problem of the space distribution of flare stars in aggre-gates. Full-drawn curves with n = 0,1, 2 and 3 are the theoretical dependences of the surface concentrations of stars $N(r)/N_O$ on the distances r for corre-sponding values of the exponent n in the law $n(x) \sim x^{-n}$. Agreement between observations and theory is obtained: for the flare stars in the Pleiades for n = 1.4 (·—·—), for the flare stars in Orion for n = 2.2 (broken line), for the Hα stars in Orion for n ≈ 3.

Orion: stars with
  Hα emission   n = 3 (full-drawn line)

Orion: flare
  stars   n ≈ 2.2 (broken line)

Pleiades: flare
  stars   n ≈ 1.4 (·—·—)

These results show that the largest value of n (= 3), and hence the strongest cen-tral concentration, is found for the Hα stars of Orion, i.e. for the youngest stars, and the lowest (n = 1.4) for the flare stars in the Pleiades, i.e. for the

oldest of these three groups of stars. There is a definite relation between the age of the stars and the degree of their concentration within the aggregate, namely, the younger the stars, the stronger they are concentrated in the aggregate. With age, or in the course of the evolution of the stars their space distribution approach-es a uniform distribution.

What conclusions can we draw from this? First, this has certainly a direct relation to the space distribution of the primary substratum from which the stars are born--discrete prestellar bodies of unknown nature or a gas-dust cloud with the possi-bility of local condensations; in either case we shall have for that substratum n > 3. Furthermore, does it mean that the decrease of n with the evolution of the stars is the result of expansion of the aggregate (association)? Evidently not, since, in the first place we do not know what percentage of the proto-stars, i.e. stars with Hα emission, are transformed into flare stars, and secondly at what rate and in what proportions the flare activity stops in stars with different abso-lute luminosities (different spectral classes). It should be noted, however, that the spherical-concentric model with n > 3 for the prestellar substratum may have a natural likeness in the case of a gas-dust cloud.

The conclusions drawn from the analysis of the space distributions of flare and Hα stars in aggregates are only preliminary. In particular, no account was taken of the influence of field stars, of the influence of interstellar absorption, the absorption by diffuse matter in the aggregate itself, etc. At the same time the above analysis is promising and should be extended to other aggregates.

13. THE PROBABLE NUMBER OF FLARE STARS IN AGGREGATES

By far the largest number of flare stars are, as we have seen, concentrated in stellar associations and aggregates. Repeated flares indicate that the total number of flare stars is large, but remains within reasonable limits. Hence the problem arises of estimating the probable number of flare stars in a certain aggregate. V. A. Ambartsumian [7] proposed the following simple method to estimate the total number of flare stars.

The basis of this method is the assumption that a star flare is an accidental occurrence and that its sequence is an event following a Poisson law. It is also assumed that the average flare frequency ν is constant for the various stars. And even more, it is assumed that ν is the same for all flare stars in a given aggregate. If all these assumptions hold, the probability of observing k flares during the effective observing time t can be written as:

$$P_k = e^{-\nu t}\frac{(\nu t)^k}{k!}.$$ (11.4)

If N is the total number of all flare stars in a given aggregate, the mathematical expectation for the number of stars n_k, showing k flares during the time t will be:

$$n_k = Ne^{-\nu t}\frac{(\nu t)^k}{k!}.$$ (11.5)

If N is sufficiently large, n_k will be, with sufficient approximation, the number of stars with k flares during time t. Writing the expression (11.5) for the cases k = 0,1,2,3 and 4, and letting n_0 be the number of flare stars with no flares during the time t, we find:

$$n_0 = \frac{n_1^2}{2n_2}$$ (11.6)

$$n_0 = \frac{2n_2^3}{9n_3^2}$$ (11.7)

$$n_0 = \frac{3}{32}\frac{n_3^4}{n_4^3},$$ (11.8)

where n_1, n_2, n_3 and n_4 are the numbers of flare stars which have in time t one, two, three and four flares, respectively. In order to determine the total number of flare stars N in a given aggregate one should add to the value of n_0 the number of flare stars already recorded.

Evidently, if in a real aggregate the assumptions made do hold, then each of the formulae (11.6), (11.7) and (11.8) must give the same value for n_0. However, as we shall see below, this does not happen.

Let us begin with Orion. We have from Table 11.3: $\Sigma n = 325$, $n_1 = 236$ and $n_2 = 60$. Using formula (11.6) we find: $n_0 = 465$, and for the total number of flare stars $N = \Sigma n + n_0 = 790$, i.e. the total number of flare stars in Orion must be close to 800.

However, completely different numbers are found for n_0 if we use formulae (11.7) and (11.8). Taking the quantities n_3 and n_4 for Orion from Table 11.3 we find for n_0 the values 110 and 284, respectively, i.e. a difference of a factor four.

Still worse is the state of affairs in the case of the Pleiades. Taking n_k from Table 11.3, we find from (11.6): $n_0 = 400$ and $N = 847$, i.e. the total number of flare stars in the Pleiades must be of the order of 1000. However, the formulae (11.7) and (11.8) give for n_0 the values 44 and 39--ten (!) times less than we found in the first case. The conclusion is clear: the requirement that the average flare frequency in the given aggregate is constant is not fulfilled either in Orion nor in the Pleiades and evidently there are direct data showing that the average flare frequency in a system can vary within wide limits. In the Pleiades, for instance, besides stars showing only one or two flares, there occur stars as well with 10, 20, 45 and in one case even up to 78 (!) flares during the same interval of time. Also a subgroup of stars was discovered with a flare frequency of 0.008 fl/hour, more than an order of magnitude higher than the average for the entire aggregate.

There is still another reason leading to such a strong discrepancy between the observed probability of detecting repeated flares and that given by a Pouasson analysis. The factor of <u>chance</u> in detecting flares on photographic plates stops working in practice, since an observer, when looking at a star field on a photographic plate, may give special attention to already known flare stars.

From all this it follows that the "method of repeated flares" for finding the true number of flare stars in a given aggregate does not lead to convincing results. The assumption of a Pouasson distribution may be in error and perhaps should be replaced by some other formula following from a fundamentally different approach to the problem.

14. DO ALL STARS IN AN AGGREGATE SHOW FLARES?

In connection with the contradictory results obtained in the preceding section as regards the total number of flare stars in a given aggregate, the question arises: how large is the possible number of flare stars as compared with the total number of physical members of the aggregate? In particular, could it be possible that <u>all</u> stars in the aggregate are flare stars and that it is only a matter of time to discover them as such?

The available data give a negative answer to this question. As an example we shall choose a group of 78 stars given in the list by Hertzsprung and others [40] for physical members of the Pleiades, with photographic magnitudes in the interval 14.50-16.05. It is found that out of this number of stars 31 already are flare stars [7]. It is interesting to note that for this group of stars the distribution of the number of observed flares is represented quite satisfactorily by a Pouasson distribution, with a suitable choice of parameters. And it appears that the expected number of flare stars for the group mentioned must be 39, roughly one half! The remaining stars have no flare activity, at least during the period of observation of the Pleiades region (1963-1975).

Finally it must be thought premature to draw far-reaching conclusions from these facts, in particular conclusions of evolutionary character. It is quite evident that application of more sensitive methods (for instance, the photoelectric method) for the detection of flares certainly will bring to light some flare activity in the second half of the group of physical members of the Pleiades. In any case, the assumption that part of the stars in an aggregate, where the processes of stellar formation and the initial stage of their evolution are taking place, need not at all pass through a phase with flare activity, apperas to be highly improbable. The Pleiades stars are no longer very young; therefore it would be quite natural to expect a decrease in flare activity in the older stars, whereas stars which have appeared later are probably at the peak of their flare activity.

In addition, if one half of all physical members of an aggregate are now flare stars, then it must be assumed that this is almost impossible to think of conditions in which the remaining half of the members of the aggregate could not be suspected of being flare stars at some earlier time.

These conclusions, however, should be confirmed by new observations. It is possible that the assumption that all stars in aggregates pass through a phase of flare activity in the first 10-100 millions of years after their formation may appear already quite plausible, but as yet we have no irrefutable and direct data in favor of this assumption.

Analysis of the data referring to flare stars of UV Cet type leads us to the conclusion that the flare frequency rapidly increases with decreasing intrinsic luminosity of the star (Chapter 1, Fig. 1.3). On the other hand, we have already convinced ourselves of the identity of the physical processes leading to a flare of UV Cet type stars on the one hand and flares of stars in aggregates on the other hand. Therefore such a rapid increase of flare frequency for fainter objects in any aggregate is certainly to be expected. A small amount of energy which is impulsively emitted by a star can easily be observed as a flare in an intrinsically faint star, and may remain completely unnoticed in an intrinsically bright star. Hence we must not be astonished when in the future observations may confirm the rapid increase of flare frequency towards fainter stars (fainter than 16^m in the Pleiades).

15. ON THE PERIODICITY OF FLARES

It has been remarked that there exists a periodicity in the flare activity of some flare stars in the Pleiades [7]. Thus, for instance, in one of the first flare stars discovered in the Pleiades by the photographic method, the star T14 = H II 906, during the year 1963 alone, four flares were observed with amplitudes larger than 1^m. The following flare of this star was recorded after two years--in the end of 1965. Moreover no flares were observed in the next two years; only in the end of 1968 one flare with amplitude $2^m_{\cdot}2$ was recorded. After 1968 two flares were observed at the end of 1972 and in the beginning of 1973. Finally, after another two years two more flares were observed--at the end of 1974. A time diagram of these flares is shown schematically in Fig. 11.10. One gets the impression that the flare activity of this star is repeated periodically with a period of 2 to 2.5 years. This regularity appears to be disturbed by the absence of flares in 1971, but perhaps they have simply not been recorded because at the time no observations were made.

The star T22 shows a periodicity in the flare phenomenon which is striking in its accuracy; during ten years of observations of this star three powerful flares were recorded (with ampliutdes of about 2.5, 4.7 and 9.0 magnitudes in U light) with intervals of five years (Fig. 11.10).

Another kind of periodicity in flare activity is shown by the stars T157 = H II 2144 and T102. During many years, starting in 1963, not a single flare was observed in these stars. And then in a period less than one year 4 to 5 flares were recorded for each of these stars. After that again for many years no flares appeared. Evidently, the period after which the flare activity of these stars repeats itself must be longer than 6 to 7 years (Fig. 11.10).

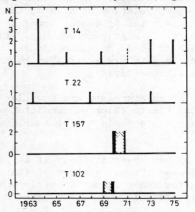

Fig. 11.10. Time curves of flares of some stars in the Pleiades. The flare phenomenon occurs in cycles: for the star T14, of about 2-2.5 years (with a probably unrecorded flare in 1971); for T22, of 5 years; for T157 and T102, more than 6-7 years.

In general, the periodicity of flare activity is quite a difficult matter; there are many reasons which make it hard to perform continuous and quite homogeneous observations during many years of a single star. Thus the total number of stars for which the flare activity is really periodic must be considerably larger than indicated from the above examples.

The possible existence of a periodicity in the flare activity should be considered as evidence that in these cases we have double stellar systems. The phenomenon itself is directly related to that which has been described in §7 of Chapter 4. In the double system one star produces the fast electrons and the second the infrared photons. During another flare the components of the system could interchange their roles. The following seems important: in double systems the efficiency of the transformation of kinetic energy of the fast electrons into optical radiation reaches its maximum value; this happens at a certain well-defined phase of the system (with reference to the observer). Herein lies the cause of the appearance of periodicity in the phenomenon considered above.

Depending on the structure of the double system itself the period of the maxima (or minima) of the flare activity may be a few years to tens of years or more.

16. THE DEPENDENCE OF THE STRENGTH OF THE OPTICAL FLARE ON THE STAR'S LUMINOSITY

If a flare star is at the same time a variable star, the amount of energy emitted in a flare will result in an increase in brightness which will be larger when the star is near minimum brightness, and vice versa. But the possibility is not excluded that the absolute amount of the energy emitted in the flare will also vary in one or the other direction depending on the absolute luminosity of the star. The answer to this question can be found only if data were available about the strength of the flares, i.e. about the mean amplitude of the flares for one and the same star at different brightnesses.

Unfortunately such data do not exist. But one can try to find the answer to the question by making use of the fact that all flare stars in a given aggregate are at the same distance from us, and hence one can either consider all these stars together with their different absolute luminosities, or the various states of one flare star with variable intrinsic luminosity.

Practical application of this simple idea is made difficult by the fact that there is a certain selection in the observational material. The selection occurs because in faint stars flares with small amplitudes may not be recorded, whereas in bright stars both faint as well as strong flares can easily be observed. As a result the mean values of the observed flare amplitudes will increase rapidly, as seen from the table below, towards the faint stars:

m(U) =	15–16	16–17	17–18	18–19	19–20	20–21	21–22
$\overline{\Delta U}$ =	1.10	1.45	1.55	2.60	4.30	4.80	6.50

The limit of the 30-inch Schmidt telescope of the Tonantzintla Observatory is $17\overset{m}{.}5$ in U light. This means that a flare amplitude $\lesssim 2\overset{m}{.}5$ in U light may be observed without any selection for all stars brighter than $\sim 20^m$ in U light. Or, amplitudes $\lesssim 1\overset{m}{.}5$ will be observed in all stars brighter than 19^m, etc.

Restricting ourselves in the analysis of the data only to flares with amplitudes larger than $1\overset{m}{.}5$, and stars brighter than 19^m (in U light), we can judge on the real dependence of the average strength of the flare on the absolute luminosity of the star. The results are given in Table 11.6.

TABLE 11.6 Dependence of the Average Amplitude $\overline{\Delta U}$ on Apparent Magnitude m(U) for Flare Stars in Orion and the Pleiades. In Parentheses the Number of Stars (Flares) is Shown, for Which m(U) $\leq$ 19 and $\Delta U \geq 1.5$.

m(U)	14–15	15–16	16–17	17–18	18–19
Orion	$1\overset{m}{.}65$ (2)	$1\overset{m}{.}87$ (6)	$1\overset{m}{.}92$ (6)	$2\overset{m}{.}14$ (15)	$2\overset{m}{.}71$ (41)
Pleiades	1.70 (4)	2.40 (3)	2.51 (10)	2.23 (21)	3.18 (37)

As seen from these data the average amplitude $\overline{\Delta U}$, and therefore the average strength of the flare (at maximum) increases, at first slowly, then more rapidly towards the stars of lower intrinsic luminosity. In other words, the specific energy production in the form of a flare is larger in the intrinsically faint stars than in the bright ones. Qualitatively this is the same as we have seen in the case of the UV Cet type stars (Fig. 6.39).

Moreover, we can find the dependence of the average flare strength at maximum $I_f(U)$ on M_U quantitatively by means of the following relation:

$$I_f(U) \sim 10^{-0.4M_U}(10^{0.4\overline{\Delta U}}-1), \qquad (11.9)$$

where the values $\overline{\Delta U}$ are taken from Table 11.6 for each interval of m(U) separately, and M_U = m(U) - 8.5 for Orion (r = 470 pc), and M_U = m(U) - 5.5 for the Pleiades (r = 125 pc). In the computations m(U) has been taken as 14.5, ..., 18.5 for successive magnitude intervals. The results of the computations for Orion and the Pleiades are shown in Table 11.7, which gives the dependence of the relative values of $I_f(U)$ on M_U, with $I_f(U)$ = 1 for M_U = 9. This dependence is also shown graphically in Fig. 11.11, where the abscissae are given in absolute magnitude in U light, M_U, and on top in V light, M_V.

It is seen from these results that the average value $I_f(U)$ of the energy produced during one flare in U light quite clearly increases with increasing absolute luminosity of the star. The result is found to be exactly the same as in the case of the UV Cet type stars (Fig. 6.39). Besides, the dependence of Log $I_f(U)$ on M_U (or M_V) for flare stars in aggregates in the range of absolute magnitudes from $M_U \sim 13$ to $M_U \sim 6$ (or from $M_V \sim 10$ to $M_V \sim 5$) is also linear, as in the case of the UV Cet type stars.

TABLE 11.7 Dependence of the Average Energy $I_f(U)$, Produced in U Light During One
Flare of a Star in Orion and the Pleiades, on the Absolute Magnitude of the Star
M_U. $I_f(U)$ Has Been Taken Equal to 1 for $M_U = 9$.

m(U)	Orion		Pleiades	
	M_U	$I_f(U)$	M_U	$I_f(U)$
14–15	6	9.4	9	1
15–16	7	4.8	10	0.86
16–17	8	2.0	11	0.38
17–18	9	1	12	0.126
18–19	10	0.72	13	0.117

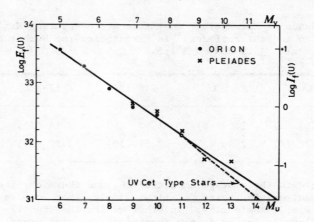

Fig. 11.11. Relation between average energy strength of one
flare in U rays $E_f(U)$ (in ergs) and the absolute magnitude of
the star M_U (or M_V) for flare stars in Orion (dots) and in
the Pleiades (crosses), normalized at $M_U = 9$ (scale of relative flare strength $I_f(U)$ at the right-hand side). Broken
line: part of this relation, taken from Fig. 6.39, for stars
of UV Cet type (the circle at $M_V \approx 8.4$ corresponds with the
position of YY Gem).

For a comparison of the gradients $d[\text{Log } I_f(U)]/d\,M_V$ for these two groups of flare
stars part of the graph in Fig. 6.39 is reproduced in Fig. 11.11 (broken line),
the curves have been superposed from $M_V = 8.5$ (the limiting point (YY Gem)) in
Fig. 6.39--towards the higher luminosities.

Figure 11.11 shows that the slope of the curve for the UV Cet stars differs slightly
from that of the flare stars in aggregates. However, we are not convinced that
this difference is real and therefore we may assert that the dependence of $I_f(U)$ on
M_V in the huge interval of absolute magnitudes from 17^m (Fig. 6.39) to 5^m (Fig.
11.11) encompassing the UV Cet stars as well as the flare stars associated with
aggregates can be represented by one law. Here, in particular, we obtain one more
proof that the two categories of flare stars are identical as regards the identity
of the mechanism of generation of the flares. The graph showing $E_f(U)$ versus M_V
for the UV Cet type stars (Fig. 6.39) was constructed in absolute units. It is
interesting to see in which conditions the $I_f(U)$ vs. M_V relation can be transformed

into the graph for the average total energy $E_f(U)$ of a flare in absolute units for the case of flare stars in aggregates. We have by definition:

$$E_f(U) = I_*(U)P(\overline{\Delta U}) , \qquad (11.10)$$

where $I_*(U)$ is, as before, the light-emission of the star in U rays in normal conditions, i.e. outside flares, and $P(\overline{\Delta U})$ is the "equivalent time" of the flare.

For the last point (left) in Fig. 6.39, corresponding to $M_V \approx 8.5$, we have, for YY Gem: $E_f(U) \approx 1.5 \times 10^{32}$ ergs and $I_*(U) = 1.61 \times 10^{30}$ ergs/sec (Table 6.8). With these data we find from (11.10): $P(\overline{\Delta U}) \approx 100$ seconds, i.e. a quantity of the same order as found earlier for UV Cet stars. This coincidence appears to be fortuitous, for the observations of flares of UV Cet type stars and of stars found in aggregates are usually made by completely different methods (photoelectric and photographic) and with entirely different time resolutions.

In Fig. 11.11 the ordinates (left) show $E_f(U)$ in absolute energy units. Combining this with Fig. 6.39 we may get a clear idea of the actual range of the energy $E_f(U)$ produced on the average during one flare for all categories of flare stars: this energy (in U rays) ranges from $\sim 6 \times 10^{28}$ ergs (CN Leo) to $\sim 4 \times 10^{33}$ ergs (the intrinsically brightest stars in Orion).

17. THE AVERAGE STRENGTH OF THE FLARE

For an understanding of the whole energetics of flares in stars connected with aggregates the following two questions are particularly interesting:

a) Which is the predominant form of energy losses of the star--weak, but very frequent flares, or powerful, but very rare flares?

b) How large is the average strength of one flare in a given aggregate?

The power of a flare is determined by the total number of fast electrons, N_e, appearing during the flare. But N_e determines in its turn the quantity τ--the optical thickness of the medium for processes of Thomson scattering. Hence it is practical to use as a measure of the strength of the flare the quantity τ.

Let us consider the first question. For weak flares τ is small, for strong flares it is large. But the relative number of weak flares is much larger than that of the strong flares. Evidently the quantity $\tau_*(\Delta U)$, given by the relation:

$$\tau_*(\Delta U) = \tau(\Delta U)F(\Delta U) , \qquad (11.11)$$

where $F(\Delta U)$, the distribution function of the flares with respect to amplitudes, can be characterized by the specific strength of the flare for a given amplitude. The problem lies in finding out for which ΔU the function $\tau_*(\Delta U)$ reaches it maximum value.

Numerical values of the function $F(\Delta U)$ for UV Cet type stars are given in Table 1.8. There the values of $F(\Delta U)$ for the flare stars in Orion and the Pleiades have been given as well, only flare stars brighter than $17^m.5$ (in photographic light) have been taken into account. The numerical values of $\tau(\Delta U)$ have been taken by assuming a Gaussian distribution of the fast electrons and for given ΔU. The numerical values of $\tau_*(\Delta U)$ found with these data and by means of (11.11) for the flare stars of Orion and the Pleiades, and also for four stars of UV Cet type and for the star H II 2411, are shown in Table 11.8.

It is seen that the function $\tau_*(\Delta U)$ reaches its maximum for $\Delta U \approx 3$ in the case of

TABLE 11.8 Dependence of the Specific Strength of a Flare, $\tau_*(\Delta U)$, on Flare Amplitude for Flare Stars in Aggregates and of UV Cet type

ΔU	0-1	1-2	2-3	3-4	4-5	5-6
Orion	0.00002	0.00015	0.00051	<u>0.00056</u>	0.00012	0.00010
Pleiades	0.00004	0.00017	0.00027	<u>0.00030</u>	0.00018	0.00015
UV Cet	0.00005	0.00017	0.00016	<u>0.00022</u>	0.00010	0.00008
CN Leo	0.00003	0.00021	<u>0.00037</u>	0.00024	0.00006	--
YZ CMi	0.00007	0.00010	<u>0.00011</u>	0.00004	0.00006	--
AD Leo	0.00007	0.00008	<u>0.00016</u>	--	0.00006	--
H II 2411	0.00007	0.00008	<u>0.00010</u>	0.00008	--	--

the stars in Orion, the Pleiades and for UV Cet, and for $\Delta U \approx 2.5$ in the case of the remaining stars. It is also seen that the brighter the star is intrinsically, the steeper is the function $\tau_*(\Delta U)$.

The results indicate that the flare stars lose energy--in the form of kinetic energy of fast electrons--mainly as a result of average and strong flares--with amplitudes up to $\Delta U \sim 4$. Weak, but very frequent flares do not play a significant role.

Let us now turn to the second question, on the average strength of a flare $\bar{\tau}$. Evidently we have for $\bar{\tau}$:

$$\bar{\tau} = \sum_{\Delta U} \tau_*(\Delta U) \ . \tag{11.12}$$

For flare stars of UV Cet type the numerical values of $\bar{\tau}$ have been found earlier (Table 6.8). For Orion and the Pleiades we found the quantity $\bar{\tau}$ by simply adding the numbers in the first two lines of Table 11.8. Thus we have:

$$\bar{\tau} = 0.0015 \text{ for Orion}$$
$$\bar{\tau} = 0.0011 \text{ for the Pleiades}$$

i.e. the average flare strengths $\bar{\tau}$ for these two groups of flare stars are nearly equal. Moreover, these flare strengths are of the same order as in the case of UV Cet and CN Leo, but are several times larger than for the stars YZ CMi and AD Leo ($\tau \sim 0.0004$). Evidently this difference is not real, but is due to selection of the observational material; flares with amplitudes smaller than 0^m6-0^m7 in aggregates cannot be recorded.

18. HEATING AND EXPANSION OF THE INTERSTELLAR MATTER IN AGGREGATES

The total number of flare stars in Orion must be, according to the estimate made above, of the order of 1000. If a flare is due to the ejection of fast electrons, for a period of $\sim 10^6$ years, then a huge quantity of energy will be set free, which will be transmitted to the interstellar medium in the immediate surroundings of the system of flare stars. This may lead to heating and expansion of the interstellar matter which is present in the aggregate considered. The possibility of heating of interstellar matter by particles of high energy has been discussed theoretically quite often [28,]9].

Menon [30] found, by detailed study of the Orion region in the line of monochromatic hydrogen radiation (21 cm) the radial expansion of a huge mass (M = 58000 $M_\odot$) in Orion with velocity 10 km/s. The question arises if this expansion may be due to heating of the interstellar matter by fast electrons ejected by flare stars?

For a frequency of n flares per year for one star, the total energy E_o, produced by N flare stars in t years in the form of kinetic energy of fast electrons will be

$$E_o = P N n t \text{ ergs} \tag{11.13}$$

where P is the total energy of the fast electrons produced by the star in one flare, given by (6.6). Thus we find for $\bar{\tau}$:

$$\bar{\tau} = \frac{E_o \sigma_s}{4\pi R^2 \varepsilon N n t} , \tag{11.14}$$

where ε is the energy of one electron ($\sim 1.5 \times 10^6$ eV), R is the radius of the star (or rather, of the fast electron envelope).

Estimating the kinetic energy of the interstellar matter in Orion, which is in a state of expansion, Menon found the value $E_o = 5.8 \times 10^{49}$ ergs. Taking N = 1000, n = 300 flares per year, t = 2×10^6 years, R $\sim 10^{11}$ cm, we find from (11.14) for the average strength of one flare:

$$\bar{\tau} \sim 0.002 .$$

This is of the same order of magnitude as found above for Orion. But for this value of $\bar{\tau}$ the value taken above for the flare frequency n must be thought to be too high. At the same time it should be remembered that in our considerations no account was taken of the role of stars with emission lines which also provide fast electrons. The vulnerability of this analysis is the large number of uncertain parameters in formula (11.14).

Notwithstanding the deficiencies, the notion that fast electrons may be the cause of heating and expansion of the diffuse interstellar matter in Orion may be real.

REFERENCES

1.   A. H. Joy, Ap. J. 110, 424, 1949.
2.   G. Haro, Ap. J. 117, 73, 1953.
3.   G. Haro, Flare Stars. Stars and Stellar Systems, Vol. VII, Nebulae and Interstellar Matter, Ed. B. M. Middlehurst and L. H. Aller, 1968, p. 141.
4.   G. Haro, E. Chavira, Bol. Obs. Tonantz. y Tacub. 31, 23, 1969; 32, 59, 1969; 34, 181, 1970.
5.   G. Haro, Bol. Inst. Tonantzintla 2, 3, 1976.
6.   L. Rosino, L. Pigatto, Contr. Asiago Obs. No. 189, 1966; No. 231, 1969.
7.   V. A. Ambarzumian, L. V. Mirzoyan, E. S. Parsamyan, H. S. Chavushian, L. K. Erastova, Astrofizika 6, 7, 1970; 7, 319, 1971; 8, 485, 1972; 9, 461, 1973.
8.   E. S. Parsamyan, E. Chavira, Bol. Obs. Tonantz. y Tacub. 31, 35, 1969.
9.   G. Haro, G. Gonzales, Bol. Obs. Tonantz. y Tacub. 34, 191, 1970.
10. L. V. Mirzoyan, E. S. Parsamyan, N. Kallogyan, Astr. Circ. USSR, No. 485, 1968.
11. I. Jankovich, D. Thesis, 1975.
12. M. K. Tsvetkov, H. S. Chavushian, K. P. Tsvetkova, L. K. Erastova, IBVS No. 909, 938, 1974; 1002, 1975.

13. M. K. Tsvetkov, Astrofizika 11, 579, 1975; D. Thesis, 1975.
14. G. Haro, E. Chavira, IBVS No. 624, 1973.
15. G. Haro, W. W. Morgan, Ap. J. 118, 16, 1953.
16. H. L. Johnson, R. I. Mitchell, Ap. J. 127, 510, 1958.
17. G. H. Herbig, Ap. J. 135, 736, 1962.
18. G. Haro, IAU-URSI Symposium No. 20, Canberra, 1962. Ed. F. J. Kerr and A. W. Rodgers. Canberra: Australian Academy of Science, 1969, p. 83.
19. G. Haro, E. Chavira, ONR Symposium, Flagstaff, Arizona, 1964.
20. R. P. Kraft, J. L. Greenstein, Low Luminosity Stars, Ed. S. S. Kummar. Gordon & Breach Sci. Publ., 1969, p. 65.
21. M. F. McCarthy, Low Luminosity Stars, Ed. S. S. Kummar. Gordon & Breach Sci. Publ., 1969, p. 83.
22. O. C. Wilson, Ap. J. 138, 832, 1963.
23. G. A. Gurzadyan, Bol. Obs. Tonantz. y Tacub. 35, 263, 1970.
24. A. D. Andrews, Bol. Obs. Tonantz. y Tacub. 34, 209, 1970.
25. L. G. Gasparian, Reprint of Garny Space Astronomy Lab. No. 1, 1975.
26. H. S. Chavushian, A. T. Gharibjanian, Astrofizika 11, 567, 1975.
27. G. W. Allen, Astrophysical Quantities, 3rd Ed. The Athlone Press, London, 1973.
28. S. P. Pikelner, Ap. J. Lett. 155, L149, 1969.
29. G. B. Field, D. W. Goldsmith, H. J. Habing, Ap. J. Lett. 155, L149, 1969.
30. T. K. Menon, Ap. J. 127, 28, 1958.
31. F. P. J. Van Den Heuvel, PASP 81, 815, 1969.
32. G. Haro, E. Chavira, IBVS No. 624, 1972.
33. G. Velin, Astr. Astrophys. 12, 312, 1971.
34. P. W. Merrill, C. G. Burwell, Ap. J. 110, 387, 1949; 112, 72, 1950.
35. W. P. Bidelman, Ap. J. Suppl. 1, 175, 1954.
36. G. H. Herbig, Ap. J. 128, 259, 1958.
37. M. Petit, Contr. Asiago Obs. No. 95, 29, 1958.
38. M. Rodono, Variable Stars and Stellar Evolution, IAU Symposium No. 67. Ed. V. I. Sherwood, L. Plaut., 1975, p. 69.
39. G. Haro, E. S. Parsamyan, Bol. Obs. Tonant. y Tacub. 5, 45, 1969.
40. E. Hertzsprung, C. Sanders, C. J. Kooreman et al., Ann. Leiden Obs. 19, No. Ia, 1947.
41. O. Struve, M. Rudkjobing, Ap. J. 109, 92, 1949.
42. G. Haro, Astron. J. 54, 188, 1949.
43. G. Haro, E. Chavira, Obs. Inst. Tonantzintla 1, 189, 1974.

CHAPTER 12

Transition Radiation and Peculiar Objects

1. THE PROBLEM

The large infrared excess and the very high luminosity of the T Tau type stars suggest that they must be surrounded by dense clouds of gas and dust. Probably the source of energy heating the particles, although connected with the central star, is the kinetic energy of the corpuscular stream--the fast electrons appearing in the outer regions of the star--above its atmosphere.

The fact that dust particles and electrically charged high energy particles co-exist makes it quite probable that <u>transition radiation</u> is generated as a result of their electrodynamic interaction. Matter in the form of dust is a constant component of many galactic and extragalactic objects, the nucleus of the Galaxy, the nuclei of Seyfert galaxies, cometary nebulae, Herbig-Haro objects, nebulae of the Barnard 10 type, supernova remnants, diffuse and possibly planetary nebulae, very small dust nebulae, or gas-dust clouds, FUORs, etc. On the other hand, there are direct or indirect data indicating that in all these objects there may exist (at least part of the time) high-energy charged particles as well--cosmic rays, relativistic or fast electrons, etc. Interaction of the high-energy electrons with dust particles may lead to transition radiation in these objects too.

The radiation--in optical and ultraviolet light, in the x-rays and radio ranges, in emission lines, etc.--in the objects mentioned above often has an abnormal character and sometimes it is simply impossible to understand it with the usual notions about thermal and non-thermal processes. Transition radiation may play some role in explaining the abnormal emissions. The importance of this problem is certainly not restricted to the T Tau type stars only, but is of a more general character. Here we shall first discuss the main properties of the transition radiation, and then we shall adapt it to astrophysical problems.

The essence of the transition radiation is as follows: When a charged particle passes from one medium to another, i.e. when it crosses the boundary between media with different dielectric properties, the electromagnetic field made by the particle is deformed or restructured, and as a result part of the field is "torn away" from the particle in the form of radiation.

If empty space and a dust particle are considered as two different media, then transition radiation will appear twice for each time the charged particle--an electron--passes through a dust particle, the first time, when the electron passes from the vacuum into the particle, and the second time when it leaves the particle and returns to empty space. This model has recently been applied to astrophysics [1-8]. Although in each of these interaction processes an extremely small amount of energy is generated, on the other hand the total number of these interactions--encounters of electrons with dust particles--is very large.

One of the remarkable properties of the transition radiation is that it can arise

in the interaction of dust particles not only with a relativistic electron, but
with a non-relativistic electron as well. In the latter case a lower limit to the
energy of the electron is determined mainly by the condition that it penetrates
into the particle to a certain depth. In practice this corresponds with an electron
energy of the order of or higher than a few Kev.

The transition radiation produced by non-relativistic electrons differs in character from the radiation due to relativistic electrons. For instance, in the case
of non-relativistic electrons the transition radiation can have the same direction
as the motion of the electron or the opposite one. But in the case of relativistic
electrons the radiation is nearly entirely in the direction of the motion of the
electron and close to it, as in the case of synchrotron radiation, within the limits
$\Theta \approx mc^2/E$. In the following sections of the present chapter the main points of the
theory of the transition radiation and its applications to astrophysics will be
discussed. It will be shown, in particular, that the transition radiation may have
a certain relation to objects which are rich in dust particles and which contain
fast electrons--such as peculiar nebulae, stars of T Tau type, FUORs, Herbig-Haro
objects. This radiation may be an important source of ionizing radiation and the
means of excitation of emission lines in these unstable objects. Also it is predicted that it may be possible to detect the relatively strong radiation of "transition" origin in the region 1000-3000 Å in T Tau type stars. These same stars may
also be sources of x-ray radiation. It is also found that the concept of transition
radiation can explain some properties of the abnormally low degree of ionization in
Herbig-Haro objects.

The "accumulation effect" may be very important for understanding the nature of
these objects. This is the fact that in certain cases the fast electrons generating
the transition radiation may be contained by magnetic fields ("magnetic trap") in
gas-dust envelopes. This radiation may be sufficiently large to give a more or less
stable emission during long periods of time. This idea of the "magnetic trap" is
especially promising for the FUORs.

2. THE THEORY OF THE TRANSITION RADIATION

The transition radiation of a charged particle was predicted theoretically by V. L.
Ginzburg and I. M. Frank in 1946 [9]. They were the first to treat this phenomenon
quantitatively. However, only a few results are represented in the form of relations
which are practical for astrophysical application [10,11,12]. We are here interested
in the properties of this radiation, in its spectral distribution over the entire
range of wavelengths, and also for all values of the energy of the electron--from
non-relativistic to ultrarelativistic.

The theory gives the following quite general expression for the loss of energy
$J_\omega(\eta)$ in the form of transition radiation, depending on the frequency ω and the
direction Θ, when an electron passes from vacuum into a particle or vice versa:

$$J_\omega(\eta)d\omega d\Omega = \frac{e^2\beta^2}{\pi^2 c}\frac{\eta^2(1-\eta^2)}{(1-\beta^2\eta^2)^2}\left[\frac{(\varepsilon-1)(1-\beta^2)\mp\sqrt{\varepsilon-1+\eta^2})}{(\varepsilon\eta-\sqrt{\varepsilon-1+\eta^2})(1\mp\beta\sqrt{\varepsilon-1+\eta^2})}\right]d\omega d\Omega \qquad (12.1)$$

where $\eta = \cos\Theta$, $d\Omega = 2\pi\sin\Theta d\Theta$, $\beta = v/c$, v is the velocity of the electron, and
$\varepsilon = \varepsilon(\omega)$ is the dielectric constant of the dust particle. The minus sign in (12.1)
corresponds to a direction of the radiation in the same direction as the motion
of the electron, the plus sign in the opposite direction.

In the general case the dielectric constant is a complex quantity and can be represented in the form:

$$\varepsilon(\omega) = \varepsilon_1(\omega) + i\varepsilon_2(\omega) \ , \tag{12.2}$$

where ε_1 and ε_2 are the real and the imaginary parts of the dielectric constant. Their values as a function of frequency are usually determined by direct measurements for a certain kind of matter or element. Here ε_1 as well as ε_2 can have positive or negative signs in some parts of the frequency range of interest.

Usually the curves for the functions $\varepsilon_1(\omega)$ and $\varepsilon_2(\omega)$ found experimentally have a rather complicated form and cannot be represented by a simple dependence on ω. However, for simplicity in certain cases one can start from some model of the dust particles for which these functions can be represented in the form:

$$\varepsilon_1(\omega) = 1 - \frac{\omega_o^2 \tau^2}{1 + \omega^2 \tau^2} \ , \tag{12.3}$$

$$\varepsilon_2(\omega) = \frac{1}{\omega\tau} \frac{\omega_o^2 \tau^2}{1 + \omega^2 \tau^2} \ , \tag{12.4}$$

where here τ is the relaxation time of the electron in the dust particle, ω_o is the plasma frequency, and depends only on the charge e and the mass m_e of the electron and the particle density N:

$$\omega_o = \left(\frac{4\pi N e^2}{m_e}\right)^{\frac{1}{2}} \ . \tag{12.5}$$

Substituting (12.2) in (12.1) and performing the necessary transformations we find the expression for the amount of energy of the transition radiation $J_\omega(\eta)$ as a function of the frequency ω and its direction Θ, and also for arbitrary values of the energy of the electron $\gamma = E/mc^2 = (1 - \beta^2)^{-\frac{1}{2}}$.

In the most general case $J_\omega(\eta)$ can be represented in the following form:

$$J_\omega(\eta) = \frac{e^2 \beta^2}{\pi^2 c} \frac{\eta^2(1 - \eta^2)}{(1 - \beta^2\eta^2)^2} P_\omega(\eta) \ , \tag{12.6}$$

where the function $P_\omega(\eta)$ depends on the signs of ε_1 and ε_2, and also on the condition $\varepsilon_1 \lessgtr 1 - \eta^2$. The following combinations are possible:

1. $\varepsilon_1 < 0$, $\varepsilon_2 > 0$. For most kinds of materials this corresponds to the optical spectral range [i.e. $\tau\omega \ll 1$ in (12.3)]. In this case we have [8]:

$$P_\omega(\eta) = \frac{(\varepsilon_1 - 1)^2 + \varepsilon_2^2}{1 \mp 2\beta\Phi\sin\phi + \beta^2\Phi^2} \frac{(1 - \beta^2)^2 \mp 2\beta(1 - \beta^2)\Phi\sin\phi + \beta^2\Phi^2}{\eta^2(\varepsilon_1^2 + \varepsilon_2^2) + \Phi^2 + 2\eta\Phi(\varepsilon_1\sin\phi + \varepsilon_2\cos\phi)} \ . \tag{12.7}$$

This expression is valid for all values of $(1 - \eta^2)$.

2. $\varepsilon_1 > 0$, $\varepsilon_2 > 0$. This corresponds to ultraviolet wavelength, and x-rays [$\omega\tau \gg 1$ in (12.3)]. In this case the problem has two solutions, depending on the value of $(1 - \eta^2)$ as compared with ε_1.

In the case where $\varepsilon_1 > 1 - \eta^2$ we have [8]:

$$P_\omega'(\eta) = \frac{(\varepsilon_1 - 1)^2 + \varepsilon_2^2}{1 \mp 2\beta\Phi\cos\phi + \beta^2\Phi^2} \frac{(1 - \beta^2)^2 \mp 2\beta(1 - \beta^2)\Phi\cos\phi + \beta^2\Phi^2}{\eta^2(\varepsilon_1^2 + \varepsilon_2^2) + \Phi^2 + 2\eta\Phi(\varepsilon_1\cos\phi - \varepsilon_2\sin\phi)} \ . \tag{12.8}$$

In the case where $\varepsilon_1 < 1 - \eta^2$ we have:

$$P''_\omega(\eta) = \frac{(\varepsilon_1 - 1)^2 + \varepsilon_2^2}{1 \mp 2\beta\Phi\sin\phi + \beta^2\Phi^2} \frac{(1 - \beta^2)^2 \mp 2\beta(1 - \beta^2)\Phi\sin\phi + \beta^2\Phi^2}{\eta^2(\varepsilon_1^2 + \varepsilon_2^2) + \Phi^2 + 2\eta\Phi(\varepsilon_2\sin\phi - \varepsilon_1\cos\phi)} \; . \tag{12.9}$$

3. $\varepsilon_1 < 0$, $\varepsilon_2 < 0$. These conditions usually correspond to the metals. In this case we have [12]:

$$P_\omega(\eta) = \frac{(\varepsilon_1 - 1)^2 + \varepsilon_2^2}{1 \mp 2\beta\Phi\sin\phi + \beta^2\Phi^2} \frac{(1 - \beta^2)^2 \mp 2\beta(1 - \beta^2)\Phi\sin\phi + \beta^2\Phi^2}{\eta^2(\varepsilon_1^2 + \varepsilon_2^2) + \Phi^2 + 2\eta\Phi(\varepsilon_2\cos\phi - \varepsilon_1\sin\phi)} \; . \tag{12.10}$$

This expression is analogous to (12.7) and holds for all values of $(1 - \eta^2)$.

4. $\varepsilon_1 > 0$, $\varepsilon_2 < 0$. As in case 2, the problem has again two solutions depending on the value of $(1 - \eta^2)$.

In the case where $\varepsilon_1 > (1 - \eta^2)$ we have:

$$P'_\omega(\eta) = \frac{(\varepsilon_1 - 1)^2 + \varepsilon_2^2}{1 \mp 2\beta\Phi\cos\phi + \beta^2\Phi^2} \frac{(1 - \beta^2)^2 \mp 2\beta(1 - \beta^2)\Phi\cos\phi + \beta^2\Phi^2}{\eta^2(\varepsilon_1^2 + \varepsilon_2^2) + \Phi^2 + 2\eta\Phi(\varepsilon_1\cos\phi + \varepsilon_2\sin\phi)} \; . \tag{12.11}$$

And in the case $\varepsilon_1 < (1 - \eta^2)$ the function $P''_\omega(\eta)$ is the same as $P_\omega(\eta)$ given by (12.7).

In all of these formulae:

$$\Phi = [(\varepsilon_1 - 1 + \eta^2)^2 + \varepsilon_2^2]^{\frac{1}{4}} \; , \tag{12.12}$$

$$\tan 2\phi = \left| \frac{\varepsilon}{\varepsilon_1 - 1 + \eta^2} \right| \; . \tag{12.13}$$

For $\varepsilon_1 = 1 - \eta^2$ we have $\varepsilon_2^{\frac{1}{2}} = \Phi$ and $\phi = \pi/4$; and the formulae (12.8) and (12.9) on the one hand, and (12.11) and (12.7) on the other hand are equal.

In the case of fast electrons, i.e. for $\gamma \sim 3$, the radiation has practically the same direction as the motion of the electron and hence, for finding the spectral distribution of the transition radiation, one needs to integrate only over the "outer" hemisphere, i.e.

$$J_\omega d\omega = d\omega \int_{(2\pi)} J_\omega(\eta) d\Omega = 2\pi d\omega \int_0^1 J_\omega(\eta) d\eta \tag{12.14}$$

or, considering the relation (12.6), we find (per unit frequency interval)

$$J_\omega = \frac{2e^2}{\pi c} \beta^2 L_\omega(\beta) \; , \tag{12.15}$$

where

$$L_\omega(\beta) = \int_0^1 \frac{\eta^2(1-\eta^2)}{(1-\eta^2\beta^2)^2} \, P_\omega(\eta)d\eta \tag{12.16}$$

in the case $\varepsilon_1 < 0$ and

$$L_\omega(\beta) = \int_{\sqrt{1-\varepsilon_1^2}}^1 \frac{\eta^2(1-\eta^2)}{(1-\eta^2\beta^2)^2} \, P'_\omega(\eta)d\eta + \int_0^{\sqrt{1-\varepsilon_1^2}} \frac{\eta^2(1-\eta^2)}{(1-\eta^2\beta^2)^2} \, P''_\omega(\eta)d\eta \tag{12.17}$$

in the case $\varepsilon_1 > 0$, and the functions P'_ω and P''_ω are taken from (12.8) and (12.9) or (12.11) and (12.7), depending on the sign of ε_2.

It should be noted that all these formulae are valid for any form and any value of the functions $\varepsilon_1(\omega)$ and $\varepsilon_2(\omega)$, independent of whether these functions have been found experimentally for a certain material, or whether they are represented by a relation of the type (12.3) and (12.4). Finally, integrating (12.15) over all frequencies, we find for the total amount of transition radiation in one interaction:

$$J = \frac{2e^2}{\pi c} \beta^2 \int_0^\infty L_\omega(\beta)d\omega \quad . \tag{12.18}$$

In the special case that the electron is relativistic ($\beta \to 1$) and $\omega \gg \omega_o$, which usually corresponds with the region of hard ultraviolet and x-rays, we have: $\varepsilon_1 = 1 - (\omega_o/\omega)^2$ and $\varepsilon_2 \ll \varepsilon_1$. Then we find from (12.15) and (12.18) [10]:

$$J_\omega = \frac{e^2}{\pi c} \left[\left(1 + 2 \, \frac{\omega^2}{\omega_o^2\gamma^2} \right) \text{Ln} \left[1 + \frac{\omega_o^2\gamma^2}{\omega^2} \right] - 2 \right] \quad , \tag{12.19}$$

$$J = \frac{2e^2}{\pi c} \omega_o\gamma \; . \tag{12.20}$$

In essence formula (12.20) is the total loss of energy by the electron in the form of transition radiation; it is found to be proportional to γ, i.e. the first power of the electron energy.

In comparison, for the inverse Compton effect the energy loss of the electron (relativistic only) in one encounter with a photon is proportional to γ^2. The loss in the case of synchrotron radiation is also proportional to γ^2 (on the average). Hence transition radiation seems a "longer lasting" mechanism than the inverse Compton effect.

The absolute amount of energy liberated by the electron in the form of transition radiation when it crosses the boundary vacuum-particle or vice versa, it is very small and equal to (for $E \sim 1.5$ MeV, $\omega_o \sim 10^{16}$ s^{-1}): $J \approx 10^{-7} \, E \approx 0.15$ eV = 2.4×10^{-13} ergs.

3. PROPERTIES OF THE TRANSITION RADIATION

The main properties of the transition radiation can be discovered by analyzing the formulae given above. For relatively small energies of the electron the resulting transition radiation will be directed not only in the same direction as the motion

of the electron, but also in the opposite direction. With increasing energy of the
electron the radiation itself gets progressively a more directed character. Because
of its directed character the transition radiation must be polarized, this has been
confirmed in numerous experiments. The degree of polarization is more than 10%,
and it increases rapidly with increasing electron energy, reaching up to 90% and
sometimes nearly 100% in the region above 50 keV. However, in a homogeneous medium
consisting of randomly oriented dust particles (non-spherical in form) and with no
preferential directions for the motions of the fast electrons, the total transition
radiation from the entire volume of the medium then may not be polarized.

From an astrophysical point of view the spectral distribution of the transition
radiation in the optical and near ultra-
violet regions is especially interesting.
Figure 12.1 gives curves of the spectral
distribution of the transition radiation
in the optical range (3000-6400 Å) constructed by means of formulae (12.7),
(12.15) and (12.16) for values of the
energy of the fast electrons from 0.1 MeV
to 10 MeV. In the computations ω_o was
taken to be 10^{16} s^{-1} and $\tau = 10^{-15}$s, (the
relaxation time of the electron) which
corresponds to the case $\omega_o^2\tau^2/(1 + \omega^2\tau^2) > 1$,
and hence $\varepsilon_1 < 0$ and $\varepsilon_2 > 0$. In the wavelength region considered the radiating
power of the transition radiation increases
towards the short wavelengths. This means
that in the optical region the transition
radiation must have a blue color (in these
and the following diagrams the radiative
power of the transition radiation is given
in the wavelength scale and in arbitrary
units).

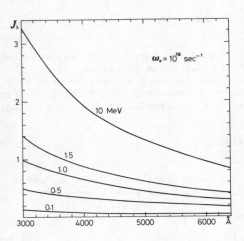

Fig. 12.1. Curves of spectral distribution of transition radiation in the
optical region for electron energies
from E = 0.1 MeV to E = 10 MeV
($\omega_o = 10^{16}$ s^{-1} and $\tau = 10^{-15}$ s).

How does the transition radiation behave
in the ultraviolet wavelength region (λ
shorter than 3000 Å)? What is the influence
of the plasma frequency on the character
of the radiative power at various wavelengths?

Figure 12.2 gives an answer to these two questions; the curves in this figure have
been computed for three values of ω_o: 0.5, 1 and 2 (in units of 10^{16} s^{-1}), and for
an electron energy 1.5 MeV. The jumps in the curves correspond with the wave-
lengths where ε_1 changes its sign, i.e. where ε_1 (ω_{cr}) = 0. The long wavelength
parts of the curves, lying at the right of the jumps ($\varepsilon_1 < 0$) are constructed by
means of formulae (12.15), (12.16) and (12.7), and the short-wavelength parts
($\varepsilon_1 > 0$), by means of (12.15), (12.17), (12.8) and (12.9).

Thus the spectral distribution of the transition radiation in a certain wavelength
region can change strongly depending on the dielectric properties of the material.

The possibility of jumps and sharp maxima occurring in the spectral distribution
of the transition radiation is confirmed by the results of computations made for
certain materials, which are thought to most probably occur in the interstellar
medium. Some of these results are shown in Figs. 12.3, 12.4 and 12.5; they refer
to solid carbon, graphite and silicon oxide (SiO). In these computations the
derived laboratory results for the dependence of ε_1 and ε_2 on ω were used (for

details, see [8]).

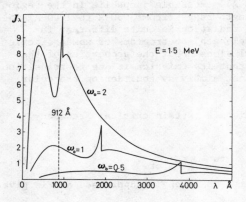

Fig. 12.2. Curves of spectral distribution of transition radiation for electron energies E = 1.5 MeV, $\tau = 10^{-15}$ s and three values of the plasma frequency, ω_0: 0.5, 1 and 2 (in units 10^{16} s^{-1}). The curves have been computed: up to the jump (long wavelength side) with formulae (12.15) and (12.16), after the jump with formulae (12.15) and (12.17).

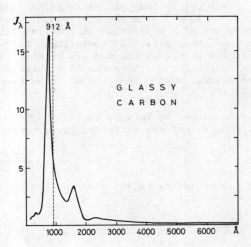

Fig. 12.3. Spectral distribution of the transition radiation of solid carbon for interaction with electrons with energy E = 1.5 MeV.

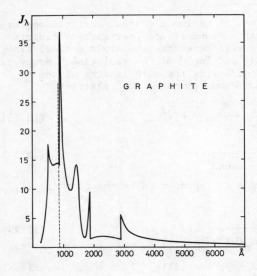

Fig. 12.4. Spectral distribution of transition radiation of graphite, for E = 1.5 MeV.

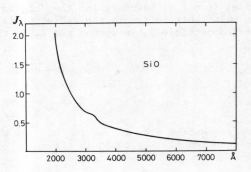

Fig. 12.5. Spectral distribution of transition radiation of SiO, for E = 1.5 MeV.

The spectral distributions of the transition radiation in these cases show no jumps, at least not in the optical range ($\lambda > 3000$ Å). Possible jumps lie in the region with $\lambda < 3000$ Å. In the ultraviolet wavelength range (from 3000 Å to 1000 Å) the structure of the spectrum of the transition radiation is quite different for various substances. By comparing short-wavelength spectrograms of some galactic object with theory one may try to draw conclusions about the properties of the dust particles on the assumption, however, that its emission is due to transition radiation. Such a comparison can be made most simply by considering the color indices.

Let us return to formula (12.19). There exists a certain critical frequency, $\omega_{cr} = \omega_o \gamma$, and the radiative power becomes:

$$J_\omega = \frac{2e^2}{\pi c} \left[\ell n \frac{\omega_{cr}}{\omega} - 1 \right] \quad \text{for } \omega \ll \omega_{cr} \, . \tag{12.21}$$

In the frequency region $\omega \gg \omega_{cr}$ the radiation is strongly suppressed and is given by the formula

$$J_\omega = \frac{e^2}{6\pi c} \left(\frac{\omega_{cr}}{\omega} \right)^4 \, . \tag{12.22}$$

From these relations it follows that the short-wavelength limit of the radiation lies at frequency $\omega \sim \omega_{cr}$, while there is a shift of this region towards the very hard photons proportional to the energy of the electrons. On the wavelength scale the transition radiation has its maximum at about $\lambda' = \lambda \exp(1 + \lambda_{cr}/2)$, where $\lambda_{cr} = 2\pi c/\omega_{cr}$. These considerations are valid as long as $\varepsilon_2 \ll \varepsilon_1$, and ε_1 is represented in the form (12.3).

4. THE EFFECT OF THE "FORMATION ZONE"

During the passage from empty space to a material medium a charged particle begins to radiate long before it crosses the geometric boundary and continues to radiate for some distance after crossing this boundary. There are no sharply defined distances from this boundary for the "beginning" and "end" of the radiation. However, the main contribution to the radiation is made during the path lengths of the charged particle Z_v in vacuum and Z_m in the medium, given by the relations:

$$Z_v = \frac{c}{\omega} \frac{1}{1 - \beta\cos\theta} \approx \frac{c}{\omega} \frac{1}{1 - \beta^2} = \frac{c}{\omega} \gamma^2 \, , \tag{12.23}$$

$$Z_m = \frac{c}{\omega} \frac{1}{1 - \beta\sqrt{\varepsilon - \sin^2\theta}} \, . \tag{12.24}$$

For frequencies $\omega \gg \omega_o$ and in the case when ε is represented in the form (12.3), the equation (12.24) takes the form

$$Z_m = \frac{c}{\omega} \frac{\gamma^2}{1 + \frac{1}{2} (\omega_o/\omega)^2 \gamma^2} \, . \tag{12.25}$$

The validity of this formula has been proved experimentally. From these equations it follows that the transition radiation will be emitted only in the case where the dimension d (diameter) of the dust particle is equal to or is larger than the dimensions of the "formation zone", i.e. when

$$d \overset{\sim}{>} Z \, . \tag{12.26}$$

All relations given in the preceding sections hold if this condition is satisfied.

If this condition is not fulfilled, the process of generation of the transition radiation will not be efficient and this will lead to a decrease of the total amount of energy produced. This is the essence of the effect of the "formation zone". The "formation zone" of this radiation can also be defined in a different way, namely as the minimum distance necessary for the electromagnetic field of the electron to reach its new equilibrium configuration--when the electron passes from the medium into the vacuum; or the distance where the equilibrium configuration breaks down when the electron passes from vacuum into the medium. The smaller the dimensions of this zone, the sooner the equilibrium configuration will be reached, and the smaller will be the amount of energy radiated. The transition radiation is a collective effect of the electrons in the material medium; therefore the radiative power tends to become zero only when the dimensions of the radiating particles approach zero.

However, it is difficult to compute the influence of the "formation zone" on the radiative power of the medium (such an attempt has been made [4]). In estimating the role of the transition radiation in astrophysics it is sufficient to determine the wavelength limit (λ_{max}) beyond which no transition radiation can be generated for the given dimensions of the particle and the given energy of the electrons.

In the case of the production of transition radiation in vacuum we have from (12.23) and (12.26):

$$\lambda_{max} = 2\pi \frac{d}{\gamma^2} . \tag{12.27}$$

TABLE 12.1 Maximum Wavelength λ_{max} of Transition Radiation Generated by Electrons with Energy E and for Particle Size d

E (MeV)	λ_{max} (Å)	
	$d = 10^{-5}$ cm	$d = 10^{-4}$ cm
0.5	6300	63000
1.5	700	7000
5	63	630
50	0.6	6.3
500	0.006	0.06

In Table 12.1 the values of λ_{max} found from this equation are given for a number of values of the electron energy E and particle size d. From these data the following conclusions can be made.

1. For particle sizes which ordinarily occur in the interstellar medium and in peculiar objects (d $\sim 10^{-5}$-10^{-4} cm) the transition radiation in the optical region --λ shorter than 6000 Å--can be generated only by electrons with energy of the order 1-2 MeV or less.

2. X-ray radiation in the range of astrophysical interest (1-10 Å) can be induced by the transition mechanism only by electrons with energies from 10-100 MeV.

3. Ultrarelativistic electrons with energies of the order of one GeV or more can generate transition radiation only in the region of gamma rays ($\lambda < 0.01$ Å).

4. In principle infrared radiation (1-10 mμ) can be induced by the transition

mechanism, too, but then the energy of the electrons must be of the order of 0.5 MeV or less.

These conclusions have been drawn from Eq. (12.27), which is valid for the generation of transition radiation in vacuum. An analogous relation can be derived also for the case of generation of radiation in a medium, i.e. using Eq. (12.24). The analysis shows that Z_m is not smaller than Z_v, and therefore the above conclusions hold for this case as well.

The transition radiation in the region $\lambda \gtrsim 1000$ Å, produced at the inner boundary of the dust particles, will be absorbed within the particle itself and cannot escape. This effect should be taken into account during a quantitative treatment of the theory.

5. THE RADIATIVE POWER OF THE MEDIUM

Let us start with the simplest case, where the electron-dust medium consists of electrons which all have the same energy, with a density n_e, and equal dust particles of spherical form, with diameter d, and a density n_p.

The number of passages of a fast electron through a dust particle during one second will be $\sigma_p n_p v \approx \sigma_p n_p c$, where $v \approx c$ is the velocity of the fast electron, σ_p is the geometrical effective cross section for the encounter of the electron with a dust particle, and to a first approximation we may take $\sigma_p \approx d^2$. Then we can write for the amount of electromagnetic energy produced in the form of transition radiation per unit frequency interval, per unit volume and per second (the volume coefficient of the radiation):

$$\varepsilon_\omega = J_\omega d^2 n_p n_e c \quad \text{ergs/cm}^3\text{s} \tag{12.28}$$

where J_ω is given by (12.15) or (12.19). Also, for the volume radiation coefficient integrated over all frequencies we have

$$\varepsilon = J d^2 n_p n_e c \quad \text{ergs/cm}^3\text{s} \tag{12.29}$$

where J is given by (12.20). These equations are suitable for making general quantitative estimates which can be applied to various objects. However, we must have information on the values of n_p, n_e and d.

One should start from the size distribution of the dust particles, from energy spectrum of the electrons, and also taking into account the effects of the "formation zone" [8].

6. COLOR INDICES FOR THE TRANSITION RADIATION

In Table 12.2 theoretical values are given of the color indices for the transition radiation in the UBV system, computed for three values of the energy of the fast electrons, namely 0.5, 1.5 and 10 MeV, and $\omega_o = 10^{16}$ s^{-1}, and for an optically thin dust medium ($\tau_\lambda \ll 1$) as well as for a thick one ($\tau_\lambda \gg 1$); in the latter case the law of selective absorption has been taken as $\sim \lambda^{-1}$. The values of the color indices for electron energies more than one MeV are practically insensitive to the value of their energy.

We note that the theoretical values of U − B and B − V given in Table 12.2 for the transition radiation are close to those given by the observations for peculiar

TABLE 12.2 Theoretical Color Indices of the Transition Radiation for an Optically Thin ($\tau_\lambda \ll 1$) and an Optically Thick ($\tau_\lambda \gg 1$) Medium

E (MeV)	$\tau_\lambda \ll 1$		$\tau_\lambda \gg 1$	
	U – B	B – V	U – B	B – V
0.5	$-1^m_.16$	$+0^m_.13$	$-0^m_.93$	$+0^m_.43$
1.5	−1.24	+0.13	−0.99	+0.39
10	−1.30	+0.12	−1.04	+0.36

nebulae, which have color indices very different from those for ordinary reflection nebulae [8]. For comparison Table 12.3 gives a list of color indices of radiation generated by four processes of non-thermal nature--transition (for four types of substances), synchrotron, bremsstrahlung (non-thermal electrons), and inverse Compton effect.

TABLE 12.3 Theoretical Color Indices of Radiation Generated by Different Non-Thermal Processes

Type of radiation	U – B	B – V
Transition radiation (for $\gamma = 3$):		
Silicon oxide (SiO)	$-1^m_.30$	$+0^m_.08$
Solid carbon	−1.37	+0.06
Graphite	−1.53	+0.08
Particles with $\omega_0 = 10^{16}$ s^{-1}, $\tau = 10^{-15}$ s	−1.24	+0.13
Synchrotron radiation:		
Monoenergetic electrons ($\gamma = 3$)	−0.90	+0.04
Electrons with $\sim E^{-\alpha}$ ($\alpha = 3$)	−0.60	+0.40
Bremsstrahlung of non-thermal electrons (E > 1 MeV)	−1.33	+0.04
Inverse Compton effect ($\gamma = 3$)	−1.80	−0.38

7. TRANSITION RADIATION IN T TAU TYPE STARS

Stars of T Tau type may be potential generators of transition radiation. In order to obtain some general estimate of the transition radiation for these stars we compute the transition radiation integrated over the entire volume of the gas-dust cloud (envelope) around the star when fast electrons pass through this cloud [6].

Let $V = (4\pi/3)R^3$ be the volume of the dust cloud, whose radius R is much larger than the radius R_* of the star. Then the total amount of energy produced by the cloud in the form of transition radiation induced by monoenergetic fast electrons will be:

$$\varepsilon_t = \varepsilon V = Jd^2 cn_e N_p \quad \text{ergs/s} \tag{12.30}$$

where ε is the volume radiation coefficient, taken from (12.29), and $N_p = n_p V$ is the total number of dust particles in the cloud. With the assumption that the radiation of the dust particles follows Planck's law for their effective temperature T_{eff}, we can write for the total luminosity of the entire cloud:

$$L = \kappa L_{\odot} = \pi d^2 \sigma T_{eff}^4 N_p \quad \text{ergs/s} .$$ (12.31)

Substituting from this the value of $d^2 N_p$ into (12.30), we find

$$\varepsilon_t = \frac{2e^2}{\pi^2 \sigma} \omega_o \gamma \frac{n_e}{T_{eff}^4} \kappa L_{\odot} \quad \text{ergs/s}$$ (12.32)

where the value of J from (12.20) has been substituted as well. We note that the equation does not contain the effect of self-absorption of the radiation in the cloud. Neither does the particle size d enter in it; however, it is replaced by the "specific luminosity" κ, the value of which can be found directly from observations. For some stars κ has the following values [13]:

Star	$\kappa = L/L_{\odot}$
RW Aur	5
SU Aur	20
RY Tau	23
T Tau	40
GW Ori	50
V 380 Ori	270
FU Ori	685
R Mon	830

Clearly the observed values of κ have wide limits, from 5 to 800. Further model computations of a T Tau type star will be made with $\kappa = 50$.

The observed characteristics of T Tau type stars can be represented directly by means of τ--the optical thickness of processes of Thomson scattering of the cloud consisting of dust and fast electrons. We have $n_e = \tau/\sigma_s \Delta R$, where ΔR is the linear extension of the cloud (the medium of fast electrons). Then (12.32) can be written in the form:

$$\varepsilon_t = \frac{2e^2}{\pi^2 \sigma \sigma_s} \frac{\kappa \tau}{\Delta R T_{eff}^4} \omega_o \gamma L_o .$$ (12.33)

The curves of the spectral distribution of the radiation for T Tau type stars show maxima in the range 1–5 mμ, corresponding to an effective temperature of the order $T_{eff} = 1000$ K. We know little about the value of ΔR; from analogy with flare stars ΔR can be one to two orders larger than the radius of the star itself. Tentatively we take $\Delta R \approx 10^{12}$–10^{13} cm. Finally, for most of the T Tau type stars $\tau \sim 0.001$. With these data we find (for $\omega_o = 10^{16}$ s^{-1} and $\gamma = 3$):

$$\varepsilon_t = 2 \quad L_{\odot} = 0.04 \quad L \quad \text{for } \Delta R = 10^{12} \text{ cm}$$
$$\varepsilon_t = 0.2 \quad L_{\odot} = 0.004 \quad L \quad \text{for } \Delta R = 10^{13} \text{ cm}$$

Thus the total energy produced by the dust envelope around the star in the form of transition radiation is of the order of one percent of the total energy it emits in the form of thermal radiation. Hence, the transition radiation does not play a decisive role.

However, the situation is different when we consider the spectral dependence of the transition radiation. Instead of (12.33) then we have for the total energy radiated by the cloud in the form of transition radiation at wavelength λ, per unit wavelength interval:

$$\varepsilon_t(\lambda) = \frac{2e^2 c}{\pi^2 \sigma \sigma_s} \frac{\kappa \tau}{\Delta R T_{eff}^4} L_\odot Q_\lambda(\gamma) , \qquad (12.34)$$

where

$$Q_\lambda(\gamma) = \lambda^{-2} \left[\left(1 + 2 \frac{\lambda_{cr}^2}{\lambda^2} \right) \ln \left(1 + \frac{\lambda^2}{\lambda_{cr}^2} \right) - 2 \right] , \qquad (12.35)$$

$$\lambda_{cr} = 2\pi \frac{c}{\omega_o} \frac{1}{\gamma} . \qquad (12.36)$$

The spectrum of the transition radiation found by means of this formula is shown in Fig. 12.6. For comparison the Planck distribution for $T_o = 4000$ K ($B_\lambda = 4\pi R_*^2 B_\lambda (T_o)$, where R_* is the radius of the star), and the energy distribution due to inverse Compton effects are shown. Figure 12.6 shows that the transition radiation in T Tau type stars is concentrated in the short wavelength spectral region, with λ shorter than 3000 Å. The ratio of the transition radiation to the Planck radiation with wavelength has the following form:

Wavelength, Å	1000	1500	2000	2500	3000	3500	6000
$\dfrac{E_\lambda \text{ (transition)}}{B_\lambda \text{ (Planck)}}$	10^9	10^4	10^2	10	2.5	0.8	<0.1

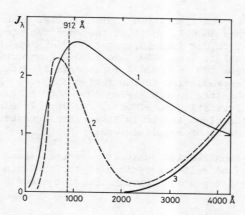

Fig. 12.6. Curves of relative spectral distribution of energy for various mechanisms of radiation (model T Tau type stars):
1. "Transition radiation" ($\tau = 0.001$, $\gamma = 3$, $T_o = 1000$ K, $\Delta R = 10^{13}$ cm, $\kappa = 50$).
2. "Compton radiation" (inverse Compton effect, $\tau = 0.001$, $\gamma = 3$, $T_o = 4000$ K, $R_* = 1\ R_\odot$).
3. "Planck radiation" ($T_o = 4000$ K, $R_* = 1\ R_\odot$).

In the visual part of the spectrum the fraction of the transition radiation is about 1% and in the infrared region it is totally negligible. It should be stressed that the relative fraction of the transition radiation will be the larger, the larger the integral luminosity of the star. This means that the transition radiation must be quite strong in the stars V 380 Ori, FU Ori, and in particular R Mon; and about an order weaker in SU Aur, RY Tau, and should be practically absent in RW Aur. A comparison of the transition radiation with the "Compton" radiation is interesting. First, in both cases the intensity of the radiation is proportional to n_e, and their ratio will not depend on the density of the fast electrons but on the value of the infrared excess. Hence for a given degree of instability of a star, characterized by the total amount of ejected fast electrons, the transition radiation will predominate for stars with large infrared excess. For stars with small infrared excess their spectrum at short-wavelengths will be primarily determined by the "Compton" radiation. However, even for $\kappa = 50$, corresponding to a rather large value of the infrared excess, transition radiation will be important mainly in the near ultraviolet, from 3000 Å to about 1500 Å; in the region $\lambda \leq 1000$ Å the two types of radiation are of the same order.

In Chapter 10 a particular group of T Tau type stars was considered, the stars of NX Mon type, known for their exceptionally strong emission in the ultraviolet part of their spectra, shorter than 3800 Å. The question arises whether this emission is somehow related to the transition radiation. If the source of the ultraviolet emission lies between the photosphere and the dust cloud, this emission will be strongly suppressed by absorption in the cloud itself. However, when it is observed, it means that either the absorption of ultraviolet photons in the cloud itself is small, or that the observed ultraviolet emission comes from the outermost parts of the cloud. In the latter case there is a good probability that this emission may be due to the transition mechanism. To answer this question, it is necessary to make complex observations of NX Mon type stars in the far infrared as well as in the near ultraviolet region.

Given that the main part of the transition radiation lies in the region of the spectrum with λ shorter than 3000 Å, and that the spectrum itself in the ultraviolet has a rather interesting form, it is important to perform special observations of T Tau stars in the ultraviolet, from 3000 Å to 1000 Å from outside the Earth's atmosphere. The region from 2000 to 1000 Å is especially interesting. The expected intensity of the transition radiation in the region 3000-2000 Å is of the order of the Planck radiation of the star in the region 4000-5000 Å, i.e. it is within the limits of detectability of existing receivers. However, if the absorption of the transition radiation in the cloud itself is taken into account, the expected flux may become more than one order of magnitude weaker.

The main weakness of the quantitative analysis performed above has been to neglect the effect of absorption (and self-absorption) of the radiation--transition, "Compton", Planck--during its passage through the dust cloud.

8. X-RAY RADIATION IN T TAU STARS

One of the interesting properties of the transition radiation is that in certain cases it leads to the emission of x-ray photons. Moreover, attempts have been made [1,2] to explain the observed x-ray background of the Galaxy, the x-ray radiation of the Crab Nebula, of the supernova remnants, and the Seyfert galaxies by the "transition" mechanism. However, more careful considerations [3,4,5] led to the conclusion that the transition radiation cannot be an important source of generation of x-ray radiation in these objects. It is interesting to ask whether x-ray radiation is important in T Tau stars.

Let us recall that the x-ray region is very far from the maximum of the transition radiation (Fig. 12.6). The radiation intensity falls quickly, following the law:

$$J_\lambda \sim (\lambda/\lambda_{cr})^2 \,, \tag{12.37}$$

where $\lambda_{cr} = (c/2\pi\omega_0)\gamma = 600\text{A}^\circ$ and $\lambda/\lambda_{cr} \ll 1$. For the ratio of the total radiation in the x-ray region ($\lambda_0 \leq 60$ Å) and the total radiation of transition origin we have, for $\gamma = 3$:

$$\frac{E_x}{\varepsilon_t} = \frac{3c}{2\omega_0\gamma} \int_0^{\lambda_0} Q_\lambda(\gamma)d\lambda \approx 0.01 \,. \tag{12.38}$$

Earlier we found that $\varepsilon_t \sim L_\odot$. Thus the total amount of energy radiated by a T Tau type star in the x-ray region will be $\sim 10^{31}$ ergs/s or 10^{40} photons/s ($\overline{\lambda} \approx 40$ Å); this value is four orders of magnitude larger than for the Sun. If the star is at a distance of 100 parsecs from the Sun, the intensity of the x-ray photons on the Earth will be ~ 0.01 photon/cm^{2}s. T Tau type stars must be sources of x-ray radiation also because of the contribution of the non-thermal bremsstrahlung of the fastest electrons.

Comparing the main characteristics of flares in cool stars with the color characteristics of the T Tau type stars we can give an estimate of the intensity of the expected x-ray radiation; it is found that it is also of the order 0.01 photons/cm^{2}s. This estimated value is within the limits of sensitivity of present-day orbiting x-ray telescopes.

Thus T Tau type stars may be potential sources of cosmic x-ray radiation. The contributions of the two mechanisms for generating x-rays--the "transition" and the "bremsstrahlung" mechanisms--are about equal.

9. EXCITATION OF THE CHROMOSPHERE OF T TAU TYPE STARS

In all the cases considered above--particles of graphite, solid carbon, silicon oxide with $\omega_0 = 10^{16}$ s^{-1} and $\tau = 10^{-15}$ s--the maximum of the spectrum for the transition radiation is in the region with $\lambda < 1000$ Å (Figs. 12.2 to 12.5). This means that besides the L_c radiation of "Compton" origin, the L_c radiation of "transition" origin can serve as an important source of excitation of emission lines of hydrogen, oxygen, possibly helium and other elements and ions in the chromosphere of T Tau type stars. In this process the L_c radiation of "transition" origin falls on the chromosphere from outside as well, i.e. from the gas-dust cloud surrounding the star, through which the fast electrons pass.

According to our model, fast electrons with energies ~ 1.5 MeV cannot reach the chromosphere from the outside because the magnetic field of the outer regions of the star turns them back. In other words, the chromosphere is not ionized by collisions of atoms with fast electrons, and we have pure photoionization (although the ionization energy is really taken from the kinetic energy of the fast electrons, and comes in the form of "transition" or "Compton" radiation).

For an electron energy of the order 1.5 MeV the spectrum of the transition radiation extends up to frequencies capable of ionizing hydrogen, oxygen, calcium, sulphur, and also helium, i.e. the elements of which emission lines are observed in the spectra of the T Tau stars.

However, the energy of the fast electrons should not be very large either; otherwise, emission lines of many times ionized atoms should appear. But these are not observed. Judging from all the data one may conclude that there must be practically no ionizing radiation in the wavelength region shorter than 600-500 Å in the chromospheres of T Tau type stars.

Thus the fact that emission lines of hydrogen, helium, etc. occur in the spectra of
T Tau type stars provides the possibility of finding a lower limit of the energy
of the fast electrons, and the fact that no lines belonging to ions with high ionization potential occur puts an upper limit to their energy. From these conditions the
probable value of the energy of the fast electrons can be found. For that purpose
a number of curves have been drawn, in Fig. 12.7, corresponding to the spectral
distribution of the transition radiation for various values of the electron energy
--from 0.1 MeV to 20 MeV. We see that the required value of the energy of the fast
electrons must be considerably larger than 100 KeV, but lower than 10 or even 5 MeV.
The probable value is of the order of 1.5 MeV ($\gamma \sim 3$). The same value for the energy
of the fast electrons holds for the case of flare stars as well.

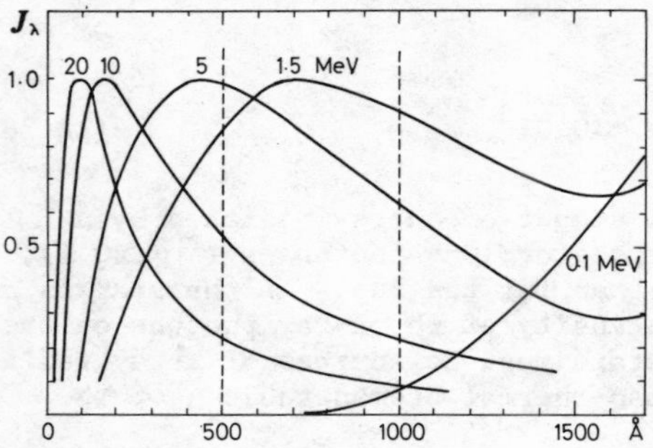

Fig. 12.7. Curves of relative spectral
distribution of transition radiation
for electron energies 0.1 to 20 MeV
and in the spectral region with λ
shorter than 1500 Å ($\omega_o = 10^{16}$ s^{-1},
$\tau = 10^{-15}$ s). In all these cases
the radiation intensity at maximum
has been taken equal to unity.

The question which is the primary mechanism
for the excitation of the chromosphere--
radiation of "transition" origin or non-
thermal bremsstrahlung--also depends on the
density of the dust particles n_p and the
density of the electrons n_e in the medium.

For the ratio of the volume emission
coefficients $\varepsilon^t(\lambda)/\varepsilon^{br}(\lambda)$ we have:

$$\frac{\varepsilon^t(\lambda)}{\varepsilon^{br}(\lambda)} = 0.015 \frac{Q_\lambda}{[\omega f(\omega,\gamma)]} \frac{d^2}{h\nu} \frac{n_p}{n_e} , \tag{12.39}$$

where the functions Q_λ and $[\omega f(\omega,\gamma)]$ have been taken from (12.35) and Table 14.1,
and n_e is the density of the electrons, thermal electrons or fast electrons, which-
ever is largest.

For the spectral region of hydrogen ionization ($\lambda \sim 1000$ Å) we have from (12.39)

$$\frac{\varepsilon^t}{\varepsilon^{br}} \approx 10^{23} \frac{d^2 n_p}{n_e} , \tag{12.40}$$

or, taking for the diameter of the dust particles $d \sim 10^{-5}$ cm,

$$\frac{\varepsilon^t}{\varepsilon^{br}} \approx 10^{13} \frac{n_p}{n_e} . \tag{12.41}$$

It should be noted that in the spectral interval 1000-5000 Å this result is almost
independent of wavelength.

The condition for the ionizing radiation of "transition" origin to predominate over
non-thermal bremsstrahlung is:

$$\frac{n_p}{n_e} > 10^{-13} . \tag{12.42}$$

For applications of this condition in individual cases--gas and dust envelopes of
T Tau type stars, Herbig-Haro objects, peculiar nebulae, the interstellar medium,
etc.--the values of these parameters must be known.

10. THE EFFECT OF ACCUMULATION OF FAST ELECTRONS

When the problem of the interaction of fast electrons with a cloud surrounding a
star is considered, the possibility that a fast electron is not destroyed by the
cloud should not be completely rejected. Let us assume that each fast electron
leaving the central star is retained for a long time in the cloud surrounding the
star. Magnetic fields may act as traps where fast electrons may accumulate for long
periods of time. This accumulation effect, in the course of time, is an enormous
supply of energy in the form of kinetic energy of fast electrons. Nearly all of
this energy is transformed into radiation. During this time every electron passes
through the dust particles many times. This effect can work in gas-dust clouds
surrounding T Tau type stars, in FUORs, in peculiar nebulae and also in Herbig-Haro
objects.

11. HERBIG-HARO OBJECTS

There is little doubt of the evolutionary significance of the Herbig-Haro objects
(H-H). Precisely, in these objects we witness the process of star formation itself,
evidently as a result of direct condensation of a gas-dust cloud. Here we are
dealing with the process of star formation, a process which has started, but has
not yet finished; however, the star must "hatch" any moment (see Plate I).

The first objects of this type were discovered in 1949 by Haro [14,15] and Herbig
[16] in the region of the Orion Nebula by spectroscopic observations of their
unusual emission spectra. The most interesting such objects, which later were
called H-H1 and H-H2, are in the vicinity of the well-known nebula NGC 1999.

The H-H objects consist of a dense nebulous structure, slightly diffuse, of very
small dimensions and rather faint, fainter than 16-17^m. All attempts to discover
or detect the image of a star within these tiny structures have failed (in the wavelength region from the ultraviolet (U) to the near infrared ($\sim$1 mμ) [17], and to the
far infrared ($\sim$3 mμ) [18]). To the limit of 21^m (from observations with the 200"
telescope) no central star has been discovered in the H-H objects.

However, when a picture was obtained with a long-focus telescope of one of these
objects, H-H1 (which on ordinary plates is a compact slightly diffuse object), a whole
"cluster" of objects were found, consisting of at least seven very small condensations, imbedded in a faint nebula. And even then no individual stars were discovered.

The spectral characteristics of the H-H objects proved to be very interesting. The
first spectrograms obtained by Herbig [16] showed a large number of strong emission
lines on a background of a very faint continuous spectrum.

A first impression is that this spectrum resembles that of a planetary nebula; it
contains the well known forbidden lines N_1 [OIII], N_2 [OIII], and also 3727 [OII],
etc. However, more careful inspection reveals a characteristic peculiarity: the
unusually large intensity of the forbidden lines 6300 [OI] and 6363 [OI], and also
of 6717 [SII] and 6731 [SII]. The strength of these pairs of lines is about equal
to that of Hα. However, no planetary or diffuse nebula is known with such extraordinary strength of the lines of neutral oxygen.

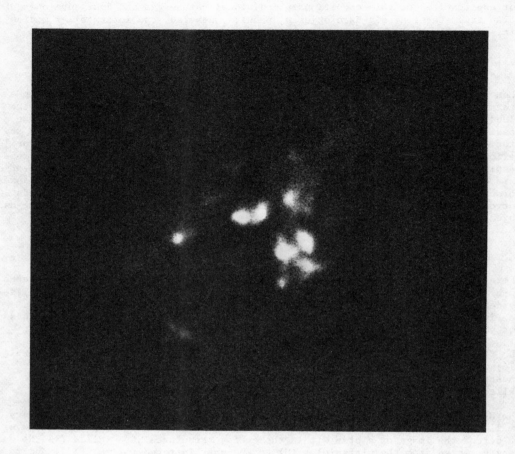

Plate I. Object Herbig-Haro No. 2, photographed with the
120-inch telescope of the Lick Observatory in red light on
December 6, 1959. Scale: 6.8" per 10 mm. The entire system
of ten condensations lies within a circle with diameter about
30". The brightness of the brightest of these condensations
is about 17^m in photographic light. (Kindly made available by
G. H. Herbig)

TABLE 12.4 Emission Lines in Herbig-Haro Object No. 1

Lines (Å)		Intensities		
		1955 [25]	1969 [26]	1969 [28]
3726	[OII]	192	73	186.1
3729	[OII]		56	
3835	H9	8	6	--
3869	[NeIII]	9	8	12.2
3889	H8	17	9	5.6
3933	CaII	12	16	12.0
4047	FeIII	--	4	
4069	[SII]	59	51	66.8
4076	[SII]		18	
4101	Hδ	19	23	24.1
4287	[FeII]	10	6	9.1
4340	Hγ	42	36	37.7
4359	[FeII]	13	10	--
4414	[FeII]	8	17	18.7
4416	[FeII]			
4571	MgI	6	7	--
4658	[FeIII]	4	5	12.7
4815	[FeIII]	--	1	27.8
4861	Hβ	100	100	100
4959	[OIII]	14	13	12.7
5006	[OIII]	40	46	39.4
5159	[FeII]	--	19	14.9
5199	[NI]	8	19	4.5
5201	[NI]			
5877	HeI	--	--	7.2
6300	[OI]	289	152	115.3
6364	[OI]		51	45.7
6548	[NII]		55	37.1
6563	Hα	580	284	265.8
6584	[NII]		130	114.2
6717	[SII]	353	77	82.7
6731	[SII]		117	107.8
10318	[SII]	--	--	10.5
10336	[SII]	--	--	11.7
10830	HeI	--	--	133.2

Furthermore, as indicated in Table 12.4, from data by Böhm et al. [25-27], the emission lines N_1 and N_2 of [OIII] and even 3727 of [OII], which are the most characteristic for ordinary nebulae, are considerably fainter than the Hβ line, whereas even in faint nebulae lines N_1 and N_2 are several times stronger than Hβ. In the spectra of the H-H objects the lines of [OIII] are much weaker than the lines of [OI]. They also show no lines of ionized helium or lines of ions with high ionization potential.

We may conclude that H-H objects are primarily emission objects with a low degree of ionization.

New interesting H-H objects have been discovered in other regions of the sky than Orion [19,20]. For all these objects it is characteristic that they are associated with dark and luminous nebulae of small dimensions, and also with T Tau type objects. They all look "star-like," with a tendency to form small clusters in the shape of compact condensations.

The fact that H-H objects are associated with dust resulted in the suggestion that their luminosity is due to reflection by a dust cloud surrounding a star. This was also suggested by Strom et al. [20,21]. The authors state the following facts and considerations in favor of their point of view: The discovery of a number of infrared sources near some H-H objects; the resemblance of the distribution of energy in the spectra of the infrared objects with that found in the spectra of T Tau type stars, in the first case the star being immersed in a medium with strong absorption, 10 to 30 stellar magnitudes; the assumption that the luminosity of the H-H objects is caused by these infrared stars, which are surrounded by an emission cloud; the assumption that there are transparent paths in the dust cloud through which the light from the central star may reach the outer regions of the cloud; the fact that the light of some H-H objects have a measurable polarization, etc. We note that the authors evade the important problem of the mechanism of excitation of the emission lines.

However, the concept of a reflection was strongly criticised by Haro, who gave a series of convincing arguments against it [22]. Other authors have developed "emission" models of the H-H objects [23,24].

Obviously we still do not understand the nature of the H-H objects. Observations from space should be made in the region of the far ultraviolet--λ shorter than 3000 Å and down to 1000 Å. Photography of these objects is also needed with exceedingly high angular resolution. The available data already suggest that the processes taking place in these objects are too complicated and cannot be explained by a simple "reflection" model.

If we agree that the H-H objects are extremely young systems, of an age $\lesssim 10^5$ years, where stellar birth and formation takes place in them, then clearly the conventional processes need not be adequate to explain their observed spectra. The excitation of the emission lines in these objects may be due perhaps to a completely different mechanism which exists only in these objects.

The "reflection" model of H-H objects requires a specific geometry of the "cloud-star" system. If the model of transparent paths in the diffuse objects is correct, then, as Haro [22] remarks quite correctly, in at least one of the numerous cases, these paths or "corridors" may be pointed at the observer and then we should be able to see the central star. But we have never observed a central star. Besides, it is unlikely, from elementary physical considerations, that the photons emitted by the central star reach the outer layers of the cloud by a very large number of scattering processes on dust particles.

As noted above, the continuous spectrum of H-H objects is very weak. Böhm et al. [27] have been able to observe the energy distribution in the wavelength range from 3300 Å to 8000 Å for a few objects; these distributions for H-H1 and H-H2 are shown in Fig. 12.8. It is seen that these spectra do not resemble the spectra of T Tau type stars. The continuous spectra of the H-H objects get stronger towards short wavelengths. This property is especially clear in the case of H-H1. Here it is assumed that the observed continuous spectrum is not the ordinary recombination radiation of hydrogen. The level of the observed continuous spectrum was found to be 3 to 5 times higher than the theoretical continuum (Fig. 12.8) due to recombination radiation (Paschen and Balmer and two-photon radiation). However, these differences may not be real, due to the low accuracy of the photometric measurements,

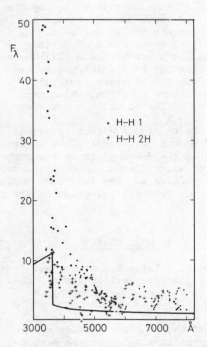

Fig. 12.8. Observed energy distribution, F_λ, in the continuous spectrum
of two Herbig-Haro objects, H-H1 (dots)
and H-H2 (crosses) in the wavelength
region 3300-8000 Å. The full-drawn
curve gives the theoretical recombination spectrum (continua of the Balmer
and Paschen series and two-photon
radiation) for T_e = 10000K, and the
observed value of the flux in the Hβ
line.

and due to model approximations (in particular, no account was taken of the absorption
by the dust component of the cloud). However it cannot be excluded that the strong
increase of the continuous spectrum in the
case of H-H1 from $\lambda \gtrsim 4000$ Å is due to the
Balmer continuum, though as we shall see
below, in principle a different interpretation of this fact is possible.

Although the nature of the origin of the
strong ultraviolet radiation, in the region
$\lambda < 4000$ Å, is not entirely clear there are
reasons to expect an even stronger increase
of the continuous spectrum in the region
shorter than 3000 Å. Judging from the fact
that the total intensity of all emission
lines, including those of hydrogen, the
forbidden lines of oxygen, sulphur, etc.
is a substantial part of the entire radiation of H-H objects, their continuous
spectrum in the region $\lambda < 900$ Å must be
very strong. Even if we assume that the
source of the ionizing radiation is the
central star, then the extremely high
opacity of the cloud will make it impossible for this radiation to reach the outer
regions of the cloud where emission lines
can form. And there is no doubt that we
observe emission lines from the outermost
regions of the cloud.

It seems that these are convincing arguments against the "reflection" model.
The question of the source of the short-
wavelength energy needed for the excitation of the emission lines remains open.

Thus we think that the observed spectra of
the H-H objects cannot be caused by short-
wavelength photon radiation from a central star. The sources of ionization must be
close to where the emission lines and the continuous spectrum are excited, i.e. in
the outer regions of the cloud.

There is no convincing explanation for the origin of the continuous spectrum of the
H-H objects; this has been pointed out by a number of investigators [27,23]. With
regard to the emission lines, the attempt [25,28] to interpret them on the basis of
thermal processes leads to postulating the presence of some hot star of type B0 in
the center of the cloud which has the required combination of the main parameters
of the medium, n_e and T_e (in particular, for various observing periods of H-H1 the
following values have been found: n_e = 1.6×10^4 to 3.2×10^4 cm^{-3}, T_e = 7500 to 10200
K). At the same time, according to Herbig [33], the central "star" must be a dwarf
of class K or M. This additional discrepancy makes us think of a non-thermal way
of exciting the emission lines, namely, inelastic collisions of high-energy charged
particles with atoms and ions [30,31] or interaction of the cloud material with a
stellar "wind" [23].

Returning to the evolutionary significance of the H-H objects, we must stress once

more that there must be conditions and processes which characterize these objects
at the earliest stage in the sequence of formation and evolution of stars. The
next phase of development for the H-H objects is the state of "active stars" of T
Tau type or stars of "NX Mon type" (cf. Ch. 10, §4), then the state of ordinary T
Tau stars, etc. The fact that the H-H objects themselves are variable [29] suggests
that the central "stars" undergo rapid changes which may be transmitted to the
cloud surrounding them. We probably see only the "echoes" of these "central"
processes in the form of emission lines and the continuous spectrum, which are
formed, as secondary phenomena, in the outer regions of the cloud.

But as soon as we are convinced that it is impossible to give a direct relation
between the photon radiation of the central "star" on the one hand and the excitation of the emission spectrum in the outer regions of the H-H objects on the other
hand, then a possibility for such a relation must be sought, by means of processes
of non-photon nature, i.e. by means of high-energy particles. The following section
is devoted to a quantitative consideration of one such possibility, based on the
theory of the transition radiation.

12.   THE IONIZATION EQUILIBRIUM OF THE HERBIG-HARO OBJECTS

As we have seen above, the Herbig-Haro objects contain a large amount of dust
particles, this must be considered to be their most characteristic feature. The
cloud itself is practically opaque with regard to a star or star-like object
immersed in it.

The observations also prove that the contribution of the photon radiation of the
central "star" is extremely insignificant in the total luminosity of the Herbig-
Haro object itself. We assume that the central "star," of the type of the "very
early stage" of T Tau, is not the direct source of ionizing radiation for the
material in the cloud. But that it is the original source of fast electrons, which
can liberate their kinetic energy in the form of radiation as a result of various
kinds of processes. The fact that there is much dust in the cloud makes it very
probable that in these objects transition radiation is generated as well. This
radiation can be partly responsible for the strange continuous spectrum observed
from these objects (Fig. 12.8). In particular, as we have seen above, in the spectrum of the transition radiation quite arbitrary changes occur, sudden jumps and
sharp peaks (cf., Fig. 12.2, and in particular Figs. 12.3 and 12.4). The rapid and
strong increase of the continuous spectrum in the region $\lambda < 4000$ Å, and also the
twisted form of the continuous spectrum in the region 4000-8000 Å in the case of
H-H1 can in principle be explained by transition radiation. Quantitative verification of this assumption is prevented by the fact that there are no reliably measured
continuous spectra, and hence there is no convincing proof that the observed increase
of the spectrum at $\lambda \sim 4000$ Å has no relation with the Balmer continuum.

The transition radiation may turn out to be an important source of generation of
radiation ionizing hydrogen and other elements, especially in the outer regions of
the cloud, where the fast electrons can reach practically without loss of energy,
although, on their way from the central star to the outer cloud regions they interact with dust particles as well as with particles like themselves, i.e., with fast
electrons.

The transition radiation created in the medium, particularly in the infrared spectral
region, can interact with the fast electrons (inverse Compton effect), and as a
result the continuous spectrum in the optical region becomes even more complex. In
general, the problem of the radiation field in a medium where transition radiation
is generated, and where it interacts inelastically with fast electrons, is at least
theoretically of interest.

Finally, we assume that the effect of accumulation of fast electrons in gas-dust clouds, considered in the preceding section, may also be valid in the Herbig-Haro objects.

Here we discuss the ionization of hydrogen and other elements in Herbig-Haro objects under the influence of ionizing radiation of "transition" origin [7].

Let us start by considering the L_c radiation ionizing hydrogen.   The total optical depth τ_0 at the frequencies of the L_c radiation is very large--of the order of 1000--in Herbig-Haro objects; this follows from an analysis of the emission line intensities [25].   The total optical depth t_0 of the same cloud at the optical wavelengths is determined exclusively by absorption by dust particles and appears to be $\tilde{\sim}$ 5. Thus, in Herbig-Haro objects the condition $\tau_0 \gg t_0$ is fulfilled. This result should be kept in mind when comparing and solving the transfer equations for the ionizing and for the optical radiations. We assume that the sources of energy generation are distributed homogeneously in the cloud. Then the solution of the transfer equation gives for the average intensity $J_\nu(\gamma)$ of the L_c radiation the following expression (in the Eddington approximation and for a model with plane-parallel layers):

$$J_\nu(\gamma) = \frac{1}{4\pi} \frac{\varepsilon_\nu}{\kappa_\nu N_1} \left[\frac{3}{2} (\tau_0^2 - t^2) + 2\tau_0 \right] , \tag{12.43}$$

where $\tau = \kappa_\nu N_1 r$, $\tau_0 = \kappa_\nu N_1 R$, κ_ν and N_1 are the coefficient of continuous absorption and the density of the neutral hydrogen atoms, respectively, R is the cloud's radius, and ε_ν is given by Eq. (12.28). In the derivation of (12.43) it was assumed that $\tau_0 \gg 1$, and the following boundary conditions were used: in the center of the cloud $\tau = 0$ and $H_\nu(0) = 0$; at the outer boundary $\tau = \tau_0$ and $H_\nu(\tau_0) = 2\pi J_\nu(\tau_0)$, where H_ν is the flux of L_c radiation.

Since $t_0 \ll \tau_0$, the optical radiation of transition origin as well as the hydrogen emission lines can propagate nearly unimpeded even from the middle parts of the cloud where $t < 1$, but τ is still sufficiently large. Therefore for these parts of the cloud, where still $\tau < \tau_0$, we have:

$$J_\nu = \frac{3}{8\pi} \frac{\varepsilon_\nu}{\kappa_\nu N_1} \tau_0^2 = \frac{3}{8\pi} \varepsilon_\nu \kappa_\nu N_1 R^2 , \tag{12.44}$$

i.e., J_ν is constant and does not depend on τ up to the boundary $\tau \sim \tau_0$. The relation (12.44) is the required expression for the average intensity of the radiation ionizing hydrogen.

We now consider the field of the transition radiation in the optical region. For the intensity I_0 of the radiation leaving the cloud we have: $I_0 = I_0' + I_0''$, where I_0' is the "direct" component equal to

$$I_0' = \frac{\varepsilon_0}{\sigma_* n_p} \left(1 - e^{-t_0} \right) \tilde{\sim} \frac{\varepsilon_0}{\sigma_* n_p} , \tag{12.45}$$

where $t_0 = \sigma_* n_p R > 1$.

The "diffuse" component I_0'' appears in the case where the albedo of the particle p differs from zero. Solution of the corresponding transfer equation gives (for $t_0 > 1$):

$$I_0'' = \frac{3}{\ell(3+2\ell)} \frac{\varepsilon_0}{\sigma_* n_p} , \tag{12.46}$$

where $\ell = \sqrt{3(1-p)}$. For $p \sim 0.5$ we find from these equations

$$I_o'' \approx 1.5 \frac{\varepsilon_o}{\sigma_* n_p} \quad .$$

(12.47)

In these expressions ε_o is the energy of the transition radiation emitted in the optical region per unit   volume per second and is equal to

$$\varepsilon_o = \frac{3}{2} e^2 \omega_o \gamma \delta_o n_p n_e \quad ,$$

(12.48)

where δ_o is the fraction of the energy radiated by the particle in the optical wavelength region.

The total luminosity of the cloud in optical wavelengths is equal to $L = 4\pi R^2 I_o$. With (12.48) and (12.47) we deduce the following for the density of the fast electrons in the cloud:

$$n_e = (2e^2 \omega_o \gamma \delta_o)^{-1} \frac{L}{R} \approx 2.5 \times 10^2 \frac{L}{R} \; cm^{-3} \quad ,$$

(12.49)

where we assume $\omega_o = 10^{16} \; s^{-1}$, $\gamma = 3$ and $\delta_o \approx 0.3$; here we have taken into account that part of the L_c radiation is re-emitted in the emission lines, whose intensities constitute a considerable fraction of the total brightness of the object itself. The numerical values for L and R are taken from the observations.

The condition of ionization equilibrium for the hydrogen atoms is written in the form:

$$N_1 \int_{\nu_*}^{\infty} \kappa_{1\nu} \frac{J_\nu}{h\nu} \, d\nu = N^+ N_e C(T_e) \quad ,$$

(12.50)

where ν_* is the ionization frequency of hydrogen, N^+ and N_e are the densities of ions and thermal electrons, and $C(T_e)$ is the total recombination coefficient of electrons with protons. Substituting the value of J_ν from (12.44) and also $\kappa_{1\nu} = \kappa_o (\nu_*/\nu)^3$, we find from (12.50), taking $N^+ \approx N_e$:

$$\frac{N^+}{N_1} = \left[\frac{e^2 \kappa_o^2}{16\pi^3 h} \frac{q(\gamma)}{C(T_e)} \right]^{\frac{1}{2}} (Rt_o n_e)^{\frac{1}{2}} \quad ,$$

(12.51)

where

$$q(\gamma) = \left[\left(1 + 2\frac{\nu_*^2}{\nu_o^2 \gamma^2} \right) Ln \left(1 + \frac{\nu_o^2 \gamma^2}{\nu_*^2} \right) - 2 \right] \quad ,$$

(12.52)

with $\nu_o = 2\pi\omega_o$, $\nu_* = 3.27 \times 10^{15} \; cm^{-1}$. This is also the formula for hydrogen ionization in Herbig-Haro objects in the case when the ionizing radiation is of transition origin and is generated uniformly at all points in the cloud.

Substituting in (12.51) the value of n_e from (12.49) and also taking $\kappa_o = 0.6 \times 10^{-17}$ cm^2, $C(T_e) = 5 \times 10^{-13} \; cm^{-3} s^{-1}$ with $T_e = 7500$ K [15], $\omega_o = 10^{16} \; s^{-1}$ and $\gamma = 3$ we find:

$$\frac{N^+}{N_1} \approx 10^{-8} \left(t_o \frac{L}{R} \right)^{\frac{1}{2}} \quad .$$

(12.53)

We apply this formula to H-H1 for which $R = 2 \times 10^{16}$ cm, $L \approx 10^{32}$ ergs/s. Taking also

$t_o \approx 5$, we find:

$$\frac{N^+}{N_1} \sim 1 \, .$$

This is 2 to 3 orders of magnitude smaller than the degree of ionization we usually have in planetary nebulae. However this result is in complete agreement with the observations. In fact, using the known intensities of the emission lines of [OII] and [OI] for H-H No 1, Osterbrock [17] found for the degree of ionization of oxygen $N(O^+)/N(O) \approx 0.5$. The first ionization potentials of oxygen and hydrogen are nearly equal. Therefore the degree of ionization of hydrogen will be of the same order as that of oxygen, hence $N^+ \approx N_1$. (Plate II shows a slit spectrogram of H-H1, and Plate III other spectra for comparison.)

For H-H No 1 we have $N_e = 3.7 \times 10^3$ cm^{-3} [15,17], therefore $N_1 \approx 3.7 \times 10^3$ cm^{-3} [15,17]. The total optical depth of the cloud at the frequencies of the L_c radiation is then $\tau_o = \kappa_o N_1 R \approx 400$, i.e. two to three orders of magnitude larger than in planetary nebulae.

The fact that t_o and, in particular, τ_o are considerably larger than one, is a very strong argument that in the case of the Herbig-Haro objects the source of ionizing radiation does not lie at the center of the cloud in the form of a high-temperature star. In these objects we have radiation sources at many points, scattered over the entire volume of the cloud or, rather, at all the points where we have fast electrons and dust particles.

With the aid of the relations derived above we find for H-H1: $n_e = 50$ cm^{-3}, the total energy of the fast-electron cloud $E_e \approx 2 \times 10^{45}$ ergs, the density of the dust particles $n_p \approx 4 \times 10^{-7}$ cm^{-3} for $d = 10^{-5}$ cm, and $n_p \approx 6 \times 10^{-8}$ cm^{-3} for $d = 5 \times 10^{-5}$ cm.

The value of the ratio $N(O^{++})/N(O^+)$ depends on the spectral distribution of the ionizing radiation (the energy γ of the fast electrons). On the other hand, the numerical value of this ratio may be found from the observations from the known intensities of the lines of [OIII] and [OII]. Hence some estimate can be obtained of the probable value of the energy of the fast electrons.

Taking the recombination coefficients of once and twice ionized oxygen to be of the same order, we may write:

$$\frac{N(O^{++})}{N(O^+)} = \frac{\displaystyle\int_{\nu_2}^{\infty} \alpha_\nu(O^+) J_\nu(\gamma) \, d\nu/h\nu}{\displaystyle\int_{\nu_1}^{\infty} \alpha_\nu(O) J_\nu(\gamma) \, d\nu/h\nu} \, , \tag{12.54}$$

where $\alpha_\nu(O)$ and $\alpha_\nu(O^+)$ are the coefficients of continuous absorption of neutral and ionized oxygen [19], ν_1 and ν_2 are the ionization frequencies of O and O^+.

Substituting the value of the function $J_\nu(\gamma)$ from (12.19) and performing the integration we find from (12.54) the quantity $N(O^{++})/N(O^+)$ for a number of values of γ. The results are given in Table 12.5 for two values of the plasma frequency: $\omega_o = 1 \times 10^{16}$ s^{-1} and $\omega_o = 2 \times 10^{16}$ s^{-1}.

For the object H-H1 Böhm has found from observations that $N(O^{++})/N(O^+) = 5.6 \times 10^{-2}$. Comparing this with Table 12.5, we find for the energy of the fast electrons $\gamma \sim 2$ to 4, i.e. on the average $\gamma \sim 3$. This estimate is similar to the value found by other means for flare stars and T Tau type stars.

Plate II. Slit spectrogram of Object Herbig–Haro No. 1 in the blue spectral region, obtained at the prime focus of the 120-inch telescope at Lick Observatory. Note the faintness of the lines N₁ and N₂ of [O III], indicating a low degree of excitation of the gas. (Kindly made available by G. Herbig)

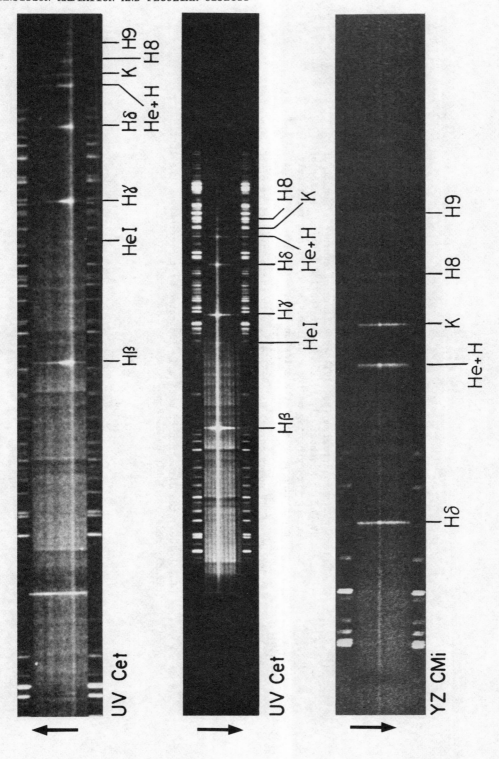

Plate III.

Top: Slit spectrogram of a very strong ($\Delta U = 5\overset{m}{.}2$) flare of UV Cet on
        October 14, 1972 in the wavelength range 6000-3800 Å. Dispersion
        150 Å mm^{-1}. A strong strengthening of the continuous spectrum,
        in particular in the short wavelength part, is seen at flare
        maximum.

Center: Slit spectrogram (No. 791) of a flare of average strength of
        UV Cet on November 11, 1971. Dispersion 150 Å mm^{-1}. A case of
        strong strengthening of the continuous spectrum and relatively
        rapid disappearing of the emission lines.

Bottom: Slit spectrogram (No. 993) of a very short-lived flare of YZ CMi
        on March 7, 1972. Dispersion 25 Å mm^{-1}. A case of practically
        equal duration of flare radiation in the continuous spectrum and
        in the emission lines.

All three spectrograms have been obtained with the Struve 82-inch telescope
at the MacDonald Observatory. The spectrograms have been taken with a moving
plateholder; the direction of motion is indicated by the arrows at the left.
(Kindly made available by B. Bopp)

TABLE 12.5 Theoretical Dependence of $N(O^{++})/N(O^{+})$ on the Energy γ of Fast Electrons (in $mc^2 = 0.51$ MeV units)

γ	$N(O^{++})/N(O^{+})$	
	$\omega_o = 1 \times 10^{16}$ s^{-1}	$\omega_o = 2 \times 10^{16}$ s^{-1}
1	2.5×10^{-2}	3.0×10^{-2}
2	3.0	5.3
3	4.1	8.1
4	5.3	10.9
5	6.7	13.2
10	13.2	21.5
20	21.6	29.0

The analysis given above enabled us to derive the probable spectrum of the ionizing radiation and to understand the conditions of ionization in the Herbig-Haro objects. However, this analysis can be applied to other emission lines as well, in particular those of [SII], [NII], [FeII], [FeIII], [NeIII], etc. One of the possible results might be to find the relative abundances of the elements in those objects, which certainly are of great evolutionary interest.

In classical nebulae--planetary and diffuse nebulae--the ionization of the atoms is due to the radiation from the <u>central</u> stars. The difference in the nature of the ionizing radiation--thermal in the case of the nebulae, and non-thermal in the case of the Herbig-Haro objects--must have decisive significance. Therefore the question arises: Is there a clear difference between nebulae and Herbig-Haro objects due to the geometric structure of the sources of ionizing radiation--namely a central point (star) in the case of the nebulae, and many points, not centralized (charged particles) in the case of the Herbig-Haro objects? If this is so, then the low degree of ionization can, to a certain degree, be used as an independent parameter, characterizing not only the nature and the strength of the ionizating radiation but also its space-geometric structure.

13. THE FUOR PHENOMENON

The observational data do not give evidence that the FUOR phenomenon--a strong, more than a hundredfold increase in the brightness of a T Tau type star lasting for months or a year--is caused by expansion, or a blowing up of the star's photosphere. The expected low expansion velocities do not agree with the observed large negative velocities of the absorption lines (~ 100 to 400 km/s), and the FUOR phenomenon does not seem to be the result of a sudden dissipation of the gas-dust cloud itself [20].

FUORs and the phenomena taking place in them evidently have some relation with the considerations given above. It is especially remarkable that FUORs seem to maintain maximum brightness for a very long period of time. Undoubtedly there must be a huge accumulation of energy especially in the outer parts of the star. There must also be a basic mechanism, which allows energy production from the accumulated energy at a more-or-less constant rate in the form of other kinds of energy.

We assume that T Tau type stars become FUORs, whose gas-dust clouds have the most favorable conditions for them to become <u>magnetic traps</u>, in which the process of capture and accumulation of fast electrons can continue for a long time. Hence the above mentioned accumulation of energy will be formed, in the form of kinetic energy of fast electrons. Processes like non-thermal bremsstrahlung of fast electrons and transition radiation may be the mechanisms which may enable a constant rate of liberation of the energy of the fast electrons, regulated by the average duration of encounters of one electron with another or with a dust particle.

When the process of "fuorization" of a T Tau type star begins, we mean that a process of constant and long-lasting outflow of matter from the stellar interior towards the outer layers of the star begins, where the production of fast electrons takes place--as a result of nuclear decomposition. These electrons are kept in the magnetic trap in the gas-dust cloud. As a result the density of fast electrons in the cloud must increase. Let us assume that the rate at which the fast electrons appear in the cloud is constant, or at least that it can remain constant during a considerable period of time. In other words, we assume that the density of the additional fast electrons which have appeared in the cloud, $n_e(t)$, increases according to the law:

$$n_e(t) = n_e(0)t , \qquad (12.55)$$

where the time t is measured from the moment the star is "fuorized," and $n_e(0)$ is the number of fast electrons produced per unit time and per unit volume. According to the formula (12.55) the accumulation of energy in the cloud in the form of kinetic energy of fast electrons must increase as well.

A constant rate of production of fast electrons also means a constant rate of ejection of ordinary gaseous matter, in particular, of hydrogen from the star's interior. Therefore, there are grounds for applying (12.55) also to the increase of the density $n_e^*(t)$ of the non-thermal electrons (or ions) in the cloud itself, i.e. $n_e^*(t) = n_e^*(0)t$, where $n_e^*(0)$ is the number of non-thermal electrons appearing per unit time and volume. With regard to the density of dust particles, we assume that it stays constant during the time of the fuorization of the star.

The increase of the density of the fast (and thermal) electrons in the cloud leads to an increase of the star's brightness. The increase in brightness with time L(t) will depend on the possible mechanisms for the liberation of the kinetic energy of the fast electrons. Thus, if non-thermal bremsstrahlung of the fast electrons is the most important mechanism, we shall have: $L(t) \sim n_e(t)n_e^*(t) \sim t^2$. But if the density of the dust particles, n_p, in the medium is so large that transition radiation predominates, then we shall have $L(t) \sim n_p n_e(t) \sim t$. Thus the law of increase in brightness with time for a FUOR can be written as:

$$\frac{L(t)}{L(0)} = 1 + \kappa(t/t_o)^2 \qquad (12.56)$$

in the case of non-thermal bremsstrahlung, and

$$\frac{L(t)}{L(0)} = 1 + \kappa(t/t_o) \qquad (12.57)$$

in the case of transition radiation. In these expressions t_o is the duration of the fuorization, and κ is determined from the condition that $L(t_o)/L_o = 10^{0.4\Delta m_o}$, where Δm_o is the total increase in magnitude. Quantitative analysis shows that the FUOR V 1057 Cyg belongs to the first case--formula (12.56). Figure 12.9 gives the observational points for this FUOR [21], together with the theoretical curve computed from (12.56) for the following values of the main parameters of its light-curves: $t_o = 360$ days, $\Delta m_o = 5^m.8$, or $\kappa = 209$ (from $m_{min} = 15^m.9$ to $m_{max} = 10^m.1$).

In the case of the FUOR FU Ori the situation is somewhat different. First of all, there are no possible values of the lightcurve parameters for which the observations can be represented by a single curve of the form (12.56) or (12.57). Only if it is

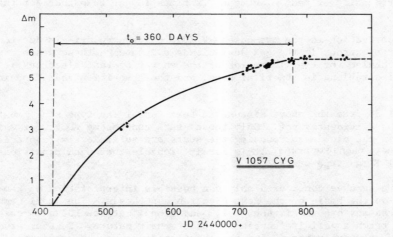

Fig. 12.9. Comparison of the theoretical lightcurve (heavy
line) for the FUOR V 1057 Cyg with the observations (dots).

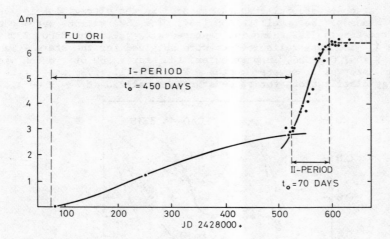

Fig. 12.10. Comparison of theoretical lightcurves (heavy
lines) computed for periods I and II of the "fuorization"
of FU Ori, with the observations.

assumed that the process of fuorization of FU Ori consists of two main periods, I
and II, with different values of t_o, Δm_o or κ, can the observations be represented
by the second law [formula (12.57)], i.e. when transition radiation is the predominating process of radiation for this FUOR.

The first period, I, of the fuorization of FU Ori is $t_o = 450$ days. There are few
observations for this period--only three. The theoretical curve constructed from
(12.57) for $\Delta m_0 = 2^m\!.8$ or $\kappa = 10.2$ is seen to agree with the observations (Fig. 12.10).

Then begins period II of the fuorization, relatively short (t_o = 70 days), but with a rate of production of fast electrons 7.2 times larger than that for the first period.

Thus, the fuorization in the two cases--V 1057 Cyg and FU Ori--is due to the appearance of fast electrons in the gas-dust clouds surrounding these stars. However, the main mechanism for the liberation of the energy of fast electrons is the non-thermal bremsstrahlung in the first case, and the transition radiation in the second case.

Can we conclude from the above discussion that each T Tau type star must finally become a FUOR? Evidently not. Only those stars can become FUORs for which the conditions in the clouds surrounding the stars are sufficiently favorable for these clouds to create magnetic traps. It is hardly probable that such conditions should exist in all T Tau type stars.

Finally, here we have considered only one possible interpretation of the FUOR phenomenon--on the basis of the fast-electron hypothesis. The small number of FUOR examples makes any hypothesis doubtful. Additional data on FUORs is essential before we can reach a definite conclusion about their nature. Further study on this problem is also very important for obtaining evidence for the existence of magnetic traps around stars.

14. THE RADIATION OF T TAU TYPE STARS IN THE ULTRAVIOLET

In 1977 the first observational results appeared in the ultraviolet for some T Tau type stars by means of the satellite ANS [34]. The observations were made with a wide-band photometer in five channels, centered at wavelengths 3300, 2500, 2200, 1800 and 1550 Å. The most reliable data were obtained for the stars T Tau, CoD -44°3318, V 1057 Cyg and, in particular, the star V 380 Ori, which was observed twice, with an interval of several months. Figure 12.11 gives the curves of the spectral energy distributions for these stars, normalized at λ = 3300 Å.

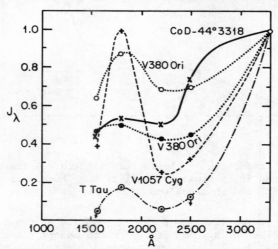

Fig. 12.11. Curves of relative spectral distributions of observed radiation fluxes in the far ultraviolet for four T Tau type stars: T Tau, V 1057 Cyg, CoD-44°3318 and V 380 Ori (two measurements). The flux of radiation at λ = 3300 Å has been taken as unity.

The most characteristic properties of these spectra are: (a) The presence of a minimum in the continuous spectrum at ⌄2200 Å in all cases; (b) The presence of a considerable radiation flux in the region ⌄2000 Å, i.e. a region where these stars should be weak considering their spectral classes. It is simply remarkable that the radiation fluxes at ⌄1800 Å, say, should be comparable with, or even be equal to, those at ⌄3300 Å.

It is also remarkable that the observed spectral curves for these stars in general resemble each other, although the distances to the stars are quite different--from 150 parsecs to 1000 parsecs, and their spectral classes range from A1 (V 380 Ori) to K - G5 (T Tau). This means that these spectral distributions are not determined by selective absorption in the interstellar medium only. The fact that they do not depend on spectral class indicates that this radiation is of "non-stellar" or non-thermal nature.

The only significant continuum in the spectral region shorter than 2000 Å can be the two-photon radiation of hydrogen. However, the radiation fluxes found from the observations are so large that they cannot be explained by the two-photon radiation. The well-known theory of the radiation of gaseous nebulae, including the two-photon (2q) radiation as well as the radiation in the emission lines of hydrogen, are derived from the same source--L_C radiation (<912 Å). Then we can find the theoretical value of the ratio $N(2q)/N(\beta)$, where $N(2q)$ is the total number of 2q-photons, emitted by the medium from 1216 Å to infinity, and $N(\beta)$ is the number of photons emitted in the Hβ line. This same ratio can be found from the observations; we shall call this $N^*(2q)/N^*(\beta)$. By comparing these two quantities we may come to some conclusions about the role of the two-photon radiation in the stars under investigation.

According to the theory of the two-photon radiation, about 1/3 of the total number of protons can, after recombination with electrons, induce 2q-transitions. Hence we have: $N(2q)/N_C = 2\times(1/3) = 2/3$, where N_C is the total number of L_C photons in the medium ($\tau_c \gtrsim 1$).

Furthermore, the total number of Balmer photons equals the total number of L_C photons. About 12% of this result is emitted in the Hβ line, i.e. $N(\beta)/N_C \approx 0.12$. Hence we have:

$$\frac{N(2q)}{N(\beta)} \approx 6 \ . \tag{12.58}$$

Omitting details (cf. [35]), we can derive for the observed value of this ratio:

$$\frac{N^*(2q)}{N^*(\beta)} = \frac{480}{W_\beta} \frac{F(\lambda)}{F(\lambda_o)} \exp[r(a_\lambda - a_{\lambda_o})] \ , \tag{12.59}$$

where W_β is the equivalent width (in Angstroms) of the emission line Hβ, $F(\lambda)$ and $F(\lambda_o)$ are the measured radiation fluxes at $\lambda = 1800$ and $\lambda_o = 3300$ Å, respectively, a_λ and a_{λ_o} are the coefficients of interstellar absorption at these same wavelengths, per kiloparsec, and r is the distance to the star.

The numerical values of the ratio $N^*(2q)/N^*(\beta)$ found from formula (12.59) are given in Table 12.6. The distances of the stars have been taken from [34] and [36], the values W_β from [37], and the dependence of a_λ on λ from [38], with a(3300) = 1.70 per kiloparsec.

The results show that the observed values of $N^*(2q)/N^*(\beta)$ are considerably larger than the theoretical value (12.58) (3 to 15 times larger). Actually this difference

TABLE 12.6 Observed Values of the Ratio of the Number of 2q-photons to the Number of Hβ Photons (last column) of Some T Tau Type Stars

Star	r pc	W_β Å	$F_\lambda = 1800$ (1)	$F_{\lambda_o} = 3300$ (1)	$\dfrac{N^*(2q)}{N^*(\beta)}$
T Tau	150	5	7	40	19
V 380 Ori	400	15	22	25	40
V 380 Ori	400	15	23	47	22
V 1057 Cyg	700	10:	5	5	90
CoD-44°3318	1000	10:	7.1	13.5	62

(1) In units 10^{-14} ergs cm^{-2} s^{-1} Å^{-1}.

must be even larger, due to a number of effects which have not been taken into account. First of all, the high values of the electron density--of the order 10^6 to 10^8 cm^{-3}--in the gas-dust clouds surrounding the T Tau type stars [35] make any 2q-radiation very improbable.

Thus the available facts exclude the possibility of two-photon radiation being generated in T Tau type stars. The radiation observed in the far ultraviolet in T Tau type stars is most probably of transition origin. The spectrum derived below for the transition radiation of the gas-dust cloud which surrounds the T Tau star shows a satisfactory resemblance with the observed spectra and at least qualitatively the assumptions made are reasonable.

Assuming that the sources of the transition radiation are distributed uniformly within the gas-dust cloud (the volume radiation coefficient ε_λ is constant everywhere), and that the absorption within the cloud is caused by dust particles only (the optical depth of the cloud $\tau_\lambda = \alpha_\lambda s$ being very large), we may write for the intensity of the transition radiation leaving the cloud and reaching the observer, taking into account the absorption in the interstellar medium:

$$F_\lambda \sim \frac{\varepsilon_\lambda}{\alpha_\lambda}\left(1 - e^{-\tau_\lambda}\right)e^{-a_\lambda r} \approx \frac{\varepsilon_\lambda}{\alpha_\lambda}\,e^{-a_\lambda r}\,. \tag{12.60}$$

Furthermore, we have to assume that with respect to their absorbing properties the particles in the cloud and the particles in interstellar space are identical, i.e. the frequency dependence of the functions α_λ and a_λ is the same.

Writing equation (12.60) for the normalized wavelength λ_o, we have the relative distribution $J_\lambda = F_\lambda/F_{\lambda_o}$ of the transition radiation as a function of wavelength:

$$J_\lambda = \frac{Q_\lambda(\gamma)}{Q_{\lambda_o}(\gamma)}\frac{\alpha_{\lambda_o}}{\alpha_\lambda}\exp\left[-r\left(a_\lambda - a_{\lambda_o}\right)\right]\,, \tag{12.61}$$

where $\varepsilon_\lambda(\gamma) = C\,Q_\lambda(\gamma)$, and the function $Q_\lambda(\gamma)$ is given by (12.35).

Using this relation curves for the dependence of J_λ on λ in the ultraviolet have been constructed for two values of the plasma frequency: $\omega_o = 10^{16}$ s^{-1} and $\omega_o = 0.5\times10^{16}$ s^{-1}, and a number of values for the total interstellar absorption at $\lambda_o = 3300$ Å, indicated next to each curve in Figs. 12.12 and 12.13.

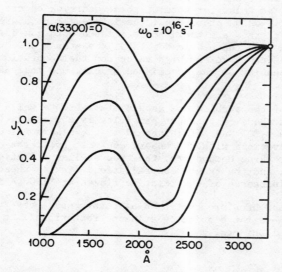

Fig. 12.12. Theoretical spectral curves of the transition radiation for a plasma frequency of the dust particles in the cloud $\omega_o = 10^{16}$ s-1, for a number of values of the total interstellar absorption a_λ at $\lambda = 3300$ Å; self-absorption in the gas-dust cloud has been taken into account. The intensity of the radiation at $\lambda = 3300$ Å has been taken as unity.

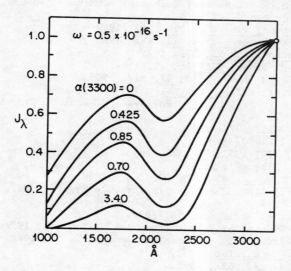

Fig. 12.13. The same as in Fig. 12.12, but for $\omega_o = 0.5 \times 10^{16}$ s^{-1}.

The general structure of the curves resembles the observations very well (Fig. 12.11). In particular, in the theoretical curves the minima at $\sim$2200 Å and the maxima at $\sim$1800Å are clearly evident.

Though the agreement of the observations with the theory appears qualitatively satisfactory, yet it is not clear if the assumed high values of the interstellar absorption can be valid, at least for some of these stars, which are relatively close to us (e.g. T Tau). On the other hand, the interstellar absorption may be unusual, since T Tau type stars are known to be related with gas-dust clouds (for instance, the star T Tau with the nebula NGC 1555, the star V 380 Ori with NGC 1999, etc.). In such cases one can have any high value for the absorption, depending on the dimensions and the density of the associated nebula, and independent of the star's distance from us.

For a better treatment of the problem one should start from a well-determined chemical composition and shape of the dust particles in the cloud, that means one should use the functions $\varepsilon_1(\omega)$ and $\varepsilon_2(\omega)$ instead of the plasma frequency ω_0. Also, when comparing the observations with the theory one must split the observed flux at $\lambda \sim 3300$ Å into at least three components: the true stellar radiation, the emission of the Balmer continuum, and the transition radiation. Taking these into account leads to still further increases of the relative fluxes at $\sim$1800 Å in Fig. 12.11.

If the conclusions reached here are confirmed by future space observations, then we shall probably obtain the first convincing evidence that transition radiation may be generated in celestial objects.

REFERENCES

1. S.A.E. Johansson, Astrophys. Lett. 9, 143, 1971.
2. I. Lerche, Ap.J. 175, 373, 1972.
3. R. D. Ramaty, R. D. Bleach, Astrophys. Lett. 11, 35, 1972.
4. G. B. Yodh, X. Artru, R. D. Ramaty, Technical Report No. 73-027, 1972.
5. L. Durand, private commun., 1972.
6. G. A. Gurzadyan, Astron. Astrophys. 20, 145, 1972.
7. G. A. Gurzadyan, Astron. Astrophys. 28, 147, 1973.
8. G. A. Gurzadyan, Astrofiziak 11, 531, 1975.
9. V. L. Ginzburg, I. M. Frank, JTEP 16, 15, 1946.
10. G. M. Garibyan, JTEP 33, 1403, 1957; 37, 527, 1959.
11. R. H. Ritchie, H. B. Eldridge, Phys. Rev. 126, 1935, 1962.
12. E. Janikova, Z. Janout, F. Lehar, P. Pavlovic, V. Zrelov, Nuc. Instr. Methods 74, 61, 1969.
13. E. E. Mendoza, Ap.J. 151, 977, 1968.
14. G. Haro, A.J. 55, 72, 1950.
15. G. Haro, Ap.J. 115, 572, 1952; 117, 73, 1953.
16. G. H. Herbig, Ap.J. 113, 697, 1951.
17. G. Haro, R. Minkowski, A.J. 65, 490, 1960.
18. G. H. Herbig, "Non-Periodic Phenomena in Variable Stars". Ed., L. Detre, Budapest, p. 75, 1969.
19. G. H. Herbig, Lick Obs. Bull. No. 658, 1974.
20. S. E. Strom, G. L. Grasdalen, K. M. Strom, Ap.J. 191, 111, 1974.
21. K. M. Strom, S. E. Strom, G. L. Grasdalen, Ap.J. 187, 83, 1974.
22. G. Haro, Bol. Inst. Tonantzint. 2, 3, 1976.
23. R. D. Schwartz, Ap.J. 195, 631, 1975.
24. G. D. Schmidt, F. J. Vrba, Ap.J. Lett. 201, L33, 1975.
25. K. H. Böhm, Ap.J. 123, 379, 1956.
26. K. H. Böhm, J. F. Perry, R. D. Schwartz, Ap.J. 179, 149, 1973.
27. K. H. Böhm, R. D. Schwartz, W. A. Siegmund, Ap.J. 193, 353, 1974.
28. K. H. Böhm, W. A. Siegmund, R. D. Schwartz, Ap.J. 203, 399, 1976.
29. G. H. Herbig, IBVS No. 832, 1973.
30. D. E. Osterbrock, PASP 70, 399, 1958.
31. G. Magnan, E. Schatzman, C.R. Acad. Sc. Paris 260, 6289, 1965.

32. M. J. Seaton, Rev. Mod. Phys. 30, 979, 1958.
33. G. H. Herbig, Vistas in Astronomy VIII, 109, 1966.
34. K. S. de Boer, Astron. Astrophys. 61, 605, 1977
35. G. A. Gurzadyan, Astrophys. Space Sci. 62, 67, 1979.
36. E. E. Mendoza, Ap.J. 151, 977, 1968.
37. I. R. Salmanov, Thesis. Shemacha Astrophys. Obs., 1975.
38. L. Nandy, G. I. Thompson, G. Jamor, A. Monfils, R. Wilson, Astron. Astrophys.
       44, 195, 1975.

CHAPTER 13
Radio Emission from Flare Stars

1. DISCOVERY OF RADIO EMISSION FROM FLARE STARS

The flux density of strong solar flares at radio frequencies, recorded on the Earth, is of the order of 10^{-19} to 10^{-20} W m^{-2} Hz^{-1}; in some cases it reaches 10^{-18} W m^{-2} Hz^{-1}, and at the time of exceptionally strong radio bursts the flux has reached 10^{-15} W m^{-2} Hz^{-1} for a few minutes at meter wavelengths. If a radio burst of such power could take place on the nearest star, which is at a distance of 5×10^5 astronomical units from us, the flux density of its radio emission on the Earth would be 10^{-26} W m^{-2} Hz^{-1}. During the last decade this quantity has already been within the limits of sensitivity of the receiving apparatus used in radio-astronomy.

The whole question is whether there are stars capable of generating such radio-waves in the neighborhood of the Sun. The most probable candidates would be unstable objects, such as flare stars.

These considerations formed the basis for the research begun in 1958 under the direction of Lovell at the Jodrell Bank Radioastronomical Observatory with the aim of discovering radio emission from flare stars [1].

The observations were made mainly with the 75-meter parabolic antenna in the meter wavelength region. At the beginning a given flare star was constantly followed and was constantly compared with the radiobackground sky. First UV Cet was chosen for observation; then four more flare stars (YZ CMi, EQ Peg, AD Leo, BD + 43°4305) were observed. The first radioflare of UV Cet was discovered on September 29, 1958. During the period from September 29, 1958 until April 14, 1960, active observations were made for 474 hours, of which 213 hours were devoted to UV Cet. During this time six radioflares of UV Cet were detected with certainty, and one flare for the other stars. The average interval between two successive flares of UV Cet was 35 hours; this was approximately equal to the then known average interval of flares from optical data. These results suggested that probably each optical flare is accompanied by a radioflare as well.

The average level of the flux from these radiobursts was found to be of the order of 6×10^{-26} W m^{-2} Hz^{-1}. We can ask if this is large or small. Lovell makes the following comparison: if UV Ceti would be at the position of the Sun, during a flare with an optical amplitude $\sim0^m5$ the strength of the radio signal reaching the Earth would be about 10^{-14} W m^{-2} Hz^{-1}. Since the radius of UV Cet is 0.08 $R_\odot$, the effective brightness temperature of such radio emission will be $\sim10^{15}$ K, i.e. 100 times higher than the effective temperature of the strongest radio burst of the Sun, and 10^7 times higher than the temperature of "ordinary" strong radio flares of the Sun.

However, for a reliable identification of radio flares from other disturbances and radio noises, simultaneous observations at radio and optical frequencies are needed.

After the first detections of radio flares a coordinated program of observations began. The following radioastronomical observatories took part in this work: Jodrell Bank, Smithsonian, Crimea, Odessa, Abastuman Astronomical Observatory,in the Northern hemisphere, and also the radio research center at Parkes (Australia) in conjunction with optical observations from the Southern hemisphere. The Baker-Nunn cameras established at various locations on the Earth were also used. Simultaneous radio and optical observations started in 1960.

2.   SIMULTANEOUS RADIO AND OPTICAL OBSERVATIONS OF FLARE STARS

During the period of simultaneous radio and optical observations in 1960-61 the results indicated [2] clearly that flare stars are radio emitters. During that period a radiotelescope constructed to work at a frequency of 240 MHz (1.25 m) observed the flare stars UV Cet, YZ CMi, and AD Leo for 727 hours; optical cameras took plates during 216 hours and the total duration of simultaneous radio and optical observations was 166 hours. As a result 23 faint flares, with amplitudes smaller than one stellar magnitude, were recorded. The average optical amplitude for these flares was found to be about $0^m.4$, and the mean flux of the radio emission was 1.8 Jansky (Jansky = Jy = 10^{-26} W m^{-2} Hz^{-1}). Thus, the average value of the radio emission per unit optical emission per unit optical amplitude (in stellar magnitudes) was 4.7 Jy.

The results obtained from this series of observations allowed us to draw the following conclusions:

a) The radioflare starts earlier than the optical flare.

b) The duration of the radioflare is somewhat longer than that of the optical flare.

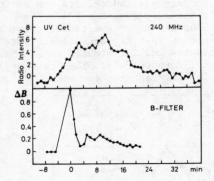

Averaged curves for three flares of UV Cet, occurring on the 13th, 16th and 19th of October 1963, are shown in Fig. 13.1. The amplitude of the optical flares according to observations at the Crimean Observatory was $\sim 1^m$, and the amplitude of the radioflares was about 6 Jy (at a frequency of 240 MHz), which gives for the radio emission 6 Jy per per 1^m. Fig. 13.2 gives analogous curves for a flare of EV Lac on August 7, 1961 [3]. The amplitude of the optical flare was equal to $0^m.6$, and in the radio region it was 4 Jy, which gives for the value of the radio emission 6.6 Jy per 1^m.

Fig. 13.1. Averaged photo-electric tracing (bottom) of three flares of UV Cet and the corresponding radio tracing at $\lambda 1.25$ m (13, 16 and 19 October 1963).

Another example of simultaneous radio and optical observations is shown in Fig. 13.3 (a flare of UV Cet, 19.X.63). Though the optical observations of this flare do not inspire confidence (they are based on naked-eye estimates), nonetheless the observations themselves are certainly interesting. In this case the radioflare lasted considerably longer than the optical flare. Though one can hardly talk about the moment the radioflare reached its maximum, its smooth and extended character seems remarkable (the same picture is observed in the case of the radioflare of EV Lac in Fig. 13.2).

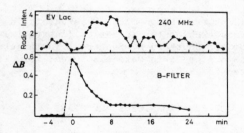

Fig. 13.2. Simultaneous radio (240 MHz) and optical (in B wavelengths) observations of a flare of EV Lac (7.VIII. 1961).

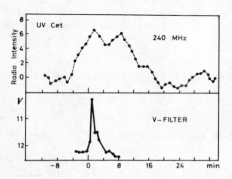

Fig. 13.3. Simultaneous radio and optical observations of a flare of UV Cet (19.IX.1963).

In these examples one more property of the radioflares is clear: the maximum of the radioflare comes after the maximum of the optical flare.

In 1963 at Jodrell Bank observations of flare stars were begun simultaneously at two radio frequencies, 240 MHz (1.25 m), 408 MHz (0.735 m). Such observations are important for discovering the dependence of the intensity of the radio emission on frequency, which is directly related to the physical mechanism of the generation of the radio emission.

Figure 13.4 shows the lightcurves of radio (at two frequencies) and optical flares of UV Cet (25.X.63) [4]. The delay of the maximum of the radioflare here is very clear. Moreover, one more property of the radio-flares appears: the delay is the larger for the lower frequency. In this case the delay was 2 minutes at 408 MHz and 3 minutes at 240 MHz. The observations of this flare made it possible for the first time to derive the character of the dependence of the radio emission on frequency (at least in the frequency interval 240-408 MHz). The frequency spectrum has the form $\omega^{-0.8}$. Fig. 13.5 shows another example of simultaneous recordings of radioflares for YZ CMi (8.II.64) at two frequencies [4]. The delays of the maxima of the radioflares with respect to that of the optical flare are 1.5 minutes and 2.5 minutes at 240 MHz and 408 MHz, respectively. The frequency spectrum in this case has the form $\nu^{-0.5}$.

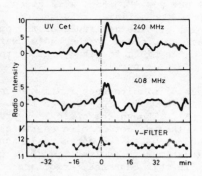

Fig. 13.4. Flare of UV Cet on 25.IX.1963. Simultaneous recordings at two radio wavelengths (1.25 m and 0.735 m) and in the optical region (in V wavelengths).

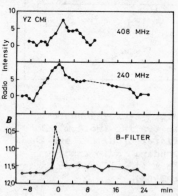

Fig. 13.5. Flare of YZ CMi on 8.II.1964. Simultaneous observations at two radio-wavelengths (1.25 m and 0.735 m) and in the optical region (in B wavelengths).

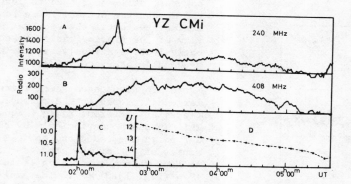

Fig. 13.6.   Strong flare of YZ CMi on 19.I.1969. Simultaneous observations at radiofrequencies 240 MHz (A), and 408 MHz (B), and in the two
optical spectral regions V and U (curves C and D).

The results of simultaneous radio and optical observations of one exceptionally
strong flare of YZ CMi, on January 19, 1969, are of particular interest. The lightcurves of this flare in V-light, and at two frequencies in the radio range are
given in Fig. 13.6 [5-6]. The maximum amplitude was determined only in visual light
as $\Delta V = 1^m_{\cdot}7$, corresponding with $\Delta B \approx 3^m_{\cdot}7$ and $\Delta U \approx 6^m_{\cdot}6$ (cf. Fig. 6.3). The maximum
flux at 240 MHz was equal to 18 Jy. Here too the delay of the maximum in the radio
emission is clearly observed; it is $\sim$15 min at $\lambda = 1.25$ m, and nearly one hour at
$\lambda = 0.735$ m. The radioflare at 1.25 m seems to have started about 15 minutes
earlier than the optical flare. Finally it is remarkable that the radioflares are
maintained for a long time, more than 1.5 hours, in the state of maximum ("flat
maximum"), though the optical flare was monotonously becoming fainter.

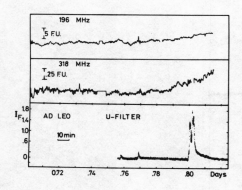

Fig. 13.7.   Flare of the star AD Leo
on 1.II.1974 (flare 12C). Simultaneous observations at two radiofrequencies (196 MHz and 318 MHz), and in
the optical region (in U wavelengths).
The radio emission continued to increase
still after the end of the optical
flare.

At the time of this flare it was discovered that the spectral index n, which
characterizes the energy distribution in
the radio emission in the frequency interval 240 to 408 MHz ($J_\nu \sim \nu^{-n}$), was not
constant. At the beginning of the flare
the index n was found to be large, about
4 to 5; then it decreased, reaching a
value about equal to one at flare maximum,
and later it increased again. On the
average $1 < n < 2.5$. In general, variations of the index n during a flare is
characteristic for nearly all flares.

The radioflares at longer wavelengths
appear considerably earlier than at shorter
wavelengths, and the latter earlier than
the optical flares. This is illustrated
best by the example of a flare of AD Leo
(12 A), the lightcurves of which are shown
in Fig. 13.7. Here the first signs of a
radio signal may be noticed at the frequency of 196 MHz nearly one hour, and at
318 MHz 40 minutes, before the moment any
optical flare appeared. The fluxes of the
radio emission at the two frequencies,

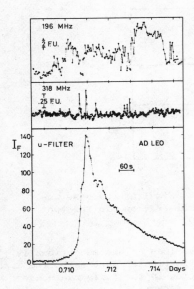

however, still continued to increase after the moment the optical flare reached its maximum and until the star was out of the view of the Arecibo radiotelescope (the optical observations were made at the MacDonald Observatory [7].

There are also cases where there is a large difference in the morphology of the flares at different radio frequencies. This is shown convincingly by the light-curves in the radio and optical spectral ranges of one of the strongest flares of AD Leo, which occurred on February 16, 1974 (Fig. 13.8) [7]. The amplitude of the flare in U wavelengths reached a record value for this star: $5^m.4$. Here the picture at 196 MHz is in general analogous to what has been described above: the radioflare starts earlier than the optical flare (about 2 minutes) and the maximum is reached considerably later than the moment of maximum of the optical flare. But the picture is completely different at 318 MHz; here only strong but very short-lived flares appear at the time of the optical flare maximum; then 2 to 3 other such flares occur during the time when the optical flare decreases rapidly.

Fig. 13.8. Very strong flare of AD Leo on 16.II.1974. Simultaneous recordings of radio emission at two frequencies (196 MHz and 318 MHz), and in the optical region (in U wavelengths).

Unique are the tracings of a radioflare (Fig. 13.9) obtained during a flare of

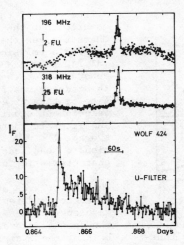

Fig. 13.9. Flare of Wolf 424 on 30.II.1974 (flare 8C). Simultaneous recordings of radio emission at two radiofrequencies (196 MHz and 318 MHz), and in U wavelengths. Case of appearance of radioflares (at about +0.867 JD) without a simultaneous optical flare.

Wolf 424 on January 30, 1974 (flare 8C) [7]. The behavior of the general level of the radio emission at frequencies 318 MHz and 196 MHz is in general the same as in the preceding case (Fig. 13.8). A characteristic peculiarity of this flare is the appearance (about 3 minutes after the maximum of the optical flare) of a strong but relatively short-lived (∿20 s) radioflare appearing in both frequencies simultaneously, but without any analogous flare detected at optical wavelengths.

This case leads us to believe that in principle radioflares can exist which are not accompanied by optical flares. The problem however is that it is very difficult to distinguish such radioflares from ordinary radio noise.

3. SHORT-LIVED RADIOFLARES

It is important and useful to make radio observations of stellar flares simultaneously with optical observations; however it should be recognized that for many reasons it is not always possible to obtain the latter. Therefore independent radio observations of flare stars should be performed anyway, which will still give very interesting results. In doing so one should take special precautions to prevent spurious identification of flare processes with other radio noise.

A series of observations have been made by means of the radiotelescope at Arecibo at meter wavelengths (196 MHz, 318 MHz and 430 MHz) of the flare stars YZ CMi, AD Leo, Wolf 424 and V 371 Ori [8]. Each of these stars was observed for about 10 to 30 hours. During this time for each of these stars (with the exception of V 371 Ori) from one to 11 flare events were recorded. More detailed data about this series of observations are given in Table 13.1.

TABLE 13.1   Data on the Radio Observations of Four Flare Stars

Star	318 MHz			439 MHz		
	Total observing time (hours)	Number of radioflares recorded	Radioflare frequency (fl/hr)	Total observing time (hours)	Number of radioflares recorded	Radioflare frequency (fl/hr)
YZ Cmi	25.9	11	0.42	13.6	1	0.074
AD Leo	26.8	5	0.19	30.6	3	0.10
Wolf 424	10.9	1	0.09	19.2	5	0.26
V 371 Ori	12.5	0	--	0	—	--

The highest flare frequency at 318 MHz was found for YZ CMi, 0.42 fl/hour, which coincides nearly exactly with the frequency of the optical flares of this star (cf. Table 1.3). The same is observed for AD Leo, with the frequency of the radioflares (at 318 MHz) being somewhat lower than the frequency of the optical flares. In both stars the frequency of flares at 430 MHz was found to be several times smaller than the frequency of the optical flares. This difference is still larger (from 4 to 10 times) in the case of Wolf 424, for which the frequency of optical flares in U wavelengths is of the order of 1 fl/hour.

We also note that Lovell [10] found a frequency of radioflares for the star UV Cet of about 0.03 fl/hr (at 240 MHz). This is nearly two orders of magnitude lower than the flare frequency of this star in U wavelengths (Table 1.3). However this estimate refers only to very powerful radioflares, corresponding with the very strong optical flares.

On November 30, 1962, a strong radioflare from the star V 371 Ori was detected at 410 MHz [9]. In this case 12.5 hours of observations at 318 MHz did not show any single flare.

The remaining characteristics of the recorded radioflares are the following: The average duration of the radioflares (in seconds) is:

	196 MHz	318 MHz	430 MHz
YZ CMi	73	23	15
AD Leo	--	12	18
Wolf 424	20	15	11

Clearly the durations are shorter than a minute and in all cases considerably less than the duration of the optical flares. There is reason to assume that these radiobursts are not accompanied by typical optical flares; they may be accompanied by optical flashes, i.e. flares lasting less than ten seconds. The above data also shows that the radioflare duration increases from short to long wavelengths.

The mean flux densities of the radioflares were found to be in the range 0.2 to 8 Jy for these stars, and the spectral index was found to be in the range +1.4 to +4.3, with an average of +2.5.

Another series of radio observations of 22 flare stars were made by Spangler [11] using the Arecibo radiotelescope at a frequency of 430 MHz, without simultaneous optical observations. For most of these stars the measured radio fluxes were less than 0.12 Jy. For four stars, AD Leo, L 1113-55, CN Leo and Ross 867 they were in the range 0.30 to 0.70 Jy. The linear dimensions of these radio sources were also estimated and were found to be of the order of 2 arc minutes or larger. This estimate however is probably not real since the size of the radio source would be a million (!) times larger than that of the star itself (with r = 5 parsecs). However, in the tracings of the radio emission real variations of the radioflux can be seen, lasting one to two seconds, which gives for the characteristic size of the radiating medium a value of the same order as the size of the star itself.

Special radio observations made of two stars, AD Leo and CN Leo, at frequencies 2695 MHz and 8085 MHz [20] show that actually there are radiosources close to these stars, but they are not connected with the flare stars AD Leo and CN Leo themselves. For instance, the radiosource near AD Leo was found to be a classical double radio-source with dimensions 5"×4" and 7"×6" (with an accuracy of ±1"). The radiosource near CN Leo turned out to be an ordinary point source with diameter even smaller than 1", and with no large proper motion.

One more series of radio observations with the Arecibo radiotelescope was aimed at discovering possible emission fluctuations of individual flare stars. The reason for investigating such a problem was the assumed role of "spots" on these stars in the generation of flare activity. These observations of the stars EQ Peg, YZ CMi and AD Leo, at 430 MHz, gave a negative result [12]. Upper limits for such fluctuations were from 0.009 Jy to 0.05 Jy.

In one case polarimetric measurements were made of the radio emission of flare stars. This was done during a strong flare of YZ CMi on April 1, 1974, when radio emission at flare maximum reached 0.52 Jy (at 430 MHz) [13]. The radio emission was found to be 56% circularly polarized at maximum, and 92% at the end of the flare. Linear polarization of 21% was observed as well at flare maximum. These facts clearly indicate that the radio emission in flare stars is of synchrotron origin.

4. THE DEPENDENCE OF THE RADIO EMISSION ON THE AMPLITUDE OF THE OPTICAL FLARE

Though the available data is scarce, there is not a clear relation between the strength (flux) of the radio emission at flare maximum and the amplitude of the flare. However, there does exist a tendency for the absolute strength of the radio emission to increase with increasing amplitude of the optical flare ΔU. Considering the large differences between the lightcurves at radio and optical wavelengths (the presence of an extended flat maximum in the radioflares, and the longer duration of the radioflares) a simple comparison of the flare amplitudes in two wavelength regions is not sufficient. It is more correct to compare the total energy in the two wavelength regions, radio and optical, integrated over the entire lightcurves.

Here we shall restrict ourselves to the following problem: how does the ratio of the fluxes of the radio and optical flare radiation behave as a function of ΔU? Let us call ΔS the flux of radio emission at a given frequency at flare maximum, and I_U the additional radiation flux in U wavelengths in units of the radiation of the undisturbed star, i.e.

$$I_U = 10^{0.4\Delta U} - I .\tag{13.1}$$

We find that the ratio $\Delta S/I_U$ is not constant with ΔU and that it varies with it in a peculiar way.

Table 13.2 shows all the cases known to us (from data up to 1976) with simultaneous measurements of fluxes of radio (ΔS) and optical (ΔU) flares for the flare stars UV Cet, EV Lac, YZ CMi, AD Leo and Wolf 424. The values ΔS are at a frequency of 240 MHz for the first six flares, and at a frequency of 196 MHz for the remaining cases. More correctly we should restrict ourselves to one frequency; however the uncertainty in determining the spectral index n in each individual case makes such an attempt meaningless.

TABLE 13.2 Ratio of Flux of Radio Emission ΔS (at 196 or 240 MHz) to Relative Flare Flux I_U (in U Wavelengths) for a Number of Flares

Star	Date or number of flare	ΔU	ΔS (Jy)	$\Delta S/I_U$	Reference
UV Cet	13. X.63	3^m3	6	0.30	1
	19.IX.63	6.0	6	0.024	2
	25. X.63	2.7	8	0.73	3
EV Lac	7.VIII.61	2.5	4	0.45	2
YZ CMi	8.II.64	3.2	10	0.56	3
	19.IX.69	6.6	18	0.04	4
	3	0.7	2.5	2.8	5
	2	0.44	(3.0)	5.0	5
	5	0.50	(1.5)	2.6	5
	7	0.10	(1.2)	12.0	5
	9C	0.29	(10.0)	33.0	5
	9D	0.29	(2.4)	8.0	5
AD Leo	2	0.10	1.85	18.5	5
	3	0.20	1.90	9.5	5
	10A	0.20	20	100	5

TABLE 13.2 (continued)

Star	Date or number of flare	ΔU	ΔS (Jy)	ΔS/I_U	Reference
AD Leo	12A	0.20	1.8	9	5
	12B	0.30	4.5	14	5
	12C	1.15	5.03:	2.6:	5
	14	5.4	12.5	0.09	5
Wolf 424	8C	1.30	3.94	1.70	5
	8E	1.20	1.80	0.90	5
	11B	1.51	2.30	0.74	5
	12	0.2 (?)	6.40	32	5
	15	0.85	1.78	1.47	5
	16	0.20	1.80	9	5

1. B. Lovell, P. F. Chugainov, Nature 203, 1213, 1964.
2. B. Lovell, Obs. 84, 191, 1964.
3. B. Lovell et al., Nature 201, 1013, 1964.
4. B. Lovell, Nature 222, 1126, 1969.
5. S. R. Spangler, T. J. Moffett, Ap.J. 203, 497, 1976.

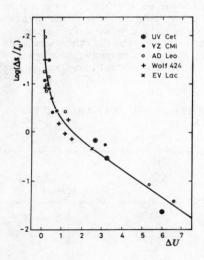

From the data in the third and fifth columns of Table 13.2 we constructed a graph of $\mathrm{Log}(\Delta S/I_U)$ versus ΔU (Fig. 13.10). The observational points are seen to lie along a well determined curve, which gives the empirical relation between $\Delta S/I_U$ and ΔU. Note that the scatter of the points around this curve is small, although the absolute luminosities of the stars used for constructing this graph differ by two orders of magnitude (from UV Cet to AD Leo).

What is the significance of the dependence of $\Delta S/I_U$ on ΔU? We cannot answer this question now, but remark only that from Fig. 13.10 the dependence of ΔS on ΔU can be derived directly. This dependence can also be represented by the following empirical formula (at ∿200 MHz):

$$\Delta S = 2.37 + 0.40\ \Delta U + 0.32\ (\Delta U)^2 \ . \qquad (13.2)$$

This formula can be applied for values $\Delta U > 0.5$. The character of the curve in Fig. 13.10 is quite uncertain in the range $0 < \Delta U < 0.5$.

Fig. 13.10. Empirical dependence of the ratio of radioflux (ΔS, at ∿200 MHz) and optical radiation in U wavelengths (I_U) at flare maximum on flare amplitude ΔU.

5. MAIN PROPERTIES OF THE RADIO EMISSION OF FLARE STARS

The following properties of the radio emission of flare stars can be stated:

1. The flux densities at radio frequencies (200-400 MHz) at flare maximum of UV Cet type stars in the neighborhood of the Sun are of the order of 2 to 10 Jy.

2. The radioflare starts earlier than the optical flare. The lower the frequency the earlier it starts.

3. The maxima of radioflares occur a few minutes after the maxima of the optical flares. The delay of the times of maximum is greater at lower frequencies.

4. The duration of the radioflare is longer than the optical flare.

5. Rise and fall of the curves of the variations of the radioflux during a flare take place more slowly than for the optical flares. Often the maxima of the radio-flares are flat and extended.

6. The dependence of the flux of radiation in the radio range on frequency can be represented in the form ν^{-n}, where the spectral index n varies not only from one flare to another, but also in the course of a given flare. The quantity n fluctuates between 0.5 and 4 or even more.

7. The frequency of radioflares is of the same order as that of optical flares (in U wavelengths). It is possible that radioflares occur more often than optical ones. Each optical flare may be accompanied by a radioflare, but the inverse need not be true.

8. The flux of radio emission of a flare follows a law represented by the curve in Fig. 13.10.

9. The radio emission of flare stars is strongly polarized.

10. In the undisturbed state of the star there is either no radio emission, or it is constant and does not show any fluctuations.

6. THE NATURE OF THE RADIO EMISSION OF FLARE STARS

The sizes and distances of flare stars are reasonably well known. It is then possible to determine the effective brightness temperatures corresponding to the radiofluxes recorded at the time of flares. For flares of UV Cet at 240 MHz the effective brightness temperature is 10^{15} K. For V 371 Ori the values of the brightness temperatures are 10^{13} K at 1410 MHz, 10^{16} K at 410 MHz, and 10^{21} K at 19.5 MHz.

These results show clearly that the radio emission of flare stars must be non-thermal. Lovell, Slee et al. come to the conclusion that the radio emission must be of synchrotron origin. It is interesting to compare the radio emission of flare stars with the Sun. In particular, it is interesting to see how the ratio of the optical flux E_0 to the radio flux E_R varies when we pass from the Sun to the flare stars. Such a comparison has been made by Lovell [3] and the results are given in Table 13.3, where the fluxes of optical radiation for a flare with amplitude $0\overset{m}{.}6-1^m$, and the radio fluxes (at 240 MHz) are given; in the last column, the ratio of the optical to the radioflux, E_0/E_R is shown.

The data in this table show that the optical radiation of flare stars is 100 times their radio emission, but for the Sun this factor is 10^5. Hence the flare stars produce 1000 times more radioenergy per unit of optical energy than the Sun does. From the data in this table it also follows that the absolute radiative power of the flare stars at radiofrequencies (their dimensions taken into account) is 10^4 to

10^6 times that for the Sun. Hence, the radiative power at radio wavelengths for "normal" flares of flare stars is considerably larger than the power of the strongest flares of the Sun. It has been known for a long time that the radio emission of the disturbed Sun is in most cases of non-thermal nature. Therefore we may accept the conclusion that the radio emission of flare stars is of non-thermal character.

TABLE 13.3 Ratio of Fluxes of Optical (E_0) and Radio (E_R) Emission for Flare Stars of UV Cet Type and for the Sun

Star	Distance (parsecs)	Flux of optical radiation, E_0 (Joules)	Flux of radio emission, E_R (Joules)	E_0/E_R
UV Cet	2.7	4.3×10^{24}	4.3×10^{22}	100
V 371 Ori	15.2	1.0×10^{26}	1.0×10^{24}	100
EV Lac	5.1	1.5×10^{26}	7.4×10^{22}	2000
YZ CMi	6.0	2.6×10^{25}	3.3×10^{22}	770
Sun	--	10^{23}	10^{18}	100000

However, the assumption made by Slee et al. that the radio emission of flare stars is synchrotron radiation due to relativistic electrons with an exponential energy spectrum appears doubtful. For explaining the observed flux of radio emission an exceedingly high density of relativistic electrons must be assumed in the atmospheres of these stars, which unavoidably leads to synchrotron self-absorption [14]. For very high densities of relativistic electrons and very small linear dimensions of the radiating object, such as the flare stars are, the radio emission will come only from a thin surface layer and the flux of radio emission will be considerably lower than the observational results.

Another mechanism for the generation of radio emission of flare stars has been suggested, namely stimulated radiation or negative absorption with synchrotron radiation [14,15]. The intensity of the radio emission J_ν of a layer of thickness ℓ is given by the following equation:

$$J_\nu = \frac{12\pi E_0}{c^2} \nu^2 \left(1 - e^{-s_\nu \ell} \right), \tag{13.3}$$

where s_ν is the coefficient of synchrotron absorption, and is approximately equal to

$$s_\nu = C n_e \frac{H_\perp}{E_0} \left(\frac{\nu}{\nu_c} \right)^{1/3}. \tag{13.4}$$

C is a constant, $H_\perp$ is the strength of the magnetic field, and E_0 is the energy of the relativistic electrons.

Computations show that in certain cases the observed intensities of the radio emission of flare stars can be explained quantitatively by stimulated emission of a monoenergetic stream of relativistic electrons with energy $E = 2 \times 10^7$ eV and a density $n_e = 10^6$ cm^{-3} in a magnetic field $H_\perp = 1$ G.

We remark that the values found for E_0, n_e and $H_\perp$, and also the type of energy spectrum of the electrons (monoenergetic) does not contradict our fundamental model of a flare star—an envelope consisting of monoenergetic fast electrons surrounding the star.

However, stimulated emission cannot be the actual mechanism for generating the radio emission in flare stars, because the frequency dependence of the radio emission differs strongly from that given by the observations. In fact, from (13.3) and (13.4) we have for the spectral index n = 2.3 for an optically thick and n = 2 for an optically thin layer. But the observations give values of n from 0.5 to 4, as we have seen above. Moreover, for a given combination of $H_\perp$ and E_0 there is negative absorption only at frequencies lower than a certain critical value, independent of the density of the relativistic electrons.

The idea that the radio emission of flare stars is purely of synchrotron origin, produced however by <u>monoenergetic</u> fast electrons with energy $\sim 10^6$ eV seems more probable. The computations made below do not exclude such a possibility.

The energy P_ν, radiated per unit of time and per unit frequency interval by one relativistic (fast) electron with energy E during the time it moves in a magnetic field with strength H is given by the expression [16]:

$$P_\nu = \frac{16e^3}{mc^2} H\, p\left(\frac{\nu}{\nu_m}\right) \qquad \text{ergs/s} \tag{13.5}$$

where the function $p(\nu/\nu_m)$ has generally a complicated form; at first it increases with increasing ν according to the law

$$p\left(\frac{\nu}{\nu_m}\right) \sim \left(\frac{\nu}{\nu_m}\right)^{1/3} \qquad \text{for } \frac{\nu}{\nu_m} \ll 1 \ . \tag{13.6}$$

Then after reaching a maximum for $\nu/\nu_m \sim 0.5$, for which p(0.5) = 0.10, it decreases rapidly with increasing ν according to the law

$$p\left(\frac{\nu}{\nu_m}\right) \sim \left(\frac{\nu}{\nu_m}\right)^2 e^{-\frac{2}{3}\frac{\nu}{\nu_m}} \qquad \text{for } \frac{\nu}{\nu_m} \gg 1 \ . \tag{13.7}$$

In these equations ν_m is the frequency at which the intensity of the synchrotron radiation reaches its maximum value; it is given by the following expression:

$$\nu_m = \frac{1}{2\pi}\left(\frac{E}{mc^2}\right)^2 \frac{eH}{mc} = 2.8 \times 10^6 \ H\gamma^2 \tag{13.8}$$

where, as before, $\gamma = E/mc^2$.

If the density of the monoenergetic electrons with energy E is called n_0, we have for the volume coefficient of radiation at radiofrequencies:

$$\varepsilon_\nu = n_0 \, P_\nu \qquad \text{ergs/cm}^3 \text{s} \ . \tag{13.9}$$

The intensity of the radio emission reaching an observer on the Earth from a flare star at distance r from us is given by the following expression:

$$J_\nu = \left(\frac{R}{r}\right)^2 \frac{\varepsilon_\nu}{\alpha_\nu}\left(1 - e^{-\alpha_\nu \Delta R}\right) , \tag{13.10}$$

where α_ν is the absorption coefficient at radiofrequencies, computed per unit path-length, ΔR is the effective linear thickness of the layer in which the magnetic field has a given strength. We shall assume for the moment that the radio emission at a given frequency ν is generated in an optically thin layer, within which the strength of the magnetic field is H. Then (13.10) can be written in the form:

$$J_\nu = \left(\frac{R}{r}\right)^2 P_\nu N_e \, , \qquad\qquad (13.11)$$

or, if P_ν is substituted from (13.5)

$$J_\nu = \left(\frac{R}{r}\right)^2 \frac{16e^3}{mc^2} N_e H \, p\left(\frac{\nu}{\nu_m}\right) \, , \qquad\qquad (13.12)$$

where $N_e = n_0 \Delta R$ is the total number of fast electrons within the radiating layer in a column 1 cm^2, R is the radius of the fast electron envelope, in which the strength of the magnetic field has the value H. To a first approximation R is of the order of several times the radius of the star's photosphere.

Equation (13.12) not only gives the value of the intensity of the radio emission, but also its frequency spectrum, which is determined by the function $p(\nu/\nu_m)$. From these results it follows that the radiospectrum must have a maximum at a certain frequency given by (13.6). Our information about the spectrum of the radio emission of flare stars is too scarce to decide whether there is such a maximum or not. It is only known, and that not quite reliably, that the intensity of the radio emission decreases with increasing frequency. This corresponds with the descending branch of the function $p(\nu/\nu_m)$, i.e. with the case $\nu/\nu_m \gg 1$. In order to simplify the computations we compare the meter region of the observed radio emission (240 MHz) with the maximum of the curve of the function $p(\nu/\nu_m)$; then we must substitute in (13.12) $p(\nu/\nu_m) = 0.10$. For an energy of the fast electrons corresponding with $\gamma^2 = 10$ and at a frequency $\sim$240 MHz, we find from (13.8) $N_e \approx 10^{17}$ cm^{-2}.

For comparison, the effective number of fast electrons required for the excitation of an optical flare with strength $\tau \sim 0.001$, is equal to $N_e \approx 10^{20}$ cm^{-2}, i.e. three orders of magnitude larger than the effective number of fast electrons for the excitation of a strong radioflare.

If we take tentatively $\Delta R \approx 10^{10}$ cm, we find for the density of the fast electrons necessary for the excitation of the radio emission, $n_0 \approx 5\times10^6$ cm^{-3}.

Thus, the synchrotron radiation of a stream of monoenergetic relativistic electrons with energy $\sim 1.5\times10^6$ eV and density 10^6 to 10^7 cm^{-3} in a magnetic field of strength ~ 10 G can explain the observed flux of radio emission during a flare. We note that the computed value of the intensity of the radio emission in the case of a hot gas (nebular hypothesis) is seven to eight orders of magnitude smaller than the observed value. We conclude that the radio emission of flare stars cannot be explained by the hot gas hypothesis.

From all this it follows that the observed radio emission can take place in those regions of the atmosphere of the flare star where two conditions are fulfilled at the same time (for an energy of the electrons $\sim 1.5 \, 10^6$ eV):

a) The strength of the magnetic field must be about 10 Gauss.

b) The density of the relativistic electrons must be smaller than a certain critical value n_{cr}; this can be determined by taking the refracting index at radio-frequencies equal to zero. The values n_{cr} for some frequencies are:

ν (MHz):	400	200	100	20
n_{cr} (cm^{-3}):	2×10^9	5×10^8	1.2×10^8	5×10^6

In regions where $n_o > n_{cr}$ generation of radio emission at the given frequency is possible, but it cannot get out from these regions.

We know nearly nothing about the magnetic fields of flare stars, but it is assumed that they may be quite strong. The existence of a region at a certain distance from the photosphere where H $\sim$ 10 Gauss, does not appear impossible. Generation of radio emission in the wavelength region of interest begins in the region where H $\sim$ 10 G only when the total column density of the relativistic electrons in that region is $N_e \sim 10^{17}$ cm^{-2}.

But with $N_e \sim 10^{17}$ cm^{-2} the optical flare cannot be observed: for this it is necessary that $N_e \sim 10^{19}$-10^{20} cm^{-2}. As the density of the fast electrons increases, it must first pass the critical value 10^{17} cm^{-2}, at which point the process of the production of radio emission begins, only after some time, when N_e reaches values 10^{19}-10^{20} cm^{-2} the optical flare can appear. In other words, the radioflare must precede the optical flare, a conclusion which is in complete agreement with the observations.

Further increase of the density of the fast electrons during the flare must lead to an increase of self-absorption, as a result of which the increase of the radio emission becomes slower. After the optical flare passes its short-lived maximum, when $N_e \sim 10^{19}$-10^{20} cm^{-2}, N_e begins to decrease to a value $N_e \sim 10^{17}$ cm^{-2}. The decrease of N_e, however, is accompanied by an increase of the radio emission, since with decreasing N_e the self-absorption gets less. Therefore in the most general case the moment the radioflare reaches its maximum may be somewhat delayed with respect to the moment the optical flare has its maximum. This conclusion is also confirmed by the observations. The self-absorption of the radiowaves, which occurs unavoidably during the time when N_e increases from 10^{17} to 10^{19} cm^{-2}, and afterwards again, when N_e falls back to 10^{17} cm^{-2}, is also the reason that the maximum of the radioflare is considerably flatter and longer than the optical maximum.

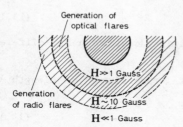

Fig. 13.11. Schematic situation of the regions around the star where the optical and the radioflare are excited.

The circumstance that N_e for the optical flare is several orders of magnitude larger than for the radioflare, and that at the time of the maximum of the optical flare the radio emission does not vanish, proves that the region where the optical flare is generated lies closer to the star than the region where the radioflare is generated. The relative positions of these two regions are shown schematically in Fig. 13.11. The boundary between the two regions may be displaced back and forth during the flare.

Further away than the region where H $\sim$ 10 G, in the outermost parts of the star's atmosphere, fast electrons with energies 10^6 eV occur; however, here no radiowaves can be produced because H $\sim$ 1 G. These outer regions can only cause absorption of the radiowaves coming from the regions where H $\sim$ 10 G. Therefore a certain dispersion in the value of the spectral index of the radio emission can be expected. Since we have relativistic electrons, the possibility exists that brief density fluctuations occur, to values 10^{18} cm^{-3}; in these cases the radiowaves may become fainter for a short time, which is sometimes observed.

V. V. Zheleznyakov [16] has also tried to explain the generation of radiowaves in flare stars by monoenergetic electrons with a moderate energy (3×10^6 eV). Here a mechanism of coherent synchrotron radiation is proposed. This mechanism requires that in the plasma with $n_e \approx 2\times10^6$ cm^{-3} in a magnetic field with a strength of a few Gauss, fast electrons appear with a density $\sim10^4$ cm^{-3} for a characteristic size of the medium $\sim10^{11}$ cm. These results do not contradict the above discussion.

We have seen earlier that in flare stars emission lines are also observed in the intervals between flares due to the fact that after the impulsive and strong increase of the ionization at the moment of the flare further processes of ionization and recombination are maintained exclusively by thermal electrons in the star's chromosphere. Here there can be no radio radiation when there are no fast electrons. Therefore the fact that there is practically no radio emission in the undisturbed state of the star should be interpreted as proof that the fast electrons have disappeared long before the following flare appears, and that the radio emission in flare stars is actually of a synchrotron nature. The contribution of the thermal electrons evidently is exceedingly small.

7. THE CONTRIBUTION OF FLARE STARS TO THE RADIO EMISSION OF THE GALAXY

Since the total number of flare stars in the Galactic system is quite large the question arises of their possible role in the general galactic radio emission. For the solution of this question it is necessary to compute the expected brightness temperature of the radio emission from flare stars and compare it with the brightness temperature of the galactic background at a given frequency. This problem was considered for the first time by Lovell [10].

Let us call E the average flux of radio emission at a given frequency which reaches us from a given flare star at a distance r_o. Let further $\Phi(M)$ be the luminosity function, i.e. the number of all stars in one cubic parsec with absolute magnitude in the interval $M\pm\frac{1}{2}$. If, finally, δ is the fraction of flare stars, we shall have for the total flux of the radio emission S reaching us from all flare stars which lie in the solid angle Ω and up to a distance R:

$$S = \Omega\delta\chi Er_o^2 \int_0^R \sum[\Phi(M)]dr , \qquad (13.13)$$

where χ is a factor which takes into account the relative duration of the different radioflares.

The brightness temperature T_S is related with the total radioflux at wavelength λ as follows:

$$T_S = \frac{\lambda^2}{2k\Omega} S , \qquad (13.14)$$

where k is the Boltzmann constant. From (13.13) and (13.14) we find:

$$T_S = \frac{\lambda^2}{2k} \delta\chi Er_o^2 \int_0^R \sum[\Phi(M)]dr . \qquad (13.15)$$

For a typical flare star (UV Cet) we have r_o = 2.6 pc, E $\sim 6\times10^{-26}$ W m^{-2}Hz^{-1} at a frequency of 240 MHz, $\chi \approx 1/30$ (for one radioflare lasting 10 minutes per 5 hours), $\delta \approx 1$ and $\Phi(M) \approx 0.18$ pc^{-3}. With these data we find from (13.15):

R, kiloparsecs	1	20	100
T_S, K	0.13	2.6	13.0

The radio maps of the Galaxy at a frequency 240 MHz [17] show a minimum value of T_S

in any direction of the order of 50 K. Out of this quantity a brightness temperature of 14 K is due to extragalactic radiosources, and 3 K to the cosmic background radiation. Then 33 K remains for the pure galactic radiobackground. Comparing these results with those obtained above, we reach the conclusion that the role of the flare stars in the total galactic radio emission is insignificant. However, this conclusion is not very reliable since most of the the calculations were made with a lot of estimates. Moreover, the role of the radio emission of flare stars in groups of stars, associations and stellar aggregates, has not been taken into account. Evidently, for definitive conclusions a more detailed analysis should be performed.

8. RADIO EMISSION OF FLARE STARS IN ASSOCIATIONS

The flare stars in the Pleiades (r = 125 pc) are about 30 times further away from us than the UV Cet type stars scattered in the neighborhood of the Sun, and those in Orion (r $\sim$ 500 pc) are about 100 times further away. Therefore the radio emission from flare stars in the Pleiades and in Orion will be about 1000 and 10000 times, respectively, weaker than that level ($\sim$1-5 Jy) which we are recording for radioflares of ordinary UV Cet type stars. This situation leaves little hope of discovering radio emission from flare stars in stellar associations and aggregates.

To a large degree this holds true also for T Tau type stars, which are permanently flaring stars. Actually, special observations performed in 1971 by Slee (cf. [18]) with the radioheliograph at Culgoora, of four selected T Tau type stars in Orion which are particularly active in the ultraviolet, gave only an upper limit for the radio emission from these stars, which was less than 0.5-0.7 Jy (at 160 MHz). Another attempt, at 5000 MHz, for a group of 12 T Tau type stars (again in Orion) gave an upper limit for the radio emission of two of these stars $\sim$0.05 Jy; for the remaining ten stars this limit was $\sim$0.02 Jy.

Attempts have been made for simultaneous optical (at three observatories) and radio (at two radio observatories) observations of flare stars in the Pleiades during October 1-6, 1972 [19]. During this period 11 optical flares were detected, and only in one case (T 53 b) an exceptionally strong optical flare (of amplitude more than 8^m in U wavelengths) was accompanied by a radioflare recorded at 170 MHz. The ratio E_0/E_R for this flare was equal to 600, in good agreement with what we found for ordinary UV Cet stars (cf. Table 13.3). Moreover, the maximum of the radioflare occurs (just as for the UV Cet type stars) after the maximum of the optical flare. The maximum itself has a rather complicated structure and essentially consists of a series of sharp radioflashes with a peak radio flux of about 35 Jy (!). This star however proved to be fainter than 22^m in U wavelengths and was not visible on the Palomar Plates. Therefore it is hard to tell if it is really a member of the Pleiades cluster. The possibility exists that this star might be a member of the Hyades cluster, which is three times closer to us (40 pc) than the Pleiades, or even that it is a member of the Ursa Major Stream, etc.

To summarize, according to the situation up to 1976, there have not been reliable cases of recording radioflares from flare stars which belong to stellar associations or young clusters, nor from T Tau type stars. However, this negative result is mostly due to observational limitations and is in no way in contradiction either with the observational facts relating to UV Cet type stars, nor with the theory. On the contrary, it indicates that those distant flare stars do not show unexpected new phenomena.

REFERENCES

1. B. Lovell, F. Whipple, L. Solomon, Nature 198, 228, 1963.
2. B. Lovell, P. F. Chugainov, Nature 203, 1213, 1964.
3. B. Lovell, Obs. 84, 191, 1964.
4. B. Lovell, F. Whipple, L. Solomon, Nature 201, 1013, 1964.
5. W. E. Kunkel, Nature 222, 1129, 1969.
6. B. Lovell, Nature 222, 1126, 1969.
7. S. R. Spangler, T. J. Moffett, Ap.J. 203, 497, 1976.
8. S. R. Spangler, S. D. Shawhan, J. M. Rankin, Ap.J. Lett. 190, L129, 1974.
9. O. B. Slee, L. Solomon, G. F. Patston, Nature 199, 991, 1963.
10. B. Lovell, Quart. J.R.A.S. 12, 98, 1971.
11. S. R. Spangler, PASP 88, 187, 1976.
12. S. R. Spangler, S. D. Shawhan, Ap.J. 205, 472, 1976.
13. S. R. Spangler, J. M. Rankin, S. D. Shawhan, Ap.J. Lett. 194, L43, 1974.
14. V. I. Slysh, Soviet Astr. 8, 830, 1965.
15. R. Q. Twiss, Austr. J. Phys. 11, 564, 1958.
16. V. V. Zheleznjakov, Soviet Astr. 11, 33, 1967.
17. R. D. Davies, C. Hazard, M.N. 124, 147, 1962.
18. G. Haro, Bol. Inst. Tonantzint. 2, 3, 1976.
19. H. M. Tovmassian, G. Haro, J. C. Webber, G. W. Swenson, K. S. Yang, K. M.
    Yoss, D. Deming, R. F. Green, Astrofizika 10, 337, 1974.
20. W. S. Gilmore, R. L. Brown, B. Zuckerman, Ap.J. 217, 716, 1977.

CHAPTER 14
X-ray Emission from Flare Stars

1. THE THEORETICAL SPECTRUM OF THE X-RAY EMISSION

During some flares of UV Cet type stars the emission line 4686 Å of ionized helium has been recorded in their spectra. This fact indicates that at the time of the flare either photons with a frequency higher than the ionization frequency of helium (i.e. λ shorter than 228 Å), or particles (electrons) with the equivalent energy are produced in the star's atmosphere. Possibly the energy spectrum of the ionizing component does not stop abruptly below the limit 228 Å; rather it may extend to the region of soft x-ray radiation--λ shorter than 100 Å. Thus at least during some flares of UV Cet type stars radiation in the x-ray wave length region must be generated as well. The question is how intensive this radiation would have to be so that it might be recorded outside the Earth's atmosphere with present-day x-ray detectors.

The first attempt to estimate the value of the expected x-ray flux from flare stars was made in 1966, on the basis of the fast electron hypothesis [1]. Later an analogous attempt was made [2] on the assumption that the radioemission of flare stars was due to magnetic bremsstrahlung. The value of the energy ($\sim$400 keV) of the electrons, their probable density ($10^{10} < n_e < 10^{15}$ cm^{-3}), and the strength of the magnetic field ($\sim$100 G) were estimated, which are compatible with the observed radio fluxes of normal flares. The expected flux of x-ray radiation due to thermal bremsstrahlung of non-relativistic electrons was, for instance, for UV Cet 2 to 3 photons/cm^{2}s on the Earth. It should be noted that in these computations the influence of self-absorption of the radio emission, which is quite large [3], was not taken into account; hence the values given in [2] for the expected flux of the x-ray radiation are clearly too low.

We have seen that the main points of the fast-electron hypothesis are in good agreement with the observations; this is also true for the radioemission. Here an analysis of the behavior of the fast electrons in the x-ray region will be made [4]. More exactly, the problem is whether x-ray radiation can be generated by <u>non-thermal bremsstrahlung</u> of fast electrons.

Let us assume that the medium where the x-ray radiation is generated consists of fast electrons and protons. Then the number of photons emitted from ω to $\omega + d\omega$ (cf. §1, Ch. 8) per unit time and per unit volume, as a result of inelastic collisions of fast electrons with fast protons (or electrons), will be:

$$P_\omega(\gamma)d\omega = n_e n_i v\sigma_\omega(\gamma)d\omega \ , \tag{14.1}$$

where n_e and n_i are the density of electrons and protons respectively, v is the velocity of motion of the electrons and the function $\sigma_\omega(\gamma)$ is given by the formulae (8.1) and (8.2). It is more convenient to give the spectrum of the photons emitted on a wavelength scale, i.e.

281

$$P_\lambda(\gamma) = P_\omega(\gamma) \frac{d\omega}{d\lambda} , \qquad (14.2)$$

where

$$\frac{d\omega}{d\lambda} = \frac{mc}{\lambda} \gamma\omega^2 . \qquad (14.3)$$

Then we have for the number of photons emitted in the interval of wavelength λ to $\lambda + d\lambda$ per unit time and per unit volume, taking $n_e = n_i$ and $v/c = 1$,

$$P_\lambda d\lambda = 4\alpha r_o^2 n_e^2 \frac{mc^2}{h} \gamma\omega f(\omega,\gamma)d\lambda . \qquad (14.4)$$

Substituting the numerical values of the constants and taking $\gamma = 3$, we find:

$$P_\lambda d\lambda = 9.1\times10^{-7} n_e^2 \omega f(\omega,\gamma)d\lambda \quad \text{photons/cm}^3\text{s} . \qquad (14.5)$$

Actually the factor $\omega f(\omega,\gamma)$ is the law of the distribution of the number of emitted photons with respect to wavelength, i.e. the spectrum of the x-ray radiation. The numerical values of the function $f(\omega,\gamma)$, computed with formula (8.2), and also the

TABLE 14.1 Values of the Functions $P_\lambda \sim \omega f(\omega,\gamma)$ and $J_\lambda \sim \omega^2 f(\omega,\gamma)$ for $\gamma = 3$

ω	λ (Å)	$f(\omega,\gamma)$	$\omega f(\omega,\gamma)$	$\omega^2 f(\omega,\gamma)$
0.55	0.0148	0.0319	0.01754	0.00965
0.50	0.0162	0.2159	0.1079	0.05395
0.45	0.0180	0.3981	0.1791	0.08059
0.40	0.0203	0.6026	0.2410	0.09640
0.35	0.0232	0.8255	0.2889	0.10111
0.30	0.0271	1.0878	0.3263	0.09789
0.25	0.0324	1.3936	0.3484	0.08710
0.20	0.0406	1.7681	0.3536	0.07072
0.15	0.0541	2.2478	0.3372	0.05058
0.10	0.0812	2.912	0.2912	0.02912
10^{-2}	0.812	6.42	6.42×10^{-2}	6.42×10^{-4}
10^{-3}	8.12	9.49	9.49×10^{-3}	9.49×10^{-6}
10^{-4}	81.2	12.56	1.26×10^{-3}	1.26×10^{-7}
10^{-5}	812	15.63	1.56×10^{-4}	1.56×10^{-9}
5×10^{-6}	1624	16.55	8.27×10^{-5}	4.13×10^{-10}
4×10^{-6}	2030	18.85	7.54×10^{-5}	3.02×10^{-10}
3×10^{-6}	2707	17.23	5.17×10^{-5}	1.55×10^{-10}
2×10^{-6}	4060	17.78	3.56×10^{-5}	0.71×10^{-10}
1×10^{-6}	8120	18.79	1.88×10^{-5}	0.19×10^{-10}

functions $\omega f(\omega,\gamma)$ and $\omega^2 f(\omega,\gamma)$ are given in Table 14.1. The first of these functions is the relative number of photons P_λ, and the second function is the relative intensity J_λ of the x-ray radiation at wavelength λ.

The x-ray spectrum due to nonthermal bremsstrahlung of fast electrons with $\gamma = 3$, i.e. the functions P_λ and J_λ are represented graphically in Fig. 14.1 (relative values are given, with $P_\lambda = J_\lambda = 1$ at $\lambda = 100$ Å). Here the maximum of the spectrum lies at $\lambda \approx 0.040$ Å for the number of photons and at $\lambda \approx 0.023$ Å for the intensity of the radiation. In the long wavelength region P_λ decreases inversely proportional to the first power of λ.

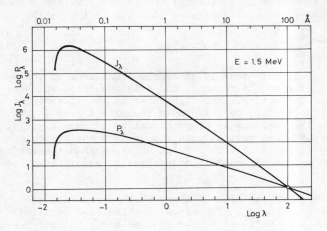

It should be pointed out that the spectrum of the x-ray radiation is not very sensitive to the value of the energy of the fast electrons. The form of this spectrum hardly changes even when the electron energy is increased by a factor of ten or twenty.

Fig. 14.1. X-ray spectrum of the flare of a star due to non-thermal bremsstrahlung of fast electrons $\gamma = 3$ (E = 1.5 MeV). P_λ is the number of photons, J_λ the intensity of the radiation; both quantities are normalized at $\lambda = 100$ Å.

The theoretical spectrum of the x-ray radiation which can be generated during a stellar flare as a result of non-thermal bremsstrahlung of fast electrons does not depend on the temperature of the star and on the flare parameters; this spectrum is the same for all stars and for all x-ray flares. Herein lies the difference between the x-ray flares and the optical flares, for which the observed spectrum depends also on the stellar temperature and on the strength of the flare itself.

2. THE LIGHTCURVE OF X-RAY FLARES

Let us call the volume of the cloud, or the envelope, near the star consisting of the fast electrons, V. Then the total number of x-ray photons generated in this volume per unit time will be:

$$N_\lambda d\lambda = P_\lambda V d\lambda \quad \text{photons/s} . \tag{14.6}$$

From this we have the flux of x-ray photons, N_λ^o, reaching the Earth:

$$N_\lambda^o d\lambda = \frac{N_\lambda d\lambda}{4\pi r_*^2} \quad \text{photons/cm}^2\text{s} \tag{14.7}$$

where r_* is the distance of the star from the observer.

In the case where the cloud of fast electrons surrounds the star more or less uniformly, forming an envelope with an outer radius $R = qR_*$ and an inner radius R_*, then the volume V of this envelope and its optical thickness τ for Thomson scattering processes can be represented by the equations (8.10) and (8.11). Then we can have from (14.7):

$$N_\lambda^o d\lambda = 6.9 \times 10^{41} \tau^2 \frac{q^3 - 1}{(q-1)^2} \frac{R_*}{r_*^2} \omega f(\omega, \gamma) d\lambda . \qquad (14.8)$$

The fraction $(q^3 - 1)/(q-1)^2$ is of the order of 10 (see §1, Ch. 8). Therefore we have approximately for the flux of x-ray photons, reaching an observer on the Earth:

$$N_\lambda^o d\lambda = 7 \times 10^{42} \tau^2 \frac{R_*}{r_*^2} \omega f(\omega, \gamma) d\lambda \quad \text{photons/cm}^2 \text{s} . \qquad (14.9)$$

This is the lightcurve of the x-ray flare, since it gives the dependence of the number of x-ray photons emitted during the flare on τ, and this quantity, which is largest at flare maximum, decreases with time. The lightcurve of the flare in the optical part of the spectrum is proportional to τ (formula (4.33) and also §4, Ch. 15). In the case of x-ray flares the spectrum is proportional to τ^2 (see (14.9)). Therefore the x-ray flare must decay after maximum much faster than the optical flare. In other words, the x-ray flare is of much shorter duration than the optical flare. This should be considered as one of the main properties of x-ray flares of UV Cet type stars.

Furthermore, it follows from (14.9) that the flux of x-ray photons reaching the Earth is proportional to the star's radius R_*. Therefore, for the same amplitude of the optical flare, i.e. for the same value of τ, the flux of x-ray photons reaching us can be different depending on the radius of the star or on its absolute luminosity.

3. EXPECTED FLUXES OF X-RAY RADIATION

Let us now estimate the expected flux of x-ray photons from a stellar flare.

We start with UV Cet--one of the intrinsically faintest flare stars. For this star we have $R_* = 0.08 R_\odot$, $r_* = 2.7$ pc. With these data we find from (14.9):

$$N_\lambda^o d\lambda = 5.6 \times 10^{14} \tau^2 \omega f(\omega, \gamma) d\lambda \quad \text{photons/cm}^2 \text{s} . \qquad (14.10)$$

For the total number of photons in the wavelength interval from λ_1 to λ_2 we have:

$$N(\lambda_1 < \lambda < \lambda_2) = \int_{\lambda_2}^{\lambda_1} N_\lambda^o d\lambda . \qquad (14.11)$$

The maximum amplitude which has been recorded for flares of UV Cet is $\sim 6^m$ in U wavelengths. This corresponds to $\tau = 0.01$. The fluxes of x-ray photons from UV Cet found from (14.10) and (14.11) for different values of $\tau \leq 0.01$ or $\Delta U \leq 6^m$ and various wavelength intervals are given in Table 14.2. The last line of the table shows the frequency of flares, $\Phi(\Delta U)$ with a given amplitude, ΔU, computed by means of equation (1.11) and from the data in Tables 1.3 and 1.8.

From the data in Table 14.2 it follows that the expected flux of x-ray photons from UV Cet is very large--from ~ 300 to ~ 3 photons/cm^{2}s in the region $\lambda < 100$ Å. In the first case one such flare is expected on the average during each day of observation, in the second case considerably more, about once every 3.5 hours of observation. The density of fast electrons at the time of flare maximum will then be 10^{10} to 10^{12} cm^{-3}. In the case of AD Leo, one of the intrinsically brightest flare stars, the picture is different. First of all, on account of the equality of the energy spectra of x-ray flares for all stars, the fluxes of x-ray photons N_1 and N_2 from two stars of the same spectral class can be expressed, according to (14.10) and (14.11), directly with their absolute luminosities L_1 and L_2 and their distances r_1

TABLE 14.2 Computed Fluxes of X-ray Radiation (photons/cm^{2}s) on the Earth from UV Cet and AD Leo for Various Strengths (τ) or Amplitudes (ΔU) of the Flare

	UV Cet			AD Leo		
τ	0.01	0.001	0.0001	0.006	0.001	0.0001
ΔU	$6^m.0$	$3^m.6$	$1^m.4$	$5^m.0$	$3^m.6$	$1^m.4$
< 0.1 Å	14	0.14	0.001	14	0.4	0.004
0.1 - 1 Å	88	0.9	0.01	100	2.8	0.03
1 - 10 Å	183	1.8	0.02	210	6.0	0.06
10 - 100 Å	270	2.7	0.03	310	8.6	0.09
Flare frequency f(ΔU) flares/hour	0.04	0.3	1.7	0.004	0.01(?)	0.07

and r_2:

$$\frac{N_1}{N_2} = \left(\frac{L_1}{L_2}\right)^{\frac{1}{2}} \left(\frac{r_2}{r_1}\right)^2 . \tag{14.12}$$

From this equation we find that the flux of x-ray radiation from AD Leo in any spectral region must be 3.2 times that of UV Cet. But the maximum flare amplitudes of AD Leo in U wavelengths recorded up till now have been smaller than 3^m; only in one case (17.II.1974) a flare was recorded with $\Delta U = 5^m.0$ [5], corresponding with $\tau \approx 0.006$. As a result the expected maximum fluxes of x-ray photons for most of the flares of AD Leo are found to be of the same order as in the case of UV Cet (Table 14.2). However, the frequency of appearances of x-ray flares in the case of AD Leo is 10 to 30 times lower than in the case of UV Cet, so that 10 days of observing time will be needed for recording one x-ray flare. The density of the fast electrons in the atmosphere of AD Leo at flare maximum will be about 10^9 to 10^{10} cm^{-3}.

Hence, UV Cet is potentially a more probable source of x-ray flares than AD Leo. Though the calculations are only estimates, it is not very probable that these conclusions can change considerably. Besides, such an analysis can be made for each of the flare stars of UV Cet type in the neighborhood of the Sun, with the aim of finding the most probable candidates for making patrol observations in the x-rays.

In principle x-ray radiation during a stellar flare can also be produced by repeated inverse Compton effects. However, estimates show that in this case the flux of x-ray photons is considerably smaller than in the case of non-thermal bremsstrahlung.

The number of x-ray photons with frequency higher than ν_o, created by n-fold scattering of a photon with fast electrons (n-fold inverse Compton effect) per unit time and per unit surface of the photosphere, can be determined from the following formula:

$$N(>\nu_o) = \frac{3}{4\pi} \left(\frac{kT}{4}\right)^3 \frac{\gamma^{2n}}{c^2} \left|\tau F_2(\tau) e^{-\tau}\right|^{n-1} J_2\left(\frac{x_o}{\gamma^{2n}}\right) . \tag{14.13}$$

The expected flux of x-ray radiation found from this formula is very small, about 0.003 photons/cm^{2}s in the region $\lambda < 100$ Å and for n = 3. Hence the expected x-ray radiation of flare stars will be practically due to non-thermal bremsstrahlung from fast electrons.

The x-ray radiation from flare stars which are members of stellar associations and clusters is expected to be exceedingly small due to their great distances from us, although in some cases this radiation may be within the limits of sensitivity of present-day x-ray detectors. As an example Table 14.3 gives estimated values of the expected x-ray fluxes in the wavelength region $\lambda < 10$ Å from a flare star which is a member of a stellar association, on the assumption that the radiation power of this star is the same as that of UV Cet during exceptionally strong flares, i.e. when the flux of x-ray radiation reaches 300 photons/cm^{2}s (on the Earth), i.e. when $\tau = 0.01$ or $\Delta U = 6^m$.

TABLE 14.3 Expected X-ray Fluxes in the Region $\lambda < 10$ Å from a Flare Star Which is a Member of a Stellar Association

Stellar group	r_* parsecs	X-ray flux on the Earth photons/cm^{2}s
Hyades	40	1.2
Pleiades	140	0.1
Orion	500	0.01

Thus, the fast-electron hypothesis also predicts the possibility of x-ray radiation being generated during stellar flares. The most characteristic properties of this radiation are the following:

a) The duration of the x-ray flare must be considerably smaller than that of the optical flare (cf. following section).

b) The frequency of x-ray flares accessible to observation must be considerably lower than the frequency of optical flares.

Are there any real possibilities of checking this theoretical prediction? Evidently, there are. The most direct and most convincing method of checking might be to send a special astrophysical observatory beyond the Earth's atmosphere. Such an observatory must have one large optical telescope, and an x-ray detector with good angular resolution. The optical axes of the two types of receivers must be parallel to each other. Simultaneous optical and x-ray observations are required.

There is also the following indirect possibility [4]. According to the above discussion, x-ray radiation will appear in flare stars only at the moment of the flare--the rest of the time there is no such radiation. Therefore, if regular observations of the sky are made in x-rays (a kind of "x-ray patrol") by means of specialized x-ray satellites, we may detect something like "x-ray flashes". Though variability of this kind in x-ray sources of cosmic origin can also be due to other causes, part of these sources may turn out to be flare stars.

4.   RESULTS OF OBSERVATIONS OF FLARE STARS IN X-RAYS

During the period 1967-1975 two attempts were made to detect x-ray radiation during flares of UV Cet type stars by direct observations outside the Earth's atmosphere. The first of these was undertaken on the orbiting solar observatory OSO-3, which was equipped with x-ray detectors sensitive to photons with energies in the range 7.7-12.5 KeV (1.6-1.0 Å). Observations were made in 1967-68 of four flare stars, UV Cet, YZ CMi, EV Lac and AD Leo. The results of these observations were published in 1975 [6].

The second attempt to discover x-ray stellar flares was undertaken with the Astronomical Satellite of the Netherlands--ANS--during October 19 to 22, 1974, and January
3-9, 1975. Two flare stars were observed, namely YZ CMi and UV Cet [7]. In this
case the observations were made with two groups of x-ray detectors, sensitive to the
energy ranges 0.2 to 0.28 KeV (62-44 Å), and 1 to 7 KeV (12.4-1.8 Å).

In both cases--OSO-3 and ANS--the x-ray observations were made with an integration
time of about 15 seconds. During the observations of OSO-3 about 80 optical flares
from the four stars mentioned above were seen. However, not a single x-ray flare
was seen in the range 1.6-1.0 Å; only statistical upper limits can be derived.

During the period of observations by ANS one flare of YZ CMi was recorded very
clearly (October 19, 1974), in two spectral ranges--62-44 Å ($\overline{\lambda} \approx 50$ Å) and 12.4-1.8
Å ($\overline{\lambda} \approx 7$ Å)--simultaneously, however, without simultaneous optical observations.
The x-ray lightcurves of these observations are shown in Fig. 14.2. The total

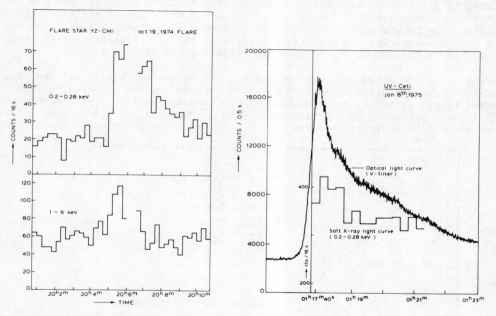

Fig. 14.2. X-ray lightcurves of a flare of
YZ CMi (19.X.1974) in two wavelength
regions: 62 to 44 Å (top), and 12.4 to
1.8 Å (bottom).

Fig. 14.3. Optical and x-ray
lightcurves, found during a flare
of UV Cet (8.I.1975).

energy liberated at the time of this flare of YZ CMi ($r_* = 6.06$ pc) was $4.2 \pm 0.3 \times 10^{31}$
ergs and $1.9 \pm 0.4 \times 10^{32}$ ergs in the spectral regions with $\overline{\lambda} = 50$ Å and $\overline{\lambda} = 7$ Å,
respectively. The intensities at the maxima of the lightcurves were $2.5 \pm 0.4 \times 10^{29}$
ergs/s and $3.6 \pm 0.7 \times 10^{30}$ ergs/s, or 1.9 photons/cm^{2}s and 1.8 photons/cm^{2}s for the
fluxes of radiation at the Earth.

In the case of UV Cet ANS recorded a strong x-ray flare on January 8, 1975. This
flare was fortunately quite strong (Fig. 14.3): its amplitude in V wavelengths was
$\sim 2^m$, which corresponds with $\Delta U = 6^m.2$ ($\tau = 0.08$, cf. Fig. 8.2). Unfortunately, the
observations of the x-ray radiation for this flare were made only in one wavelength
region, 62-44 Å. The total energy emitted in this spectral range was $2.9 \pm 0.6 \times 10^{29}$

ergs, and the x-ray luminosity at the time of flare maximum was $6.1 \pm 1.3 \times 10^{28}$ ergs/s, or 2.3 photons/cm^{2}s for the flux at the Earth.

It is interesting to note that the x-ray luminosities of UV Cet and YZ CMi at the time of these flares were about 30 times and more than 100 times, respectively, the luminosity of these stars in U wavelengths. However, the ratio of the x-ray luminosity of the flare L_x to the <u>total</u> optical luminosity of the flare L_{opt} in this case was less than one ($L_x/L_{opt} = \overline{0.03}$ for UV Cet).

Observations by ANS established upper limits for the x-ray emission of four faint optical flares of UV Cet (January 4-6, 1975) in the two spectral regions, $\overline{\lambda} \approx 50$ Å and $\overline{\lambda} \approx 7$ Å.

Although the data is clearly incomplete, x-ray radiation has been discovered at the moment of a flare in flare stars of UV Cet type. This is undoubtedly important for investigating the physical processes taking place in these objects. There is no doubt that by understanding the nature of the x-ray radiation we shall finally understand the nature of flares and other phenomena in unstable stars. From this point of view comparing the spectra of x-ray radiation of flare stars with spectra of the x-ray radiation of the coronas of ordinary stars may prove useful.

Constant x-ray radiation has been established in two stars, Capella and the Sun, in which the x-ray radiation is emitted by their coronas and has a thermal origin (thermal bremsstrahlung), corresponding with a corona temperature of the order of 2×10^6 to 10^7 K. Using the observational data for Capella [8] and the Sun [9] we can construct the averaged spectrum of the "stellar x-ray emission"; that is shown

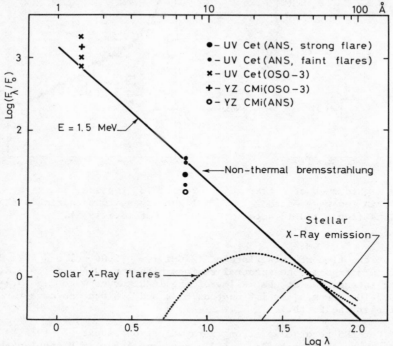

Fig. 14.4. Comparison of averaged "stellar emission" with data about an x-ray flare of YZ CMi (real measurements), and UV Cet (upper limits). The dotted line shows the spectrum of strong x-ray flares of the Sun. The full-drawn line gives the non-thermal bremsstrahlung of fast electrons for E = 1.5 MeV.

in the lower right-hand angle of Fig. 14.4, with the value of the intensity at
$\lambda = 50$ Å taken as unit. The maximum of the "stellar x-ray emission" lies at ~ 50 Å,
and in the region $\lambda < 50$ Å we observe a decrease of the intensity.

In the same figure the results of the x-ray observations of YZ CMi and UV Cet are
given. The figure shows the ratio of the intensities $F(\overline{\lambda})/F_o$ (where $F_o = F(50$ Å$)$)
for the spectral ranges $\overline{\lambda} \sim 50$ Å and $\overline{\lambda} \sim 7$ Å.

It should be pointed out that only one symbol in Fig. 14.4 (the circle for YZ CMi)
represents an actually measured ratio of radiation fluxes; the other symbols give
only upper limits for these ratios (for details cf. [10]).

This figure shows that the x-ray spectrum of a flare star (YZ CMi) continues to get
stronger in the short-wavelength region and its maximum lies in the region $\lambda < 7$ Å
(circle), i.e.   at a wavelength at least one order of magnitude shorter than the
wavelength of maximum in the case of stellar coronas. Thus we have some indication
that there exists a difference between the x-ray spectra of flare stars and of
stellar coronas.

This conclusion is also valid when applied to solar x-ray flares. The effective
temperature of solar x-ray flares is higher than 10^7 K. The observed spectrum of
strong x-ray flares of the Sun [11] is also shown in Fig. 14.4 (dotted line). In
the short-wavelength region these two spectra of an x-ray flare--of YZ CMi and of
the Sun--go in opposite directions, at least in the region 50-7 Å, i.e. there is
an increase in the first case and a decrease in the second.

In Fig. 14.4 the full-drawn line shows the theoretical spectrum of the radiation
due to non-thermal bremsstrahlung of fast electrons for E = 1.5 MeV (Table 14.1).
This spectrum is, at least qualitatively, in agreement with the observations,
specially in the case of the flare of YZ CMi. In principle it is possible to
explain the observed data for YZ CMi with thermal bremsstrahlung of thermal
electrons with an effective temperature about 5 to 10 times as high as what we have
in the case of strong solar flares. Therefore it is exceedingly difficult to draw
definite conclusions from these data about the question of whether the x-ray flares
are of thermal or non-thermal nature. Only observations made in the wavelength
region 1-2 Å or shorter will give a convincing answer to this question. As we have
seen above, the maximum of the x-ray spectrum in the case of non-thermal bremsstrahlung is expected at ~ 0.05 Å for the number of photons, or ~ 0.03 Å for the intensity.

One of the important aspects of the fast-electron theory refers to the <u>duration</u> of
flares; it must be considerably less than the duration of optical flares. The only
example thus far, where both the optical and the x-ray lightcurves are known is
the flare of UV Cet (Fig. 14.3); it confirms this prediction: The optical flare
lasted about 8 minutes, whereas the x-ray flare lasted about 50 seconds, ten times
shorter.

Most optical flares hardly last more than one minute. From this it follows that
the duration of the x-ray flares must be exceedingly short, of the order of a few
seconds. Such x-ray flares, or rather flashes, cannot be recorded with detectors
with time constants of the order of 15 seconds. The observed intensities of the
x-ray flashes with long integration times will be much lower than the expected
values. It is possible that this is the reason for the observed point for YZ CMi
in Fig. 14.4, which lies below the theoretical spectrum. Perhaps this is also the
reason for the negative results of flare observations by OSO-3 in the region 7.7-
12.5 KeV, mentioned above.

The fast-electron theory predicts very high intensities at the moment   of maximum
of x-ray flares. Unfortunately this point is not often taken into account when the

theory is compared with the observations [6,12]. The fact that the star spends an excessively short period of time at maximum results in lower observed intensities compared to theoretical predictions. The same applies for the recorded flares of UV Cet and YZ CMi. In the first case the theoretically expected value of the intensity of the radiation N_t in the region 0.2-0.28 KeV (62-44 Å) will be approximately (using the data of Table 14.2 for $\Delta U = 6^m2$ and for the wavelength range 10-100 Å): $N_t = [(62-44)/90]270 = 54$ photons/cm^2s. This value is about 25 times the observed intensity (2.3 photons/cm^2s). Evidently, this discrepancy becomes smaller with decreasing integration time of the x-ray counting system. The whole question of course deals with the duration of the maximum in the x-ray lightcurve. Some estimate of this value can be obtained from the following considerations.

Cristaldi and Rodono [13], working with a time resolution of 0.5 s, succeeded in recording the fine structure of the lightcurve at flare maximum at optical wavelengths. They discovered very sharp-pointed, needle-like lightcurve and flare maximum of very short duration. It appears that important variations of the brightness can take place during fractions of a second. But we have seen above that the duration of the x-ray flare is one order of magnitude less than that of optical flares. Therefore we come to the conclusion that the duration of the maximum in the x-ray lightcurve must be excessively short, of the order 0.01 sec. Such sharp maxima cannot be recorded with instruments which have time constants of tens of seconds.

We have seen that we can determine the relative forms of the lightcurves of a stellar flare in three fundamental ranges of the electromagnetic spectrum: the x-ray, optical and radio regions; such a comparison is shown in Fig. 14.5. The x-ray flares must be considered to be the shortest, the radio-flares the longest. There is also a fixed sequence in the appearance of the flares: first appears the radioflare, then the optical flare, and later the x-ray flare.

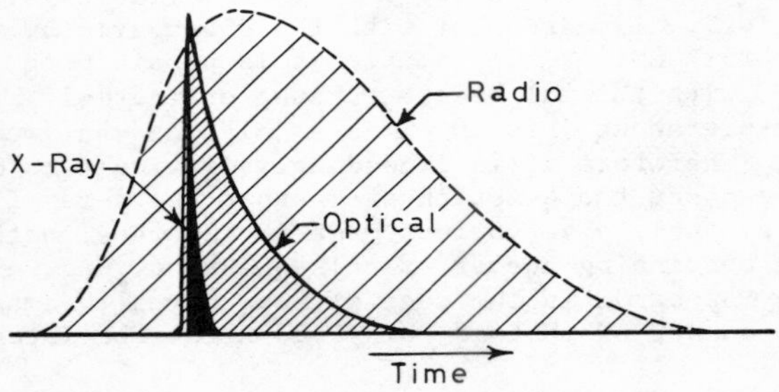

Although the available x-ray observational data is limited, we can still conclude that the hypothesis that fast electrons is the main one for the generation of all types of flares, optical, radio and x-ray, is valid.

Fig. 14.5. Lightcurves for optical, radio and x-ray flares for stars of UV Cet type (model).

5.   ON THE X-RAY BACKGROUND OF THE GALAXY

In connection with the x-ray radiation of flare stars the question arises whether this leads to the creation of a general x-ray background of the Galaxy.

The total number of flare stars in the Galaxy must be large, probably of the order 10^9 to 10^{10}, if their density is assumed to be the same everywhere in the Galaxy, and corresponding with a number of about 50 such stars within a sphere of radius 20 parsecs around the Sun. Moreover, the expected flux of the x-ray radiation during flares is quite large, of the order of several photons per square centimeter per second.

However there are factors leading to a weakening of such an assumed x-ray background.

This is the rapid decrease of the intensity of the stream of x-ray radiation of a flare star with its increasing distance from the Sun, which is inversely proportional to the square of the distance, etc. A quantitative analysis is in order to answer this question unambiguously.

Let the space density of flare stars be the same everywhere in the plane of the Galaxy and equal to $n(0)$. At distance z from the plane of the Galaxy their concentration will be:

$$n(z) = n(0)e^{-\beta z} = n(0)e^{-\beta r \sin\theta} , \qquad (14.14)$$

where r is the distance of the star from us, and θ is the position angle measured from the Galactic plane. Interstellar absorption can be neglected if we restrict ourselves to x-rays with λ shorter than 30 Å.

The number of flare stars dn per unit volume at a distance r and in the direction θ, will be:

$$dn(r,\theta) = n(0)e^{-\beta r \sin\theta} r^2 dr \, d\Omega , \qquad (14.15)$$

where $d\Omega = 2\pi\cos\theta d\theta d\phi$.

Let p_x be the probability that at a given moment the star is in the state of an x-ray flare. Then,

$$p_x = \frac{\Delta t_x}{t_x} \qquad (14.16)$$

where Δt_x is the duration of an x-ray flare, and t_x is the interval of time between two successive x-ray flares.

Assuming a constant radiative power E_* for all flare stars, we have for the flux of x-ray radiation reaching the observer from all flare stars within a volume element of the Galaxy,

$$dE(r,\theta)d\Omega = \frac{E_*}{4\pi r^2} dn(r,\theta)p_x d\Omega . \qquad (14.17)$$

The flux of x-ray radiation reaching the observer from all flare stars lying along a path from the Sun to a distance r, will be by integration:

$$\frac{E(r,\theta)}{E_*} d\Omega = \frac{d\Omega}{4\pi} \frac{n(0)p_x}{\beta\sin\theta} (1 - e^{-\beta r \sin\theta}) . \qquad (14.18)$$

Equation (14.18) gives the value of the brightness of the x-ray background of the Galaxy expressed in units of flux of the assumed x-ray radiation E_* of the nearest flare star. The distance r is also in units of the distance of the nearest flare star (i.e. in units of $\sim$10 parsecs).

With $\beta r \gg 1$ and $\sin\theta \neq 0$, we have from (14.18):

$$\frac{E(r,\theta)}{E_*} \sim \frac{1}{4\pi} \frac{p_x}{\beta} n(0) . \qquad (14.19)$$

The space density of flare stars in the neighborhood of the Sun is of the order $n(0) \approx 10^{-3}$ star per cubic parsec. There are reasons to assume that the probability of an x-ray flare is at least one order of magnitude smaller than that of an optical flare, i.e.

$$p_x \approx 0.1 \ p_{opt} = 0.1 \ \frac{\Delta t_0}{t_0} \ . \tag{14.20}$$

Taking the mean duration of an optical flare $\Delta t_0 \sim 2$ minutes and assuming their average frequency to be one flare per day (with amplitude $\Delta U \sim 2^m - 3^m$) we find $p_{opt} \sim 10^{-3}$ or $p_x \sim 10^{-4}$. With these data (14.19) gives:

$$\frac{E(r,\theta)}{E_*} \approx \begin{array}{l} 10^{-6} \text{ for } \beta \sim 10^{-2} \\ 10^{-5} \text{ for } \beta \sim 10^{-3} \end{array} , \tag{14.21}$$

which means that the x-ray background of the Galaxy due to flare stars must be weak, of the order of 10^{-5} photons/cm^{2}s sterad (with $E_* \approx 1$ photon/cm^{2}s). However, this conclusion can change qualitatively if the initial data are changed. By direct observations the value of the x-ray luminosities of flare stars can be made more accurate, and similarly their effective duration.

Considering only the plane of the Galaxy the x-ray background due to flare stars is not negligible. For this background we have:

$$E(r,\theta) \sim \frac{1}{\beta \sin\theta} (1 - e^{-\beta r \sin\theta}) \ . \tag{14.22}$$

For a given value of β the galactic x-ray brightness becomes stronger with increasing r up to a certain distance, from which the x-ray radiation can reach us without significant absorption. This limiting distance in the plane of the Galaxy increases with decreasing wavelength (the absorption of x-ray radiation in the interstellar medium is due to hydrogen atoms and is proportional to λ^{-3}). Therefore the relation (14.22) can also be formulated differently: the galactic x-ray brightness must increase from longer to shorter wavelengths.

Equation (14.18) gives a statistical mean of the value of the brightness of the x-ray background of the Galaxy. For higher values of the flux of x-ray radiation from discrete sources (flares) we must see only individual x-ray flares lasting much shorter than a minute, from relatively nearby flare stars. In other words, we must observe a kind of x-ray scintillations. The frequency of these scintillations, f, will on the average increase with decreasing intensity of the x-ray source. Actually, the total number of flares which can be observed simultaneously within a band or thickness at a distance r from us, will be $r^2 d\Omega n(r,\theta) p_x$. This at the same time is the flare frequency, i.e. $f \sim r^2$. For the average flare intensity we have $E \sim r^{-2}$. From this the following interesting relation between flare frequency in a given direction and observed flare intensity follows:

$$f \ E = \text{constant} \ . \tag{14.23}$$

Thus, if scintillating sources of cosmic x-ray radiation could be discovered, and if they are from flare stars, then it should be expected that formula (14.23) holds.

6. THE POSSIBILITY OF DETECTING INTERSTELLAR HELIUM

Interstellar space must be completely opaque for radiation in a wide wavelength range, 912 Å to at least 50 Å. The absorption is mainly due to interstellar hydrogen and it is weaker towards the short wavelengths (proportional to λ^{-3}). This seems true for ordinary stars.

The flare stars are very close to the Sun; therefore for them the opacity zone in the spectrum will be considerably narrower. Thus for UV Cet the optical depth of interstellar neutral hydrogen, t_λ, will be of the order of one at $\lambda \sim 250$ Å, if we take for the density of the hydrogen atoms $n_H = 1$ cm^{-3}; and at $\lambda \sim 550$ Å, if $n_H = 0.1$ cm^{-3}. This means that in principle a sufficiently strong flare of UV Cet can be recorded in the region of wavelengths shorter than 250 Å in the first case, and shorter than 550 Å in the second case. In connection with this an interesting possibility arises of discovering interstellar helium, making use of the fact that in this region a boundary jump occurs (at 504 Å) in the continuous absorption of neutral helium.

The continuous emission during a flare in the region ~ 500 Å is due equally to inverse Compton effects and to non-thermal bremsstrahlung. The latter, however, plays the predominant role in the region $\lambda < 550$ Å. The spectrum of this radiation is represented by the broken line 3 in Fig. 14.6.

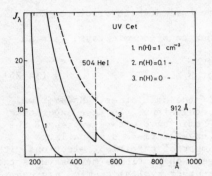

Furthermore, if it is assumed that the density of interstellar helium is 10 times less than that of hydrogen, we can write for the optical depth (in the region $\lambda < 504$ Å):

$$t_\lambda = [6.3(\frac{\lambda}{912})^3 + 0.76(\frac{\lambda}{504})^2]n_H\, 10^{-18}\, r_* \, ,$$

$$(14.24)$$

where r_* is the distance of the star from us. For UV Cet we have $r_* = 2.6$ parsecs. With these data spectra have been computed of the radiation reaching us during a flare of UV Cet for two cases: $n_H = 1$ cm^{-3} and $n_H = 0.1$ cm^{-3}. The results are shown in Fig. 14.6 (curves 1 and 2). For $n_H = 1$ cm^{-3} no jump will be visible at 504 Å; that will be seen for $n_H = 0.1$ cm^{-3}, and the value of the jump will be $J(504^+)/J(504^-) = 1.65$. Towards the short wavelengths the intensity increases rapidly, reaching a maximum in the x-ray spectral region (<100 Å).

Fig. 14.6. Theoretical structure of the continuous spectrum of UV Cet during a flare in the region 200-1000 Å. A jump in the continuous spectrum is seen at 504 Å; this is caused by absorption by interstellar neutral helium.

When we succeed in photographing the spectrum of a flare of UV Cet in the region with λ shorter than 600 Å, some absorption lines of neutral helium will be expected to appear, in particular λ 584 He I, λ 537 He I, possibly λ 522 He I, etc.

The flare star closest to us is V 645 Cen; its distance is half that of UV Cet. It will be possible to detect flares from this star even from 700 Å or shorter (with $n_H = 0.1$ cm^{-3}). With regard to the remaining flare stars of UV Cet type, it will be possible to record the flares of most of them in the region λ shorter than 300 Å. Observations of flares from outside the Earth's atmosphere, in the wavelength region shorter than 500-600 Å are certainly very important. It is expected that such observations may detect interstellar helium.

7. FLARE STARS AS CONTRIBUTORS OF PRIMARY COSMIC RAYS?

We may assume that the fast electrons find "loopholes", and escape from the magnetic fields of the stars. When this happens they leave the star practically with all of their initial energy (of the order of 1 MeV). Since the flare frequency is quite high and the total number of flare stars is not small, the question arises: are the flare stars contributors of primary cosmic rays of low energy?

First we shall try to estimate the value of the total energy transmitted to the interstellar medium in the form of kinetic energy of the fast electrons by means of stellar flares. We shall start from some average flare model, corresponding to a value $\tau \sim 0.001$. In this case the energy liberated during one flare in the form of kinetic energy of fast electrons is of the order of 10^{35} ergs. With a flare frequency of 0.1 to 1 fl/hr, the average "radiative" power of a star in the form of injection of high-energy particles will be 10^{31}–10^{32} ergs/s. If we start with a very rough estimate of the total number of flare stars in the Galaxy of the order of 10^{10}, then we have for the average power of all of the flare stars together,

$$W_{f1} \sim 10^{41} \text{ to } 10^{42} \text{ ergs/s} . \tag{14.25}$$

This estimate is very approximate and it is hard to estimate its error range.

At present supernovae are thought to be the main source of cosmic rays, with an estimated power of [15]:

$$W_{sup\ nov} \sim 10^{40} \text{ to } 10^{41} \text{ ergs/s} . \tag{14.26}$$

Comparing this with (14.25) we see that they are of the same order. However we cannot conclude that the two types of sources are equivalent. In fact supernovae inject into the interstellar medium "ready-made" cosmic rays with energies of the particles up to 10^{12} to 10^{13} eV. However, in the case of flare stars a large injection energy can only be reached due to the number of particles, electrons with energies only of 1 MeV. In the most favorable case such particles may serve as the initial "raw material" from which in some way or another typical cosmic rays can be formed. In particular, a fraction of these particles may get additional energy as a result of statistical acceleration in the interstellar medium (Fermi mechanism).

However, statistical acceleration of particles in the interstellar space in some ways does not enjoy great popularity. Discussions of its small efficiency seem to be well-founded [15]. It is assumed that in the most favorable case interstellar acceleration can increase the initial energy of the particles only by one or two orders of magnitude.

While recognizing the low efficiency of acceleration of cosmic rays, we should however take into account that in principle such an acceleration becomes efficient if there are large chaotic velocities in the gas on small scale sizes. This is more or less what we have in stellar associations and young clusters, where many flare stars occur. Therefore, contrary to the situation in the general interstellar medium, the mechanism of statistical acceleration has a chance to work within a stellar association. However, even in this case it is not very probable that the energy of the fast electrons may reach values of 10^8 eV.

Thus, there are not sufficient arguments to assume that the flare stars are injectors of primary cosmic rays which produce particles, with very high energies of the order of 10^{12} to 10^{13} eV, in the interstellar medium. However, there are reasons to believe that these stars are injectors of particles with energies of the order of 10^6 eV, and that these particles can get a large acceleration in the interstellar medium afterwards. Therefore, with regard to the problem of the origin of cosmic rays in general, we can ask what role flare stars play in the production of the low-energy end of the cosmic ray spectrum, i.e. the region of 10^6 to 10^7 eV. It seems that their role is very important indeed.

Lovell [14] has reached the same conclusions through a somewhat different way, namely he assumes that the acceleration of the particles takes place in the flare star itself. An analogy with solar flares indicates that the collapse of the magnetic field of the sunspots is the main physical process for the excitation of flares. An argument in favor of the similarity of the physical processes in the generation of flares of the Sun and flare stars seems to be that the ratio of the energy liberated during the flare at radiofrequencies ($\sim 10^3$ MHz) to the bolometric energy is the same ($\sim 10^{-5}$) for M and K type stars, and for the Sun. The important parameter which is the fraction of the total energy of the flare which is transformed in cosmic-ray energy can be estimated from solar flare data; it is ~ 0.01, i.e. only 1% of the flare energy is transformed into cosmic-ray energy.

According to Lovell the total energy liberated by flare stars in the form of cosmic rays is due to M and K type flare stars. This cosmic ray energy is in the spectral region of 10^6 to 10^8 eV, and partly 10^9 eV.

While agreeing with the main conclusions made by Lovell regarding the role of the flare stars in the production of the low-energy part of the cosmic-ray spectrum, we must remark that the possibility of additional acceleration of low-energy electrons ($\sim 10^6$ eV) in the magnetic fields of the stellar spots does not mean that the flares receive their energy from the energy of the magnetic fields. The magnetic collapse and the direct liberation of energy in the form of electromagnetic radiation is not so evident. We shall return to the question of magnetic fields in the last chapter.

Concerning the acceleration of charged particles we can discuss a stream or bunch of fast electrons moving with a velocity close to the velocity of light. If such a group of electrons catches and takes with it a certain number of protons, their energy will be (if at the same velocity as the fast electrons) about three orders of magnitude higher than the energy of the electrons. It is clear that this mechanism of proton acceleration may work most efficiently in places where there is a large concentration of fast electrons, i.e. in the medium immediately surrounding the star. This means that even before leaving the star such protons may have energies of the order of 10^9 eV. This mechanism allows the possibility that flare stars inject not only fast electrons but also protons with high energy. By additional acceleration in the interstellar medium the energy of the protons can reach up to 10^{10} eV. Considering the large number of flare stars and the very high frequency of the flares we cannot entirely exclude the possibility that flare stars are sources of primary cosmic rays (proton component).

REFERENCES

1. G. A. Gurzadyan, Doclady Academy Nauk SSSR 166, 821, 1966.
2. T. E. Grindlay, Ap.J. 162, 187, 1970.
3. V. I. Slysh, Astron. J. USSR 41, 1038, 1964.
4. G. A. Gurzadyan, Astron. Astrophys. 13, 348, 1971.
5. K. Osawa, K. Ichimura, Y. Shimizu, H. Koyano, IBVS No. 906, 1974.
6. V. Tsikoudi, H. Hudson, Astron. Astrophys. 44, 273, 1975.
7. J. Heise, A. C. Brinkman, J. Schrijver, R. Mewe, E. Gronenschild, A. Boggende, J. Grindlay, Ap. J. Lett. 202, L73, 1975.
8. R. C. Catura, L. W. Acton, H. M. Johnson, Ap.J. Lett. 196, L47, 1975.
9. T. A. Chubb, H. Friedman, R. W. Kreplin, Proc. I-Inter. Space Sci. Symposium. Ed. H. Kallman, Amsterdam, 1960, p. 695.
10. G. A. Gurzadyan, Astrophys. and Space Sci. 48, 313, 1977.
11. C. W. Allen, Astrophysical Quantities, 3rd ed. London: The Athlone Press, 1973, p. 194.
12. P. J. Edwards, Nature 234, 75, 1971.
13. S. Cristaldi, M. Rodono, Astron. Astrophys. Suppl. 2, 223, 1970.
14. B. Lovell, Phil. Trans. Roy. Soc. London 277, 489, 1974.
15. V. L. Ginzburg, S. I. Syrovatsky, The Origin of Cosmic Rays, Moscow, 1963.

CHAPTER 15
Flare Dynamics

1. INTERPRETATION OF FLARE LIGHTCURVES

Often the lightcurve of a stellar flare after maximum is approximated by a law which is exponential with respect to time:

$$J \sim e^{-\beta t} . \qquad (15.1)$$

However, only rarely the entire curve from maximum (t = 0) to the moment where the original state of the star has been completely re-established can be represented by one value of the parameter β. In nine out of ten cases the observed values of the brightness lie higher than the curve given by the above formula [1]. This led a few investigators to divide the entire lightcurve into a number of parts with different values of β. The numerical value of β is found to be lower for parts of the lightcurve further away from maximum. An exponential law of the form (15.1) assumes a priori that a certain well-defined process is taking place in the star's atmosphere, namely extinction of the radiation, and the brightness decreases after flare maximum. Thus this method of splitting up the lightcurve into pieces with different values of β makes the interpretation of the flare phenomenon somewhat artificial.

It has also been tried to represent the lightcurve of the decrease of brightness by a formula of the form:

$$J \sim e^{-\beta t^n} , \qquad (15.2)$$

which contains two parameters, β and n. However, this formula cannot represent the entire lightcurve either [1].

Below it will be shown that the lightcurve of a flare has nothing in common with extinction of the radiation, i.e. with the law (15.1), and with (15.2), and that it can be deduced, as a consequence, of the fast electron hypothesis [2].

2. THE PROBLEM. THE EFFECTS OF ENERGY LOSSES OF FAST ELECTRONS

In a qualitative interpretation of the lightcurves of flares such factors as the form of the energy spectrum of the fast electrons, or the values of their energies, or the model assumed for the atmosphere of the star do not play an important role. The fast-electron hypothesis in itself leads to an unambiguous totally determined type of lightcurve.

In our analysis we shall start from formula (4.27), which gives the relative intensity of the radiation at a given moment of the flare and which has been derived for the model of the "real photosphere" and electrons all having the same energy:

$$J_\lambda(\tau,\gamma,T) = \left[E_4(\tau) + \frac{3}{2\gamma^4}\frac{e^x-1}{e^{x/\gamma^2}-1}F_1(\tau)\right]e^{-\tau} , \qquad (15.3)$$

where $x = hc/\lambda kT$. The parameters in this formula are the electron energy γ, the effective optical depth of the envelope or layer of fast electrons above the star's photosphere, τ, and the effective temperature of the star T. The value of the intensity of the flare J_λ in (15.3) is given as a fraction of the normal radiation of the star at a given wavelength in its stable state.

In the quiet state of the star (before the flare), when t = 0, we have $E_4(0) = 1$, $F_1(0) = 0$ and $J_\lambda(0) = 1$. Evidently

$$\Delta J_\lambda = J_\lambda - 1 \qquad (15.4)$$

will be the excess intensity, or the fraction of the additional radiation at a given phase of the flare.

Equation (15.3) gives the value of the intensity for a certain phase of the flare, characterized by the values of the parameters τ, γ and T. Formally any values of J_λ will be determined by the behavior of these parameters during the flare.

The temperature of the star, immediately after the peak of the flare, practically does not vary; some increase of the temperature (judging from the drift of the star in the U – B vs. B – V diagram (Ch. 7)) may take place in the period corresponding to the second half of the lightcurve. Hence the variations of J_λ must be related-- to a first approximation--with variations of either γ, or τ or of both at the same time, in which case these parameters will be functions of time: $\gamma = \gamma(t)$, $\tau = \tau(t)$. The influence of $\gamma(t)$ on the lightcurve will be called "the effect of energy loss of the electrons", the influence of $\tau(t)$ the "effect of expansion of the envelope of the fast electrons". We must now solve the problem of determining the forms of the functions $\gamma(t)$ and $\tau(t)$, given $J_\lambda(t)$ from the observations.

First of all we remark that $\gamma(t)$ cannot be an increasing function of time, since we observe a decrease in brightness after the maximum (the time t is counted from the moment of flare maximum). Thus, we can speak about the energy loss of the fast electrons during the flare.

The fast electrons which are above the photosphere lose energy in the following ways:

(a) Losses by magnetic bremsstrahlung (synchrotron radiation).

(b) Losses by inelastic collisions with thermal photons (inverse Compton effect).

(c) Ionization losses.

(d) Radiation losses, i.e. non-thermal bremsstrahlung of gamma-rays due to inter-
     actions of fast electrons with electrons and protons.

The first two types of losses are not important in our case, the third will be considered at the end of this chapter. For the moment we shall consider only the influence of the radiation losses.

The interaction of fast electrons with fast protons and also electrons, leads to their retardation (acceleration), due to which part of the energy is liberated in the form of gamma rays. In the non-relativistic case this corresponds to the ordinary continuous radiation due to free-free transitions.

The expression for the radiation losses has the following form:

$$\gamma(t) = \gamma_o e^{-kt} \, , \tag{15.5}$$

where γ_o is the initial energy of the fast electrons, and k depends on the density n_p of the protons and very weakly on the electron energy. The most probable expression for k is the following (cf., e.g. [3]):

$$k \approx 4 \times 10^{-16} \, n_p \, \text{sec}^{-1} \, . \tag{15.6}$$

From this it follows that a considerable loss of energy can occur at the time of the flare, i.e. during a time $t \sim 100$ s, if the density of the protons is of the order $\sim 10^{14}$ cm^{-3}. For smaller densities the electron energy will be practically constant during the flare. The fall of the star's brightness after flare maximum is at least not due to energy losses of the fast electrons.

In order to check on the validity of this assumption we must first write the expression for the theoretical dependence of the brightness variations with time given Eq. (15.5). Thus, substituting (15.5) in (15.3), we have:

$$J_\lambda(t) = \left[E_4(\tau) + \frac{3}{2} \frac{e^{4kt}}{\gamma_o^4} \frac{e^x - 1}{e^{xe^{2kt}/\gamma_o^2} - 1} F_1(\tau) \right] e^{-\tau} \, . \tag{15.7}$$

This is also the theoretical lightcurve when the effects of energy loss by fast electrons are taken into account.

Let us assume for the time being that τ is constant during the flare. Then we can determine the numerical value of k by comparing the theoretical relation (15.7) with the observed lightcurves of flares. In practice this is done by means of two arbitrarily chosen points on the lightcurve. Then, introducing the value found for k in (15.7), we find the theoretical curve of the brightness variations over the entire flare. If the assumption that the energy of the electrons varies considerably during the flare is true, then the theoretical curve must agree with the observed one.

Such computations have been made for a number of flares with well-defined lightcurves (UV Cet, YZ CMi, AD Leo, etc.). The results were all negative: for values of k > 0 not a single one of the theoretical curves agreed with the observed lightcurves. There are no real values of k, other than zero, for which the computed lightcurves can be made to agree with the observed ones. In other words, k = 0.

In part this result means that the fast electrons leave the star with practically their entire original energy (Compton losses are 10^{-4} to 10^{-5} of γ_o). Also, the decrease of the brightness after flare maximum is not due to energy losses of the electrons. These conclusions can hardly be doubted.

The conclusion that during the flare the fast electrons suffer no significant losses of energy makes it possible to find an upper limit for the density of the fast protons in the same medium where the fast electrons exist. With the condition that k < 0.1, this density is $<10^{14}$ cm^{-3}.

Thus, the first of the two possible effects--energy losses by the electrons during the flare--cannot be the cause of the decrease in brightness after flare maximum.

3. THE EFFECT OF "EVAPORATION" OF THE FAST ELECTRONS

The cloud, or envelope of fast electrons, produced at the moment of the flare, expands outwards, and electrons are lost from the system. This leads to a decrease of the effective optical depth as a function of time.

Little is known about the behavior of the ensemble of the fast electrons after they appear. It can only be assumed that in general the variation of the optical depth follows the law $\tau(t) \sim t^{-n}$, where the value of the exponent n, characterizing the rate or the character of the variation of τ, can be determined from the observed lightcurves. But as soon as we assume the possibility of the fast electrons "evaporating" (practically with the velocity of light), we can find the form of the function $\tau(t)$ on the assumption that the total number of fast electrons remains constant and that they expand isotropically, i.e. the case n = 2. Then we have:

$$\tau(t) = \frac{\tau_o}{(1 + t/t_o)^2} , \tag{15.8}$$

where τ_o is the optical depth at flare maximum, t_o is a constant with dimensions of time, its numerical value is determined from the observations (although its physical significance is not quite clear). Thus the question is whether the function $\tau(t)$ of the form (15.8), when substituted in formula (15.3), agrees with the observed lightcurves.

Therefore, in the case where the decrease in brightness after flare maximum is due to variations of τ, and γ = constant, the theoretical lightcurve has the form:

$$J_\lambda(t) = \left\{ E_4[\tau(t)] + A_x(\gamma,T)F_1[\tau(t)] \right\} e^{-\tau(t)} , \tag{15.9}$$

where $\tau(t)$ is given by (15.8) and $A_x(\gamma,T)$ is given by

$$A_x(\gamma,T) = \frac{3}{2\gamma^4} \frac{e^x - 1}{e^{x/\gamma^2} - 1} . \tag{15.10}$$

For a Gaussian spectrum of the fast electrons, instead of (15.9) (cf. formula (4.50)), we have:

$$J_\lambda(t) = \left\{ E_4[\tau(t)] + A'_x(\gamma_0,\sigma,T)F_1[\tau(t)] \right\} e^{-\tau(t)} , \tag{15.11}$$

where

$$A'_x = \frac{3}{\sqrt{\pi}\sigma} (e^x - 1)\Phi_x(\gamma_0,\sigma) . \tag{15.12}$$

Equations (15.9) and (15.11) are theoretical lightcurves, where the effect of spherical expansion of the fast electrons has been taken into account (n = 2), but with a constant value γ for the energy of the electrons. We note that these formulae have been written for a lightcurve constructed strictly for a given wavelength λ. However, to a first approximation they can also be used for the construction of lightcurves in U, B and V wavelengths, when the corresponding values of the effective wavelengths (λ_U, λ_B or λ_V) are substituted in these formulae.

The procedure of comparing the theoretical lightcurves with the observed ones consists of the following:

(a) From the value of the observed amplitude at flare maximum (ΔU, ΔB or ΔV) we find τ_o, using the graphs given in Fig. 6.3 or Fig. 8.2, depending on the adopted

model of the photosphere, or on the energy spectrum of the fast electrons. With
the aid of the same graphs we find the values of τ as a function of τ.

(b) From the known values of τ_0 and $\tau(t)$ we find the numerical value of the
parameter t_0 from the following relation derived from (15.8):

$$t_0 = t \frac{(\tau/\tau_0)^{\frac{1}{2}}}{1 - (\tau/\tau_0)^{\frac{1}{2}}} . \tag{15.13}$$

(c) Evidently, if the conclusion that the fast electrons are expanding in all
directions is valid, the values of t_0 found from different points of the lightcurve
must be the same.

When the values of t_0 are more of less constant, then it is reasonable to construct
a lightcurve for a given flare using the averaged value t_0. To do this one finds
$\tau(t)$ from the value of t_0 for several times during the flare. Then ΔU (or ΔB or
ΔV) are obtained from Fig. 6.3 or Fig. 8.2. The graphical representation of the
dependence of ΔU on t or of F_U on t (intensity scale) is the required computed
(theoretical) lightcurve, in which the measured points of the observed lightcurve
can be drawn for comparison.

In the manner described above lightcurves were computed and were compared with the
observations for a large number of flares of UV Cet type stars. Some of these are
given below.

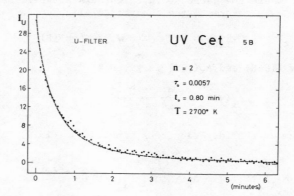

Fig. 15.1. Interpretation of the lightcurve of a flare of
UV Cet, 5B, in U wavelengths. Broken line--theoretical
lightcurve for given flare parameters τ_0, t_0, T and spherical expansion of the fast electrons (n = 2). Time is
counted from the moment of flare maximum.

In Figs. 15.1 and 15.2 the broken lines show the computed lightcurves in U wavelengths for two flares of UV Cet, 5B and 1B, observed by Moffett [4]. The corresponding values of τ_0, t_0 and T are shown on the figures. The dots indicate the
observations. As we see, the agreement of the observations with the theory is
quite satisfactory.

The flare of UV Cet numbered 53B was recorded without a light filter, with an
effective wavelength of the received light at λ_{eff} = 4450 Å [4]. The computed
lightcurve for this flare was derived from formula (15.9). Given λ_{eff} and for

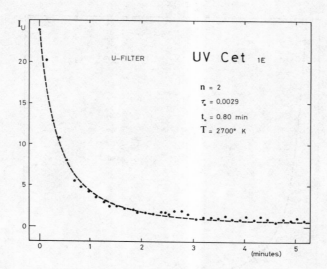

Fig. 15.2. Interpretation of the lightcurve of a flare of
UV Cet, 1E (cf. Fig. 15.1).

$T = 2700$ K we find from (15.10) $A_x = 769.4$. From the value of the observed amplitude at flare maximum $\Delta m = 1^m.19$, or $I_f = 2.76$ on the intensity scale, we find
$\tau_0 = 0.0076$, from which we find from (15.13) $t_0 = 10$ s. Since τ_0 is small, we may
take $E_4(\tau) \approx 1$, $e^{-\tau} \approx 1$. Then (15.9) can be written in the form:

$$I_\lambda = J_\lambda - 1 = 769.4\ F_1(\tau) , \qquad\qquad (15.14)$$

where the numerical values of the function $F_1(\tau)$ can be taken from Table 4.1. The
lightcurve constructed by means of (15.14) is given in Fig. 15.3 (broken line)
together with the observed dots. Here, too, the agreement of the observations with
the theory is not bad.

The results of similar comparisons for two flares of YZ CMi--No. 26 [4] and 30.XI.75
[5]--are shown in Figs. 15.4 and 15.5 (both in U wavelengths). Figures 15.6 and
15.7 give the comparisons of the computed curves with the observations for the same
flare of YZ CMi (7.XII.75) [5], but recorded in U and B wavelengths simultaneously.
As expected, t_0 is the same for both cases (7.5 minutes), but the values of τ_0 are
slightly different (0.009 for U wavelengths and 0.011 in B)! However, we see that
here is sufficiently good agreement of the observations with theory.

In all these cases the graphs giving ΔU versus time have been constructed for a
given effective temperature of the star, for the model of the "real photosphere",
for a Gaussian spectrum of the fast electrons, and with non-thermal bremsstrahlung
taken into account.

But are there cases where the observations do not agree with the theory? Yes,
there are, though relatively few. Three such cases are given in Fig. 15.8 (Wolf
424), Fig. 15.9 (EQ Peg) and Fig. 15.10 (UV Cet). Two interesting facts appear.
First, for some time following the flare maximum there is nearly complete agreement
of the observations with the theory, but in each individual case the disagreement
begins only at a certain moment in the lightcurve of the flare. Secondly, the

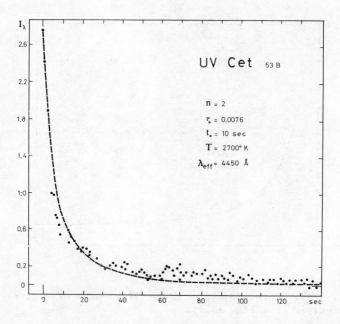

Fig. 15.3.   Interpretation of the lightcurve (without filter,
λ_{eff} = 4450 Å) of a flare of UV Cet, 53B (cf. Fig. 15.1).

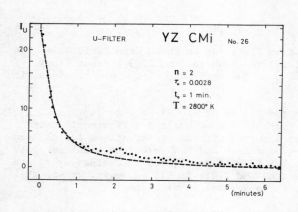

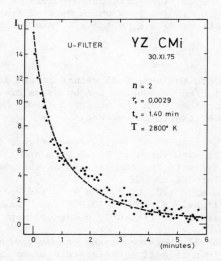

Fig. 15.4.   Interpretation of the lightcurve
in U wavelengths of a flare of YZ CMi, No.
26 (cf. Fig. 15.1).

Fig. 15.5.   Interpretation of the
lightcurve in U wavelengths of a
flare of YZ CMi, 30.XI.75 (cf.
Fig. 15.1).

observed points are higher than the theoretical curves. In other words, from the
time of disagreement the rate of the decrease in brightness is smaller than that
predicted by the theory.

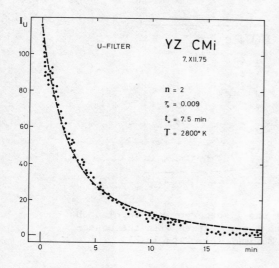

Fig. 15.6. Interpretation of the lightcurve in U wavelengths of a flare of
YZ CMi, 7.XII.75 (cf. Fig. 15.1).

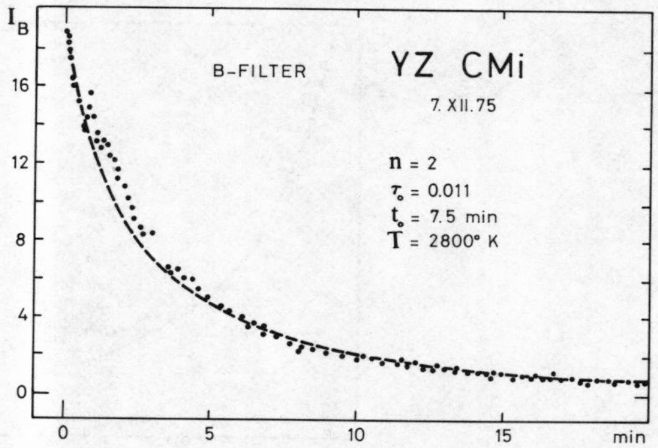

Fig. 15.7. Interpretation of the light-
curve in B wavelengths of the same flare
(Fig. 15.6) of YZ CMi, 7.XII.75 (cf.
Fig. 15.1).

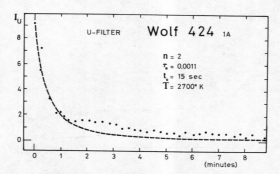

Fig. 15.8. Interpretation of the lightcurve of a flare of
Wolf 424, 1A in U wavelengths; dots--observations, broken
line--theory. The theory agrees with the observations for
1.5 minutes immediately after flare maximum (see text).

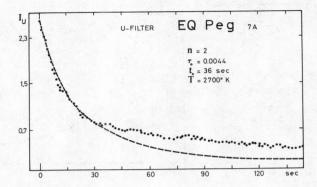

Fig. 15.9. Interpretation of the lightcurve in U wavelengths
of a flare of EQ Peg, 7A. The theory agrees with the observations for 30 seconds after maximum (cf. Fig. 15.8).

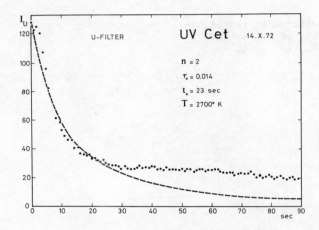

Fig. 15.10. Interpretation of the lightcurve in U wavelengths of a very strong flare of UV Cet, 14.X.72. The
theory agrees with the observations for 25 seconds after
maximum (see Fig. 15.8).

These facts find a simple explanation when the two-component structure of the lightcurve (considered in detail in Ch. 9) is accepted. According to this concept the
initial part of the lightcurve following maximum is mainly due to continuous
emission, and the final part to radiation in the emission lines. But all the
above considerations and quantitative applications refer exclusively to the continuous emission. Therefore, there seems to be no discrepancy between the observations and the theory of the fast-electron hypothesis referring to the continuous
emission.

Matters are different with respect to the behavior of the radiation in the emission
lines. Here the decrease in brightness may sometimes be slower than the decrease

of the continuous emission. If we represent the intensity variations in the light-curve, beginning from a point past the divergency point, in the form $I_\lambda \sim t^{-p}$, then for the computed lightcurves (continuous component) the quantity p_c lies in the range 1.10 to 1.25 for the three flares considered, whereas for the observed light-curves (emission lines) and for the same interval of time the quantities p_e were found to be 0.8, 0.5 and 0.3, respectively, for the flares of EQ Peg, Wolf 424 and UV Cet. The values of p_e must be somewhat different than those given above because the curves beyond the divergence point had not been corrected for the continuous component.

On the other hand, the intensity decrease of the emission lines corresponding with the region of "divergence" on the lightcurves, takes place at a rate with a value of $p_e = 0$ to 1.6, depending on the state of the chromosphere or the degree of its excitation (these quantities can easily be derived from Fig. 9.6, Ch. 9).

With this information the cause of the absence of "divergence points" on the light-curves of the flares considered above—Figs. 15.1 to 15.7—becomes clear; in these cases the rate of decrease of the radiation in the emission lines simply coincided with the rate of decrease of the radiation in the continuous emission, i.e. the condition $p_e \approx p_c$ holds.

Of course, in a more detailed analysis a distinct division of each lightcurve into the two components should be made. For this the flares must be recorded synchronously in the continuum and in the emission lines, a requirement which is very difficult. However, the observed lightcurves seem to behave as predicted by the fast-electron hypothesis.

We remark that our analysis covers cases of flares which differ from each other in duration by one order of magnitude (EQ Peg, 7A, and YZ CMi, 7.XII.75) and in strength by more than one order of magnitude (UV Cet, 53B and UV Cet, 14.X.72).

The good agreement of observations with theory shows that the assumptions made in the derivation of the theoretical lightcurve are essentially correct. This also refers to the relation (15.8), postulating the possibility of fast electrons flying off in all directions (though the meaning of the parameter t_o in this equation is unclear).

4. THE DECREASE OF BRIGHTNESS AFTER FLARE MAXIMUM

We have shown that the decrease in brightness after flare maximum is due to the effect of the expansion of the fast-electron cloud. Substituting therefore the value of $\tau(t)$ from (15.8) in (15.9) we find for the theoretical lightcurve (in simplified form) with $E_4(\tau) \approx 1$ and $F_1(\tau) = \tau/2$:

$$J_\lambda = \left[1 + A_x(\gamma,T) \frac{\tau_o}{2} \left(1 + \frac{t}{t_o}\right)^{-2}\right] e^{-\tau_o(1 + t/t_o)^{-2}} . \tag{15.15}$$

This formula can be applied for any point on the lightcurve between its maximum and the moment when the radiation in the emission lines begins to predominate.

For the initial part of the lightcurve, immediately after maximum, we have

$$J_\lambda \sim \left(1 + \frac{t}{t_o}\right)^{-2} e^{-\tau_o(1 + t/t_o)^{-2}} . \tag{15.16}$$

At sufficiently large distances from the maximum, where $(t/t_o) > 1$, we have for

$I_\lambda = J_\lambda - 1,$

$$I_\lambda \sim \left(\frac{t_o}{t}\right)^2 . \tag{15.17}$$

Finally, in the case of weak flares ($\tau_o < 0.01$) we find for the entire lightcurve --from maximum to the complete disappearance of the flare in the continuous spectrum

$$I_\lambda \sim \left(1 + \frac{t}{t_o}\right)^{-2} . \tag{15.18}$$

We note that these relations do not have anything in common with an exponential law of the type $e^{-\beta t}$, mentioned in the beginning of this chapter.

5. ON THE EFFECTIVE STRENGTH OF THE FLARE

The theoretical lightcurves of flares discussed in the preceding sections are valid for any wavelength. The dimensionless quantity τ_o, which we call the "effective strength of the flare", is constant for all wavelengths. This implies that when the inverse problem is solved, namely when τ_o is determined from the value of the flare amplitude ΔU or ΔB or ΔV, known from the observations, we must find the same value of τ_o for all cases.

However, there are many reasons for expecting some scatter in the values of τ_o, found from different photometric wavelength ranges. How large is this scatter and what can it say for the theory? In order to answer this question we have tried to find the quantity τ_o from the rather homogeneous measurements by Moffett [4] of the amplitudes ΔU and ΔB, for quite a large number of flares of UV Cet and CN Leo. In doing this the effective temperature for both stars was taken to be 2700 K, and the graphs for ΔU and ΔB versus τ were constructed for the model of the "real photosphere" and the Gaussian distribution of the fast electrons. By means of these graphs the values of $\tau_o(U)$ and $\tau_o(B)$, which in the ideal case must be equal to each other, were determined from the quantities ΔU and ΔB, which are known for each flare. For greater reliability only those flares were used for which the amplitudes in B wavelengths were found to be larger than 0^m1.

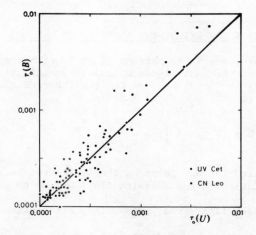

Fig. 15.11. Comparison of "flare strength" $\tau_o(U)$ in U wavelengths with "flare strength" $\tau_o(B)$ in B wavelengths, from observational data of UV Cet and CN Leo.

The results of the computations are shown in Fig. 15.11 in the form of a graph of the relation between $\tau_o(U)$ and $\tau_o(B)$. The points scatter around the line $\tau_o(U) = \tau_o(B)$, and in the range $\tau = 0.001$ to 0.0001 the scatter is more or less symmetrical. This may mean that the scatter is caused by measuring errors. This assumption becomes especially probable if we consider the great sensitivity of $\tau_o(B)$ on ΔB in particular in the τ range mentioned above. Furthermore, in the range $\tau = 0.01$ to 0.001 the scatter of the points is not only large, but asymmetrical, too; the values of $\tau_o(B)$ were found to be systematically larger than those of $\tau_o(U)$ (for the same flare). The cause of this asymmetry is quite simple: for relatively strong flares (large τ) the role of the non-thermal bremsstrahlung begins to exert its influence (Ch. 8), but this has not been taken into account in this case.

The consideration of the causes for the scatter of the points, as well as for their asymmetry, in the $\tau_o(U)$ vs. $\tau_o(B)$ diagram should continue. It is desirable to continue such a kind of analysis, in particular to find out the consequences of the difference in effective temperatures of the star in the U and B regions of the spectrum, especially if these temperatures can vary slightly from one flare to the next.

6. ON THE CLASSIFICATION OF THE FORMS OF FLARE LIGHTCURVES

The lightcurves considered above are simple in form but are not found often. As a rule the lightcurves of most flares are complex. Meanwhile these complicated lightcurves are very different from each other. Attempts have been made to make some classification of the forms of the lightcurves, assuming these differences are significant. However, one gets the impression that these differences are only apparent and have an instrumental origin, in particular due to insufficiently high time resolution of the recording apparatus.

Examples confirming this point of view have already been mentioned in Chapter 6 where we discussed the lightcurves of flares of UV Cet, obtained by the group of Cristaldi. If these are considered carefully (for instance, Figs. 6.34, 6.35) it is seen that nearly all the lightcurves (weak and strong ones) of the given series of recorded flares resemble each other. These lightcurves were obtained with an integration time of 0.5 second. It is quite clear that if such a flare is recorded with a time resolution of ∿5 seconds a smoothing effect or blending of different outbursts will take place.

The analysis made in the preceding sections leads us to the conclusion that the form of the flare lightcurve is universal; this form does not depend on the strength or the amplitude of the flare and therefore it must be the same for all flares. Therefore the "complex" form of the lightcurve must in each individual case be explained as a blend of simple flares of various strengths, i.e. of different durations and different amplitudes. As long as the interval between various flares is longer than the time constant (integration time) of the photoelectric system, such flares will be distinguished in pure form, and their lightcurves will be represented by a relation of the form (15.9) or (15.11).

7. THE POSSIBILITY OF DISTINGUISHING THERMAL AND NON-THERMAL RADIATION

The problem of distinguishing the non-thermal radiation component from the total radiation of some unstable star is quite difficult. However, in the case of flare stars it is relatively easy due to the fact that all of the additional radiation appearing at the time of the flare in the form of continuous emission and emission lines is of non-thermal origin. The fact that at infrared wavelengths the brightness of the star does not increase during the flare proves that at least the flare

is not accompanied by heating of the photosphere, i.e. an increase of its temperature.

Matters are different in the case of unstable stars for which the brightness fluctuations are caused by simultaneous fluctuations of the thermal and non-thermal components. The long-period variables, in particular, may serve as an example. The
fact that the spectral class of these stars changes in the course of time indicates
that real fluctuations of temperature take place in their photospheres. At the same
time in most of the objects of this category considerable fluctuations of the
polarization parameters have also been observed [6,7]. These variations cannot be
related to the interstellar medium, and must be due to some unstable phenomena
occurring in the atmospheres of the stars themselves. It is possible that the
fluctuations of the polarization of the light of these stars are due to dust clouds
wandering above their photospheres. However, a number of factors indicate that the
variations of the polarization parameters in long-period variables may be caused
by non-thermal processes taking place in the outer regions of their atmospheres.

The anomalous colors of long-period variables. Colorimetric observations for a
large group of long-period variables in the UBV system have been made by Smak [8]
and Landolt [9]. Their lists contain more than 60 stars, 40 are of class M; the
remaining ones belong to spectral classes N and S. The color indices found from
these observations for the stars of class
M have been drawn in the theoretical $U-B$
vs. $B-V$ diagram (Fig. 15.12). In those
cases where there is more than one observation for a given star, the points have
been connected by straight lines.

This figure shows that part of the long-
period variables lie on the main sequence
or close to it, but more than half of
them are rather far--up to $\sim1^m5$--from the
main sequence. Farthest away is R Leo, and
also R Aql, S CrB and RR Sco (the uppermost points in the diagram).

About twenty of these stars with anomalous
colors occur in the lists of Serkovski and
Zappala [6,7] and show fluctuations of the
polarization parameters. These fluctuations
are very different in character. For some
stars, for instance, the degree of polarization is observed to increase with the
increase of the star's brightness (μ Cep,
χ Cyg, S Cep, R And). For others, on the
contrary, the degree of polarization
decreases, and sometimes rapidly, with

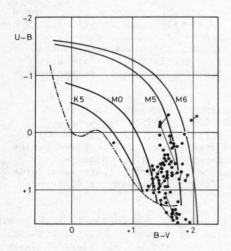

Fig. 15.12. Long-period variables in
the theoretical $U-B$ vs. $B-V$ diagram.

increasing brightness of the star (V CVn, W Peg, T Tri, U Her, Z UMa). And there
are cases where the brightness variations of the star, reaching two to three, and
sometimes six to seven stellar magnitudes, occur practically without any variations
of the polarization (W Lyr, R CVn, o Cet, S CrB).

Flares of long-period variables. There are three cases known of photographic
recordings of flares of long-period variables; one flare each for R Aql, R Tri and
RT Cyg, with approximate amplitudes of 0^m36, 0^m4 and 0^m6, respectively [10]. For
the first two stars the amplitudes in photovisual light were recorded as well;
they were 0^m15 and 0^m3, respectively. The duration of the flares was of the order
of ten minutes, and the forms of the lightcurves resemble what we observe in

ordinary flare stars.

For R Aql there are no polarimetric measurements, but its color is anomalous: $U - B = +0.03$, and $B - V = +1.85$. For another star, R Tri, there are no data about the color, but the results of the polarimetric measurements indicate that its varia- tions are unusual. Both of these stars are anomalous and it seems possible that they may show some flare activity.

With regard to RT Cyg, there are no polarimetric measurements for it, but the color indices ($U - B = +1.21$, $B - V = +1.72$) do not indicate any peculiarities.

We know of no other flares of long-period variables. Therefore the results discussed above should be considered with some caution.

Analysis of the colorimetric observations of long-period variables and the cases of the possible flares, lead to the interesting conclusion, which however should be further confirmed, that the processes occurring in the atmospheres of unstable stars may be related to the processes taking place in the atmospheres of flare stars.

In this connection, the problem arises in distinguishing the thermal and non-thermal components of the radiation on the basis of the dependence of the observed degree of polarization p_* on the brightness m of the star. It turns out that such a separa- tion can be made by means of the following formula:

$$p_* = p_0 \left[1 - \frac{1}{1+\alpha} \left(\frac{1}{1+q} + \frac{q}{1+q} 10^{-0.4\Delta m} \right) \right] \left(\frac{1}{1+q} + \frac{q}{1+q} 10^{-0.4\Delta m} \right) , \qquad (15.19)$$

where p_0 is the degree of polarization of the pure non-thermal radiation and $\Delta m = m_0 - m$ is the variation of the star's brightness. By α and q we mean

$$\alpha = \frac{J_0}{B_0} , \qquad q = \frac{\Delta J_0}{\Delta B_0} , \qquad (15.20)$$

where B_0 and ΔB_0 are the constant and the variable components of the thermal (Planck) radiation, and J_0 and ΔJ_0 the same for the non-thermal radiation. p_0, α and q are the polarization parameters. Formula (15.19) can be applied for any spectral region.

Analysis of formula (15.19) shows that, depending on the values of the polarization parameters, i.e. on the ratio of the thermal and non-thermal radiation components, a wide range of observed polarization p_* and amplitude Δm are possible. Comparing this formula with the curves p_* vs. Δm, constructed from the observational data for a certain star, we can find the fraction of the constant (α) and the variable (q) components of the non-thermal radiation. Hence, we can unambiguously separate the non-thermal component from the total radiation.

The results of such comparisons are shown in Fig. 15.13 and in Table 15.1. It should be pointed out that in most of the cases the theoretical curves agree quite well with the observational data. According to the results all stars were divided into three groups, depending on their value of q.

In the first group (Table 15.1) there are stars for which $q \to \infty$, i.e. the brightness fluctuations are wholly due to variations of the non-thermal radiation component. In the second group there are stars for which $q \sim 1$; in this case the fluctuations are due both to thermal and to non-thermal radiation. Finally, the third group contains stars for which $q \to 0$; here the fluctuations in brightness are due exclu- sively to changes in the temperature of the star.

Thus, the observed variations in the degree of polarization in long-period variables

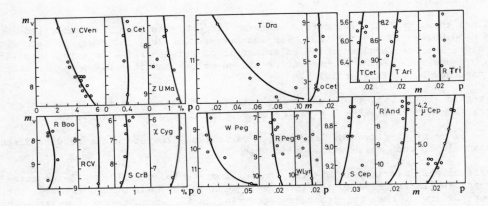

Fig. 15.13. Comparison of the observed degrees of polarization p (in percentages or in stellar magnitudes m) with the theoretical function p ∿ Δm (full-drawn lines) for some long-period variables.

TABLE 15.1 Numerical Values of the Parameters of the Non-thermal Radiation, q, α and p for Some Long-period Variables

Star	q	α	p (%)	Spectral class	Measurement of the polarization
Group I					
T Tri	∞	∞	0.9	M4e–M8e	7
T Cet	∞	1	0.9	M5e–II	7
A Ari	∞	0.9	1.1	M6e–M8e	7
S Cep	∞	0.72	1.8	N8e (C7e)	7
μ Cep	∞	0.48	2.1	M2e Ia	7
μ Cep	∞	0.24	2.6	M2e Ia	7
Group II					
R And	5.0	1.5	1.8	S6, 6e	7
o Cet	2.7	0.54	0.9	M5e–M9e	7
χ Cyg	2.25	0.66	3	S7e–I0e	6
W Lyr	2.0	1.78	0.7	M3e	6
R CVn	1.55	0.82	2.0	M6e–M8e	6
S CnB	1.20	0.27	2.1	M6e–M8e	6
R Peg	1.18	4	1.1	M6e–M9e	7
W Peg	0.57	0	3.2	M6e–M8e	7
o Cet	0.50	0.40	1.4	M5e–M9e	6
R Boo	0.50	0.30	3.0	M3e–M5e	6
V CVn	0.25	1.86	8	M4e–M6e	6
Z UMa	0.25	0.43	5	M5 IIIe	6
Group III					
T Dra	0	0	4.6	N0e (C8e)	7
U Her	0	0.47	5	M7e–M8e	6
RS Cnc	0	0.43	1.6	M61–II (S)	6
SS Vir	0	0.33	1	Vire (C6e)	6

may result from fluctuations in the thermal as well as in the non-thermal radiation components, though there are cases where these variations are only due to fluctuations of the non-thermal component ($q \sim \infty$), and cases where there is no variable non-thermal radiation component at all ($q \sim 0$). In this analysis we have assumed that the plane of polarization remains the same during the fluctuations of the star's brightness, but this is not always realistic. For a strict discussion of the problem variations of both the degree of polarization and its position angle should be taken into account simultaneously.

8. HEATING OF THE PHOTOSPHERE

The stellar photosphere can be heated at the time of a flare by the influence of the following factors:

(a) Absorption of gamma-rays, generated by the fast electrons.

(b) Penetration of fast electrons from the outer regions of the star into its photosphere.

(c) Absorption of optical radiation of Compton origin directed towards the photosphere.

Let us consider these points in more detail.

Absorption of gamma-rays. In the envelope or cloud consisting of fast electrons in principle gamma rays can be produced by non-thermal bremsstrahlung. Part of these photons will be directed towards the photosphere, where they can be absorbed. As a result the temperature of the photosphere may increase. However, as we have seen above (§2), the radiation losses during a flare can be important only if the densities of the fast electrons are of the order $\sim 10^{14}$ cm^{-3} or more. However analysis of the lightcurves led us to the conclusion that the density of the protons at the moment of flare maximum is certainly less than this value and becomes even lower after flare maximum. Therefore the total energy of the gamma rays generated under these conditions will be considerably less than the total energy of the fast electrons. The fraction of the electron energy that is transformed into energy of gamma rays can be determined from the following relation (when $k\Delta t \ll 1$):

$$\frac{\gamma_o - \gamma}{\gamma_o} \ll k\Delta t_m \sim 10^{-16} n_p \Delta t_m , \qquad (15.21)$$

where γ_o is the initial energy of the electron ($\sim 10^6$ eV), Δt_m is the duration of flare maximum, which is of the order of a second.

Here we consider a model of a strong flare, with $\tau \sim 0.01$. We take for the linear extension of the fast-electron layer $\Delta R \sim 1$ to 10 stellar radii ($R_* \sim 0.1 R_\odot \sim 10^{10}$ cm). Hence we find for the total number of electrons in a column of 1 cm^2: $N = \tau/\sigma_s \approx 10^{21}$ cm^{-2}, where σ_s has been taken equal to $\gamma^2 \sigma_T = 6.65 \times 10^{-24}$ cm^2 and $\gamma^2 = 10$. For the density of the fast electrons (protons) we have $n_e \sim n_p \sim N/\Delta R \approx 10^{10}$ to 10^{11} cm^{-3}. With these data we find from (15.21) the fraction of the energy of the fast electrons transformed into gamma rays: $(\gamma_o - \gamma)/\gamma_o \sim 10^{-5}$ to 10^{-6}.

The total energy of the fast electrons surrounding the star on all sides is $E = 4\pi R^2 N\epsilon$, where $\epsilon \sim 10^{-6}$ ergs is the energy of one electron. Substituting the respective values we find $E \approx 10^{36}$ ergs (for $\tau = 0.01$). Hence the amount of energy of all gamma rays will be: $E_{gm} \approx 10^{30}$ to 10^{31} ergs. We note that the total (bolometric) luminosity of a dwarf of class M5 ($R_* \sim 0.1 R_\odot$ and $T \sim 3000$ K) is about a thousand times less than the luminosity of the Sun, i.e. of the order $\sim 10^{30}$

ergs/s.

Thus, at the moment of flare maximum, for a few seconds the star is irradiated from outside by a stream of gamma rays. This energy is comparable with the flux of the star's normal radiation. However it is not difficult to see that even in the case where this energy is completely absorbed by the star, the equilibrium temperature of the photospheric layers will not differ much from the temperature of the star in its stable state. The gamma rays penetrating into the photosphere will undergo Thomson scattering by thermal electrons, including electrons which are connected with atoms and ions. After many scattering processes part of these photons will be reflected in the opposite direction and may leave the star forever. Of course this reduces the degree of heating the photosphere. Heating can occur only when a certain fraction of the gamma rays are absorbed. Therefore it is necessary to know the optical depth t_{gm} of the layer of the photosphere of depth ℓ for true absorption processes. We can write:

$$t_{gm} = x_{gm} n \ell, \qquad (15.22)$$

where x_{gm} is the effective cross section for absorption of the gamma rays, n is the density of the particles absorbing the gamma rays.

The efficiency of the absorption of gamma rays by hydrogen and helium, the most abundant elements in stellar atmospheres, is exceedingly small. The number of other particles--heavy nuclei and ions with many electrons--which might absorb gamma rays, is at least two to three orders of magnitude smaller than that of the hydrogen atoms. Therefore if we take $x_{gm} \approx 10^{-25}$ cm^2 we find $t_{gm} \sim 10^{-3}$ to 10^{-4} in the case that $\ell \sim 100$ km (ℓ is determined from the condition that the effective optical depth of the photosphere for Thomson scattering is of the order of unity, i.e. $n_e \sigma_T \ell \sim 1$).

Thus the fraction of gamma rays absorbed by the photosphere will be less than 1% or 10^{28} to 10^{29} ergs. This is less than the energy radiated by the star in its stable state. Under these conditions the heating of the photosphere will be insignificant and it will even be hard to detect. These calculations refer to $\tau = 0.01$, which is a rather strong flare.

Our analysis of the role of the gamma rays has been somewhat qualitative. The question of heating the photosphere of flare stars by gamma rays is quite complex and deserves further investigations.

<u>Penetration of fast electrons into the photosphere</u>. If a fast electron falls on a stellar photosphere then by ionization losses its kinetic energy will finally be converted into thermal energy. As a result the temperature of the star's outer layers may increase. However, before reaching the surface of the photosphere the electron must first pass through the magnetic field of the star. Little is known about the character and the strength of the magnetic fields of red dwarfs and, in particular, of flare stars. According to the opinion of some investigators there exist large spots on flare stars, possibly with strong magnetic fields, in the range of thousands of gauss. But even if the strength of the magnetic field is of the order of ten gauss, not a single electron with an energy of the order 10^6 eV will reach the photosphere from outside; it will be reflected by the magnetic field at considerable distances from the surface.

Thus generally it seems impossible that fast electrons penetrate from outside into the star's photosphere. Only a certain number of the electrons, captured along the magnetic lines of force of the dipole field of the star, can reach the surface from the direction of the magnetic poles. In such cases it may be possible to observe a local effect of heating of the photosphere in the polar regions of the star. An

investigation of the problem of the interaction between the dipole magnetic field
of a star and the fast-electron shell surrounding it will be extremely interesting.

<u>Absorption of optical radiation of Compton origin</u>. Part of the optical radiation
of Compton origin produced during the flare will be directed towards the photosphere.
In this process the radiation ionizing hydrogen will be absorbed in the chromosphere
and will be radiated again in the form of emission lines and the continua of the
hydrogen spectral series.

Regarding the remaining radiation of Compton origin, longer than 912 Å, it can penetrate deep into the lower layers of the chromosphere and of the photosphere as long
as the optical depth is less than unity.

However we have seen that only 10^{-5} of the energy of a fast electron is transformed
into Compton radiation. The total flux absorbed by the photosphere in the form of
an impulse will increase the normal radiating power of the star by at most a factor
of ten for $\tau = 0.01$, i.e. for very strong flares. During ordinary flares, characterized by values of $\tau = 0.001$ to 0.0001 the irradiation will be considerably less.
Nevertheless, the equilibrium of the temperature of the photosphere is somewhat
disturbed, and it is hard to tell how soon this equilibrium re-establishes itself.

<u>Ionization losses of fast electrons</u>. The medium in which the fast electrons occur
is highly ionized and therefore ionization losses do not play any important role.
Ordinarily also losses of energy by Cerenkov radiation of plasma waves must be taken
into account. These losses are described by the following relation [3]:

$$- \frac{dE}{dt} = 7.62 \times 10^{-9} \, n_p \, \left(3 \, \ln \frac{E}{mc^2} + 18.8\right) \text{ eV/s} .$$ (15.22)

From this we find the characteristic time, with $\gamma = E/mc^2 = 3$:

$$t_i \approx 1.5 \times 10^{12} \, \frac{\gamma}{n_e} \approx \frac{5 \times 10^{12}}{n_e} \text{ sec} ,$$ (15.23)

where n_e is the density of the protons (electrons) in the medium. With $\Delta R \sim 10^{10}$ cm;
$n_e = \tau/\Delta R \sigma_s \approx 10^{13} \, \tau$,   and $t_i \approx 0.5/\tau$ s. For the values $\tau = 0.0001$, 0.001 and
0.01 the characteristic time due to ionization losses is 5000, 500 and 50 seconds
respectively. For strong flares ($\tau \sim 0.01$) ΔR will be larger than the assumed
value, which leads to still a further increase of t_i.

Actually the assumed value $n_e \approx 10^{13} \, \tau$ is the instantaneous density of the fast
electrons at flare maximum only; after maximum n_e decreases very rapidly ($n_e \sim t^{-n}$).
But the expression (15.23) for t_i was derived on the assumption that $n_e = $ const.
for the duration of the entire flare. Therefore the values found above for t_i must
be considered as lower limits. For $n_e \sim t^{-2}$ these numbers should be increased by
at least one to two orders of magnitude. As a result t_i is considerably larger
than the duration of the flares. This means that the conclusions reached above,
namely that the fast electrons leave the stars with practically their initial
energy remains valid. Here we have restricted ourselves to a qualitative discussion.
The problems of the behavior of the fast-electron cloud, its stability and energy
losses, are exceedingly complex and require special study.

With regard to the heating of the stellar photosphere at the time of the flare the
following should be said. Whereas the first two sources of possible heating of the
photosphere--gamma radiation and penetration of fast electrons into the photosphere
--are to a certain degree hypothetical (since they are consequences of the fast-
electron hypothesis), the emission of a large amount of optical radiation during
the flare is an observational fact. It is also an observational fact that during
the flare the temperature of the photosphere does not increase, at least not significantly; this is indicated by the fact that there is no positive flare in the region

of infrared wavelengths. Then any mechanism for the excitation of the flare (except
the inverse Compton effect) leads inevitably to the situation that the photosphere
is irradiated by photons, at least in the optical range. In principle no flare
theory whatever can avoid this problem. Since we do not observe a significant
increase of the temperature of the photosphere, this may mean that the real time of
return to stability is considerably longer than the duration of the flare itself.
Under these conditions the increase of the temperature of the photosphere may be
insignificant. These problems however deserve a lot more consideration in the
future.

REFERENCES

1. P. F. Chugainov, Izv. Crimean Obs. 28, 150, 1962.
2. G. A. Gurzadyan, Astrofizika 5, 383, 1969.
3. V. L. Ginzburg, S. I. Syrovatskiy, Origin of Cosmic Rays, 1964.
4. T. J. Moffett, Ap. J. Suppl. No. 273, 29, 1, 1974.
5. S. Cristaldi, M. Rodono, 1975, private commun.
6. K. Serkovski, Ap. J. 144, 857, 1966; IBVS No. 141, 1966.
7. R. R. Zappala, Ap. J. Lett. 148, L81, 1967.
8. J. Smak, Ap. J. Suppl. 9, 141, 1964.
9. A. U. Landolt, PASP 78, 532, 1966; 79, 336, 1967; 80, 228. 1968.
10. T. K. Kiseleva, Astr. Zircular USSR No. 483, 1968; No. 436, 1968.

CHAPTER 16

Fast Electrons

1. THE FAST-ELECTRON HYPOTHESIS

The present monograph is devoted to the theory of stellar flares, based on the hypothesis which we have called the "fast-electron hypothesis". This hypothesis was put forward for the first time in the mid-sixties, and it has been too unusual and even heretical to be accepted easily. It is probably one of those scientific hypotheses which do not obtain quick acceptance, in spite of the supporting observational facts. It seems that some psychological notions must be overcome before people can consider radical ideas which may be the essence of some natural phenomenon.

The fast-electron hypothesis turned out to be equally fruitful with respect to stellar flares in the neighborhood of the Sun, T Tau type stars, and stars with Hα emission belonging to stellar associations and young stellar aggregates. Besides, this hypothesis can explain numerous and most diverse observational facts concerning ordinary stellar flares. The fast-electron hypothesis proved to be capable of also encompassing exceedingly rare and extreme forms of the flare phenomenon. In particular:

- The fact that the brightness of a star can increase by a factor of more than 2000 during a flare (4 or 5 such cases are already known);

- Cases where the star's brightness increases by more than a factor of ten per second during some flares;

- The unusually large fluxes of radio emission from celestial objects like a star.

These flare phenomena, representing extreme properties, constitute reliable observational facts, and may serve as "cornerstones" for testing various theories. These extraordinary properties of stellar flares seem to have been avoided by the previous "magnetic" and "thermal" theories, which have involved the "hot spot" and the "hot gas" concepts. Whereas for ordinary flares it may still be possible to "select" the required parameters of the "hot gas" or the "magnetized spot", this becomes simply impossible for flares with the extraordinary properties mentioned above.

The exceptionally large speeds of the development of the flare, and the huge supply of energy, are properties of nuclear transformations, in particular, fast electrons. To "squeeze" these properties out of a "hot gas" is in our opinion not the way to proceed.

The fast-electron hypothesis could make a number of interesting predictions, one of which—the possibility of x-ray radiation generated during a flare—has already been confirmed by direct observations. However, much remains to be done in other areas.

A comparative analysis of the fast-electron hypothesis on the one hand and the remaining hypotheses (in particular that of a "hot gas") on the other hand is not the primary aim of the present monograph. It is useful however to make one comparison: in order to induce a super-strong flare, producing a 1000-fold increase of the star's luminosity in U wavelengths, it is necessary that in the course of a very short time $\sim 10^{21}$ g of ordinary stellar matter be ejected according to the "hot-gas" hypothesis, or $\sim 10^{11}$ g of "nuclear-active" matter in the case of the fast-electron hypothesis.

Ordinarily there is more than enough gaseous matter inside the star. But there is "nuclear-active" material as well, in particular in stars which are in a state of strong convective mixing. Hence as soon as the mixture of "nuclear-active" matter is more than 10^{-10} parts of the entire amount of matter carried off during a flare, the fast-electron hypothesis will work as a non-thermal mechanism for inducing flares. Intuition suggests that such a very small amount of "nuclear-active" matter is to be expected in the matter ejected by flare stars, and by unstable stars in general. The circumstances that the fast-electron hypothesis provide a satisfactory description of many facts related with the stellar flare phenomenon can serve as evidence that this intuition is reasonably good.

According to the fast-electron hypothesis the essence of a stellar flare lies in the spontaneous appearance of β-electrons in the outer regions of the star, due to β-decay of some unstable nuclei, which are thrown out from time to time from the stellar interior. The energy spectrum of these β or fast electrons is gaussian with a small dispersion, and with a maximum at ~ 1.5 MeV, i.e. nearly monochromatic.

Thus the main or primary phenomenon during the flare is the appearance of an ensemble of fast electrons. All the other phenomena--the sharp increase in the brightness of the star due to processes of inverse Compton effects and non-thermal bremsstrahlung, the generation of radiowaves (of synchrotron origin), the emission of x-rays, etc.--are secondary and with respect to energy less essential. One of the main conclusions of this hypothesis is that elementary particles with high energies play a huge role in flares, and that the total amount of energy liberated during a flare in the form of kinetic energy of elementary particles is many times as large as the energy of the optical radiation.

An argument against the fast-electron hypothesis is that the total amount of energy of the flare is many orders of magnitude larger than the energy of the observed photon radiation of the star. To many it seems strange to believe that the observed flare radiation may be the result of secondary processes!

But no one has shown that the observed flare in optical wavelengths is the primary phenomenon. In looking for the sources of flare energy, why should we start by only requiring equilibrium for radiant energy? And on the basis of which facts, or considerations, must we assume that our observational means provide all the forms of the flare phenomenon?

It is possible that such questions touch to a large degree on the cosmogonic and even the philosophical side of the problem, but we cannot ignore them in such a universal problem as stellar flares.

It is more logical in our opinion to admit the possibility that flares are associated with higher energies. Such considerations may be related with formation and development of stars, the evolution of the sources of energy production in the nuclei of the stars, etc. The fast-electron hypothesis approaches the problem precisely from this point of view.

2. THE PROBLEM OF THE ORIGIN OF THE FAST ELECTRONS

In the above discussion the question regarding the "civil rights" of the fast-electron hypothesis was pointed out. Let us now consider the problem of the origin of the fast electrons. First of all, we stress the fact that the optical flare is accompanied by the production of powerful radio emission. This proves that, in some way, at the time of the flare, relativistic electrons appear in huge quantity. Here we certainly have to do with relativistic electrons; the flux of radio emission produced by the ordinary ionized gas during the flare is found to be seven to eight orders of magnitude less than the observed values. The density of these relativistic electrons must be very high since the volume of space around the star where the radio emission is generated has relatively small dimensions.

Such a dense, short-lived, cloud of relativistic or fast electrons must be at the same time a powerful generator of electromagnetic fluctuations in other frequencies as well, including the optical region. This is exactly the problem which is the main subject of investigation in the present monograph, where the predominating role of two processes has been established: inelastic collisions of infrared photons with fast electrons (inverse Compton effect), and the non-thermal bremsstrahlung—the radiation of fast electrons due to retardation in the Coulomb field of the elementary particles. It appears that most of the optical flares are due to the inverse Compton effect; only very strong flares may be induced by non-thermal bremsstrahlung.

Thus the fact that the flare stars emit radio radiation compells us to introduce fast electrons for the explanation of the nature of optical stellar flares. With regard to the inverse Compton effect as the main mechanism for the inducing of optical flares, this is part of the question which we shall discuss during the analysis of the energy transfer in the neighborhood of the star.

The discovery of x-ray radiation must be considered as an additional important argument to show that during stellar flares fast electrons must appear.

Nevertheless, it is necessary and important to continue to search for new facts which may clarify the role of the fast electrons in stellar flares.

At the same time a clear distinction should be made between two problems, one of which aims at explaining the entire complex of phenomena and facts connected with stellar flares, on the basis of the fast-electron hypothesis; and the other is the explanation of the production of the fast electrons themselves.

The first problem is in our opinion more important. If we are logical all the way up to the end, the second problem—that of the origin of the fast electrons—can be posed only after we are completely convinced that the phenomenon of a stellar flare can be explained in no other way than by fast electrons as the primary source of energy.

In stating the problem this way, it seems to us that there is no contradiction. Astrophysical problems have often encountered such situations. An historical example deals with the explanation of radiogalaxies and supernova remnants.

We have no reasons to believe that the problem of the generation of fast electrons in the atmospheres of flare stars, and of unstable stars in general, can be easily solved, even if we are totally convinced that they—the fast electrons—are the cause of the optical, radio and other types of flares. Here we restrict ourselves in discussing some general considerations about the possible ways in which fast electrons can be produced in the outer parts of a star.

The following ways in which fast electrons may appear in the outer regions of a star may be thought of:

I.   Ejection of fast electrons directly from the nucleus of the star.

II. Acceleration of ordinary thermal electrons in the magnetic field of the star.

III. Spontaneous nuclear β-decay of some unstable nuclei or nuclear systems ejected outwards from the nucleus of the star.

IV. Other phenomena, unknown to us, related to the internal structure of the star and the properties of the material within it, which may in some way or other lead to production of fast electrons.

Case I appears clearly unreasonable. Even if a fast electron can be ejected directly from the layers below the photosphere, its entire energy will be spent in the form of ionization losses along its way, long before it leaves the photosphere.

Case II cannot be adopted either, primarily on energetic considerations (cf. last section of this chapter).

Case IV is similar to the idea of V. A. Ambartsumian, with this essential refinement and addition that in our case the fast electrons are first separated from the material ejected from the stellar interior, and that later part of the energy of these electrons is transformed into flare energy. This idea has not been treated quantitatively, and when it appeared (in 1954) the possibility of producing fast electrons was not put forward utterly.

Case III appears to us to be the most reasonable one.

3.   BETA DECAY

What are the properties of the fast electrons responsible for optical flares? On the basis of the results obtained in the preceding chapters the following properties of the fast electrons may be considered as accepted:

(a) the electrons appear quickly, during a few seconds;

(b) the value of their energy is of the order of 10^6 eV;

(c) the energy distribution of the fast electrons resembles a Gaussian distribution.

Electrons (or positrons) liberated in β decay of some unstable nuclei have such properties. Therefore we must not exclude the possibility that the fast electrons are appearing in the outer regions of a star as a result of β-decay of unstable nuclei, ejected outwards from the star.

It is hard to find the type of nucleus responsible for the appearance of β-electrons with the properties mentioned above. Here we restrict outselves to some general remarks. For instance we may take as one of the main criteria the half-life period of the particles; it must be comparable with the duration of the most rapid flares, i.e. it must be of the order of a second.

The most attractive process appears to be the decay of the neutron, during which a β-electron, a proton and an antineutron appear. The possibility of a superdense (neutron) nucleus in the stars indicates that this may actually be a reliable source

of fast electrons after the neutrons find their way in the outer regions of the
star. However, the neutron cannot be a source of fast electrons; its half-life
period is very long, 11.7 minutes.

However, there are nuclei for which the half-life period is of the order of a
second or less. For instance, for the nucleus of He^6 it is 0.813 s, for Li^8 0.89 s,
and for B^{12} it is even less, 0.025 s [1]. The decay of nuclei of the type of He^6
is of interest. This does not mean that the short-lived isotope He^6 is ejected
directly from the stellar nucleus. A nucleus of the type of He^6 can be formed
possibly while matter flows towards the surface (the photosphere) of the star.[†]
It is desirable to consider He^6 from a purely formal point of view as model of a
hypothetical nucleus, since we think it is premature to pretend that precisely
helium should be the source of the fast electrons in the atmospheres of flare stars.

Two considerations come to mind when we speak about He^6. The first is related to
the energy spectrum of the fast electrons arising upon the β-decay of He^6. The
beta-decay of He^6 has been studied experimentally by Wu and her collaborators [1],
who give their results in the form of a Fermi-Curie graph. Using these results
we can construct the energy distribution of the
fast electrons; this is shown in Fig. 16.1, where
N_e is expressed in arbitrary units, and the energy
of the electrons γ in mc^2 units. It is interesting
that the maximum of this curve corresponds with
the value $\gamma_{o_2} \sim 3.5$ (~1.7 MeV), of $\gamma_o^2 \sim 10$. As we
have seen, $\gamma^2 \sim 10$ explains the main properties of
flares.

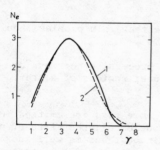

Fig. 16.1. Energy spectrum of
the electrons of beta-decay of
He^6: full-drawn line—experiment, broken line—curve
corresponding to Gaussian distribution with σ = 2 (in mc^2
units). The energy γ of the
electron is given in mc^2 as
unit, and the number of
electrons N_e is in arbitrary
units.

Furthermore, the energy distribution of the electrons arising in the decay of He^6 is rather well
represented by a Gaussian curve (broken line in
Fig. 16.1) for a value of the dispersion σ = 2 (in
mc^2 units). This is also in agreement with what
we found earlier for a Gaussian distribution of
the fast electrons.[††]

The second consideration is related to the fact
that one of the products of the decay of He^6 is
one of the isotopes of lithium, Li^6. It is known
that T Tau type stars for instance, and objects
resembling them, contain lithium in abnormally
large quantities (cf. next section).

There is a problem, however, since Li^7 is the most
abundant isotope of lithium, at least on the Earth and Sun, where the ratio Li^6/Li^7

[†]Possibly as the result of the reaction $He^4 + 2p \rightarrow He^6 + \ldots$, which does not take
place by direct encounter of three particles, but consists of the successive reactions of gamma-capture and β-decay, in the same way as the reaction $4p \rightarrow He^4 + \ldots$.
There is also the following possibility: He^6 is a fragment of a more complex,
more or less amorphous nuclear system, whose decay and appearance of He^6 nuclei
occur somewhere during the displacement of the stellar matter towards the photospheric layers of the star.

[††]It should be noted that the curve of the true distribution of the electrons
intersects the axis at a certain value of the electron energy γ_o, i.e. the number
of electrons is equal to zero in the region $\gamma \geq \gamma_o$, whereas a Gaussian curve does
not intersect this axis.

is of the order 0.08 [2]. It is hard to say anything definite about the ratio
Li^6/Li^7 in the atmospheres of T Tau stars, since we know little about the reactions
leading to the formation of Li^7 (maybe, $Li^6 + p?$). However, the possibility of an
abnormally high value of the ratio Li^6/Li^7 in the atmospheres of T Tau stars cannot
be excluded.

In connection with this problem it is interesting to remark that Herbig [3] has
found some main sequence stars of class G with ratios of Li^6/Li^7 higher than that
of the Sun.   It is possible that this difference is due to active processes of
chromospheric flares taking place intensively in the atmospheres of these stars,
and resulting into β-decay of nuclei of He^6.

We stress that the decay of He^6 has been considered only as a formal illustration.
We have no data indicating that the decay of He^6 should be the main source for the
appearance of fast electrons in the outer regions' stars. Neither can we indicate
at present the type of nucleus responsible for the appearance of fast electrons as
a result of β-decay. But there can hardly be any doubt that the decay of hypothetical nuclei--unstable, short-lived configurations unknown to us--is the most probable
source of production of fast electrons.

The role of neutron decay cannot be completely excluded either this problem should
be investigated further.

Though β-decay should be considered as the most probable process, other ways or
mechanisms for the generation of fast electrons in the outer regions of a star
should also be investigated.

Talking about β-decay there exists the possibility that gamma rays are emitted in
the decay process itself. If the classical β-decay is considered, the birth of a
gamma ray is impossible. The β-spectrum for He^6 is continuous, and is well known;
the continuity of the β-spectrum is related to the fact that in the final state
there are three particles--an electron, a neutrino and a proton. It is probable
that if gamma rays appear at all during a star flare, then they must be the result
of secondary processes (non-thermal bremsstrahlung of the fast electrons after
their production). It would be most interesting to make special observations of
flare stars in the region of gamma radiation.

4.   THE ANOMALOUS ABUNDANCE OF LITHIUM IN THE ATMOSPHERES OF T TAU STARS

The resonance doublet of neutral lithium $\lambda 6708$ Li I was discovered by Herbig [4]
in the spectra of two T tau type stars: T Tau and RY Tau. Bonsack and Greenstein
[5] confirmed the observations by Herbig, and found this strong absorption line
in three more stars of this type: SU Aur, GW Ori and RW Aur. The amount of
lithium in the photospheres of these stars was found to be 50 to 400 times that of
the Sun. There are some twenty T Tau type stars whose ratio of lithium to metals
is two orders of magnitude higher than that in the Sun. At the same time there
are no observational indications that lithium occurs in late type ordinary dwarf
stars (later than K0 [6]).

Lithium belongs to the elements which cannot exist in stellar nuclei. At temperatures higher than 3×10^6 K lithium rapidly disappears, combining with hydrogen to
form helium. Therefore the presence of lithium in stellar atmospheres is usually
considered as proof that some elements are formed directly in stellar atmospheres
as a result of certain nuclear processes. If this is correct, then the abundance
of lithium in the atmospheres of T Tau stars must be considered as evidence of
exceptional nuclear activity in them.

It can be shown that the actual number of lithium atoms in the atmospheres of T Tau stars must be much larger than the observations indicate. The absorption line λ 6708 Li I is due to neutral lithium and arises in the transition from the ground state to the nearest level. But in the atmospheres of stars which are in a permanent flare state, lithium must be mainly in the once-ionized state, since the intensity of the non-thermal radiation beyond the ionization frequency of lithium (λ shorter than 2300 Å) must be considerably higher than the intensity of the Planck radiation of a normal star of this type. This also follows from the fact that there are hydrogen emission lines in the spectra of these stars, which could not be excited if there was not a sufficiently strong radiation in the short-wavelength region of the spectrum.

In this connection it is interesting to determine the actual abundance of lithium in the atmospheres of T Tau stars as compared with the Sun [7]. We should consider the ionization of lithium under the conditions in the atmospheres of these stars. Since we have no observational data about the energy distribution in the continuous spectrum of T Tau stars in the region shorter than 3000 Å, we shall assume, as a working model, that the non-thermal radiation ionizing lithium is due to inverse Compton effects.

The initial condition is the equilibrium between the processes of photo-ionization from the ground state of neutral lithium and the recombination processes. We have:

$$n_1 \int_{\nu_*}^{\infty} x_{1\nu} \frac{H_\nu(\tau,\gamma,T)}{h\nu} \, d\nu = n^+ n_e D(T_e) \, , \tag{16.1}$$

where n_1 and n^+ are the densities of neutral and once-ionized lithium atoms, ν_* is the ionization frequency of lithium, $x_{1\nu}$ is the continuous absorption coefficient from the ground state of neutral lithium; n_e is the density of thermal electrons; $H_\nu(\tau,\gamma,T)$ is the intensity of the radiation of Compton origin ionizing lithium. The lithium is assumed to be in the upper layers of the star's photosphere and that the ionizing radiation falls upon it from outside--from the medium consisting of fast electrons.

A relation analogous to (16.1) can be written for the Sun, where the corresponding quantities are N_1, N^+ and N_e. Let us write $Z_* = n^+/n_1$--the degree of ionization of lithium in the atmosphere of the T Tau star, $Z_\odot = N^+/N_1$, the same in the atmosphere of the Sun. Then we have from (16.1):

$$Z_* = \frac{1}{n_e D(T_e)} \int_{\nu_*}^{\infty} x_{1\nu} \frac{H_\nu(\tau,\gamma,T)}{h\nu} \, d\nu \; ; \tag{16.2}$$

$$Z_\odot = \frac{1}{N_e D(T_e)} \int_{\nu_*}^{\infty} x_{1\nu} \frac{B_\nu(T_\odot)}{h\nu} \, d\nu \, . \tag{16.3}$$

The abundance of lithium in the atmospheres of T Tau stars is usually expressed as the ratio to the abundance of neutral sodium, sometimes of neutral calcium, barium, iron--elements with ionization potentials close to that of lithium. Therefore it is necessary to check if the ionization of the elements does not change from the non-excited star (the Sun) to the excited one (T Tau star). Such changes are to be expected since only the ionization potentials are equal (or nearly equal) for these elements, but the dependence of the continuous absorption coefficient on frequency is completely different for them [8]. Moreover the character of the continuous spectrum in the short-wavelength range (λ < 2300 Å) is greatly different from that for the Sun and for a T Tau star.

Computations show that the ratio of the degree of ionization of Li to the degree of ionization of Na hardly changes (comparing the Sun to a T Tau star). Therefore any deviations in the degree of ionization of lithium in T Tau stars with respect to sodium can be ascribed to a change in the actual abundance of lithium.

The recombination coefficient $D(T_e)$ is not very sensitive to the electron temperature, and hence the difference in the electron temperatures of the photosphere of the star and the photosphere of the Sun can be neglected. Then from (16.2) and (16.3) we find:

$$\frac{Z_*}{Z_\odot} = W \, \frac{N_e}{n_e} \, \frac{F_2(\tau)}{\gamma^4} \, \frac{\int_a^\infty x_{1\nu} \frac{x^2 dx}{e^{x/\gamma^2} - 1}}{\int_b^\infty x_{1\nu} \frac{x^2 dx}{e^x - 1}} \, , \tag{16.4}$$

where the value of the function $H_\nu(\tau,\gamma,T)$ is taken from (4.29), $a = h\nu/kT$, $b = h\nu/kT_\odot$, and the value of $x_{1\nu}$ is taken from [8].

Taking $T = 3600$ K for a T Tau type star, and $T_\odot = 5500$ K for the Sun, and performing the integration we find (with $\gamma^2 = 10$):

$$\frac{Z_*}{Z_\odot} \approx 10^4 \, W \, \frac{N_e}{n_e} \, \tau \, , \tag{16.5}$$

where we have substituted $F_2(\tau) \approx \tau/2$.

The most uncertain parameter is the electron density n_e in the atmospheres of T Tau stars. Though n_e increases slightly during the flare it will normally be no larger than the density of hydrogen atoms in the photosphere of the star. Taking $N_e \sim 10^{12}$ cm^{-3} (Sun) and $n_e < N_e$ we have, with $W = 0.1$ and $\tau = 0.01$:

$$\frac{Z_*}{Z_\odot} \geq 10 \, . \tag{16.6}$$

The activity of the T Tau stars is characterized by the value of $\tau \sim 0.01$. Hence we may conclude that the degree of ionization of lithium in the atmospheres of T Tau type stars must be at least one order of magnitude higher than the degree of ionization of lithium in the photosphere of the Sun. However, as discussed above, in the T Tau type stars there is about 100 times more neutral lithium than in the Sun. Hence it follows that the total number of lithium atoms in the atmospheres of T Tau stars must be 1000 times larger than that on the Sun. The last conclusion characterizes once more the unusual and powerful processes taking place in the atmospheres of T Tau type stars and similar other objects.

The lines of once-ionized lithium are in the region of soft x-ray radiation (~ 180 Å), and thus in principle ionized lithium cannot be detected in stellar spectra. There is however one weak point in the above considerations. According to our discussions above β-decay takes place at a large distance from the star's photosphere in the medium (envelope) consisting of fast electrons. Hence the lithium atoms are formed in this medium. Part of these atoms leave the star and populate the interstellar space, but another part is directed towards the star's photosphere and may, under certain circumstances, accumulate in the upper layers of the photosphere, which leads to the observed enhancement of the lithium absorption line λ 6708 Li I. But the lithium atoms, before reaching the photosphere, must pass through the chromosphere of the star. Therefore a high density of lithium should be expected not only

in the photosphere, but in the chromosphere as well. Therefore we should observe
the presence of the <u>emission</u> line $\lambda\,6708$ Li I (of chromospheric origin) in the star's
spectrum. But the observations do not confirm this.

The explanation of this discrepancy may be the following. For the excitation of
the emission line $\lambda\,6708$ Li I by fluorescence it is necessary that lithium be singly
ionized. However, there are reasons to assume that lithium in the chromosphere is
ionized twice ($\lambda_{ion} \leq 165$ Å). The density of the radiation in the region $\lambda < 200$ Å
in the chromospheres of flare stars and T Tau stars is still sufficiently large, as
indicated in some cases by the line $\lambda\,4686$ He II in the spectra of these stars. The
lines from twice-ionized lithium are in a region inaccessible to observation.

In the case of flare stars matters are more complicated. If we start from the above
considerations, then lithium can be formed only during flares. Then the total
amount of lithium accumulated in the photospheres of flare stars must be considerably
smaller than that in the photospheres of T Tau type stars in which the processes of
lithium formation have a more or less permanent character. Detection of the line
$\lambda\,6708$ Å directly during the flare is therefore more difficult due to the fact
that lithium is in its second ionization state. Therefore in UV Cet type flare
stars the line $\lambda\,6708$ Li I may be absent. Evidently, this is actually the case; for
AD Leo, for instance, the lithium abundance is even smaller than that on the Sun
[9].

The above conclusions about the high lithium abundance in the atmospheres of T Tau
type stars may be specially interesting in relation to the abnormally high abundance
of light elements, including lithium, in cosmic rays. The ratio Li/H for the Sun
is of the order 10^{-11} [4], whereas for cosmic rays it is of the order 10^{-3}. Usually
it is thought that lithium is a fragment due to splitting of heavy nuclei taking
place in the interstellar medium by collisions with protons. This assumption,
however, requires that a considerable number of heavy nuclei should be present in
the source of cosmic rays, exceeding by one or two orders of magnitude their
natural distribution (for details, cf. [10]).

Does the anomalous lithium content in cosmic rays have any relation to the anomalous
abundance of lithium in the atmospheres of T Tau stars? Are these stars and similar
objects suppliers of cosmic lithium? It is hard to tell, particularly because it
is not clear that the lithium atoms can be accelerated under the conditions of the
interstellar medium after leaving the star.

Some light may be shed upon all these questions by an analysis of the isotopic
composition of cosmic rays. If the assumption that there is a relation between the
lithium in cosmic rays and the lithium in the atmospheres of T Tau stars is valid,
the number of Li^6 nuclei in cosmic rays must be larger than the number of Li^7
nuclei, i.e. this ratio must be the inverse of what is observed under ordinary
conditions on the Sun and on stars. The available data, though exceedingly scarce,
do not contradict this. In an experiment [11] recording cosmic-ray particles (in
the energy range 180 to 400 MeV) two tracks were recorded, one of which was found
to belong to Li^6, and the second to Li^7, which gives $Li^6/Li^7 = 1$. Let us recall
that for the Sun and the Earth $Li^6/Li^7 = 0.08$.

However, more reliable data about the isotopic composition of lithium in cosmic rays
are required before we can draw definite conclusions about the role of T Tau type
stars as suppliers of the lithium in cosmic rays. It would be more than desirable
to make special experiments for determining the isotopic composition of lithium in
primary cosmic rays.

In this respect the experiment performed by the artificial satellites IMP-7 and
IMP-8 [30] are of special interest. On these satellites cosmic-ray telescopes had

been installed, with the aim of finding the chemical composition of the cosmic rays. The measurements were made in the energy interval 32 to 159 MeV per nucleon. During the total observing time, i.e. 7150 hours (from January to December 1974) 77 Li^6 particles and 63 Li^7 particles were recorded, which gives Li^6/Li^7 = 1.07--in complete agreement with the value expected from our considerations explained above. It is interesting to note that the distribution over the directions of the nuclei of the lithium isotopes recorded was found to be isotropic--a clear indication of their interstellar origin (modulated only by the influence of the solar "wind" and by the interplanetary magnetic field). Therefore we have reasons to believe that the value 1.07 found above for the ratio Li^6/Li^7 refers exactly to the initial composition of the cosmic rays.

It has been stressed above that β-decay of He^6 nuclei as a source for the appearance of fast electrons in the outer parts of the star is only a model. Meanwhile this model allowed us to discover a number of properties of our hypothetical nucleus. In particular, one of the products of the decay of this nucleus must be Li^6, and not Li^7. This question is of great astrophysical interest.

5. CHROMOSPHERIC ACTIVITY--CONVECTION

The theory of gravitational condensation developed by Hayashi and his collaborators [12], particularly the theory of the convective structure of the stars, has some popularity among astrophysicists. According to this theory every star, which is in a state of gravitational contraction, can at certain periods in its evolution be in a state of complete mixing (convection). Hayashi computed tracks for the displacement of a star from the upper right-hand part of the spectrum-luminosity diagram, where they appear immediately after birth, to the main sequence. Stars with different initial masses, from 0.05 $M_\odot$ to 4 $M_\odot$ or more, evolve differently and at different rates. It turns out, for instance, that stars with the characteristics of flare stars are somewhere in the middle of these tracks, i.e. in the region corresponding to a completely convective state. Poveda [13], combining Hayashi's results with the "radiative" theory of stars, gives a boundary for the flare stars, corresponding to spectral class K1, to the right of which (towards the later spectral classes) is the region of the flare stars.

Here we do not intend to consider in detail the theory of Hayashi. The theory is mentioned only in relation to other questions: has the flare activity of a star any relation to the state of internal convection of the star from the point of view of the fast-electron hypothesis?

At the present time it is hard to answer this question definitively, but in principle the existence of such a relation does not appear improbable. Convection makes possible a transport of material, including nuclear-active matter, from a star's central parts to its surface layers. From this nuclear-active matter fast electrons are emitted by β-decay. The stronger the convection, the higher the probability that, together with gaseous material, nuclear-active matter will also be ejected. Evidently, it would not be wrong to characterize a stellar flare as the result of fluctuations of the convective structure of the star near its surface, with the condition that the convection itself encompasses the whole, or nearly the whole, star. The flare frequency of some star will be higher, with larger probability that nuclear-active matter is ejected, and this probability depends directly on the convective activity of the star.

Another important phenomenon, the chromospheric activity, must be closely related to the flare and convective activity of the star.

Generally speaking, what is a chromosphere? In its classical definition, made first

for the Sun, the chromosphere is the region in the atmosphere of a star where
emission lines arise. But emission lines arise in the corona as well. Therefore
the definition is made more precise: in the chromosphere emission lines arise
which correspond only to permitted transitions of the atoms, induced by photo-or
other types of ionization, with successive cascading transitions downwards, whereas
in the corona forbidden lines arise by inelastic electron collisions, etc.

However, the meaning of a "chromosphere" has been broadened after it had been shown
by observations that a chromosphere exists even in stars of late classes, cool
dwarfs of classes K to M. The well-known chromospheric emission lines of hydrogen
and ionized calcium, and also the ultraviolet doublet of ionized magnesium (λ 2800
Mg II) were found to be much stronger in these stars than in the Sun. The integral
brightness of the Sun's chromosphere is nearly a million times less than the brightness of its photosphere. In some cool dwarfs the chromospheric lines are easily
detected against the background of their continuous radiation. This means that the
integrated brightness of the chromospheres of these stars is not much less than the
brightness of their photospheres. Judging from the structure of the emission lines
and from the value of the Balmer jump, the electron temperature of the chromosphere
of the cold dwarfs is not much lower than the electron temperature of the solar
chromosphere.

A peculiar situation exists: it seems that the chromosphere and its physical conditions have no direct relation to the photosphere of the star. The chromosphere,
immediately adjacent to the photosphere, maintains an independent existence. The
chromosphere can be subjected to considerable variations, indicated by the fluctuations in the intensities of the emission lines. At the same time these chromospheric
variations are not accompanied by any significant changes in the photosphere of the
star itself.

Any variations of the observed activity of the chromosphere are due first of all to
fluctuations in the amount of energy exciting the chromosphere. This is the energy
of the L_c-radiation causing the photo-ionization of hydrogen, and also of the continua corresponding to the frequencies of continuous absorption by other atoms and
ions. In the preceding chapters it was shown that in the case of flare stars, and
stars with emission lines (T Tau), the radiative energy of excitation of the chromosphere comes from outside (from above, rather than from the photosphere), from the
regions where the fast electrons are produced.

The observations do not confirm the assumption that the chromospheres of cool
dwarfs have local structure, such as extensive regions of strong excitation. If
this were the case a periodicity should be expected in the appearance and disappearance of the emission lines, due to the axial rotation of the star, which is not
observed. In the cool dwarfs there exists a genuine chromosphere covering the whole
or nearly the whole surface of the star. To produce general excitation of the
chromosphere is much simpler if the sources of energy for its excitation are lying
beyond the boundaries of the chromosphere itself, in the space surrounding the star.

Thus it would be more correct to characterize the chromosphere as a layer of the
star's atmosphere which gets energy for its excitation from outside. In this case
the behavior of the chromosphere will depend directly on the behavior of the fast
electrons themselves--their appearance and disappearance. In particular, changes in
the chromosphere take place much more slowly and much less suddenly; this question
has been considered in sufficient detail in Chapter 9.

Haro has given convincing arguments [14] for the idea that the chromospheric activity
is directly related to the flare activity of a star. The flare activity must be
characterized by the convective activity of the star as well. But only the flare
and chromospheric activity produce direct observational results. Therefore it will

be better to make the inverse conclusion: stars showing signs of chromospheric
and flare activity are at the same time objects in a state of complete or nearly
complete convective activity.

Wilson [15] established an important relation between chromospheric activity and
age of a star (see Ch. 11), based on an analysis of extensive observational material:
the chromospheric activity is highest in young stars (Orion), has moderate strength
in stars of average age (Pleiades) and has nearly disappeared in old stars (Praesepe,
Hyades). On the average, the variation of the flare activity of stars in these three
age groups falls along the same sequence. This also means that the convective
activity becomes weaker when we pass from the youngest stars to the older ones.

6. STELLAR AND SOLAR FLARES

It is surprising that sometimes stellar flares are casually compared with flares
of the Sun. Such attempts are made without any basis at all. The properties of
solar flares are even used as indicators, or even as standards, in estimating the
validity of one or the other theory of stellar flares.

The attempts to identify stellar flares with flares of the Sun have no basis, primarily on cosmogonical considerations. Not all stars show flares, not even those
of the same spectral class M. Only newborn stars, which are not yet completely
formed, show flares. With increasing age of the star the flare activity declines.
The Sun does not belong to the category of young stars. Even during the most powerful solar flare the total flare energy is hardly one percent of the normal radiation
of the Sun. For the Sun the period of flare activity in its early age has long
passed away. Instead of the processes leading to strong and frequent "stellar"
flares, processes occur which can produce much less powerful flares, which we call
"solar" flares.

The above considerations to a certain degree are very general. However, one can
give more convincing arguments that stellar flares on the one hand and solar flares
on the other are generated by processes which are essentially different. We recall
the important conclusion reached in Ch. 6 (§ 22) where the absolute amount of
energy $E_f(U)$ produced by the star on the average during one flare depends strictly
on its absolute magnitude M_V. This relation between $E_f(U)$ and M_V was represented
graphically in Fig. 6.39, and also by the empirical formula (6.19), from which it
follows that the higher the absolute luminosity of the star the larger the amount
of energy emitted during the flare. The extreme point limiting the empirical
relation between $E_f(U)$ and M_V at the high-luminosity side is the star YY Gem, for
which $M_V = 8.26$. If we extrapolate this relation to the Sun, i.e. to the value
$M_V = 4.83$, we find for the expected amount of energy emitted by the Sun during one
flare a value equal to 2.5×10^{33} ergs (in U wavelengths). This is several orders of
magnitude higher than the energy recorded during even the strongest solar flares.
It follows that formula (6.19) cannot be extrapolated to the Sun, and that the
validity of this formula ends somewhere between stars of class M0 (YY Gem) and class
G2 (Sun). From this point the flare mechanism of "stellar" type begins to lose its
importance and a switch-over to the flare mechanism of "solar" type begins. Realiz-
ing how strongly the Sun differs from the results of formula (6.19), we may conclude
that these two mechanisms for inducing flares--of "stellar" and of "solar" type--
differ radically from each other. Does the flare mechanism of solar type work in
"stellar" flares? Apparently it does, though it is very hard to prove this. But
it will have no practical significance, its contribution to the energy will be at
least two to three orders of magnitude less than that of the "stellar" flare
mechanism. In reality the "stellar" mechanism first weakens and then disappears
completely, and the "solar" flare mechanism continues to be present.

Summerizing, we may say that the assumption, perhaps attractive and even trivial, that the causes of flares for all stars and the Sun are fundamentally the same is not correct. The nature of solar flares differs from the nature of stellar flares, in a way that these flares differ from outbursts of Novae.

7. THE ROLE OF MAGNETIC FIELDS

It has been suggested that stellar magnetic fields may be the cause of excitation and generation of flares. It has even been mentioned that the energy of the flare radiation is connected with the annihilation of magnetic fields. Evidently, the idea of the "magnetic fields" as the source of energy for stellar flares was introduced mainly because "there was no other hypothesis," as Parker, one of the authors of this idea, recognizes [16].

Here we do not intend to thoroughly analyze the "magnetic" hypothesis of stellar flares. We shall only deal with a few points which in our opinion do not confirm the assumption that the flare energy is taken from magnetic fields.

Magnetic fields and solar flares. The physical processes leading to the excitation of solar flares are not yet well understood [17], therefore assuming a resemblance between stellar and solar flares cannot solve the fundamental problem of the mechanism of flares in stars. The facts relating solar flares with the magnetic fields of sunspots are often quite contradictory. Thus, for instance, there exists a relation between the frequency of chromospheric flares and the average number of sunspots, the magnetic fields of the sunspots and their polarities do not appear to have any relation to the chromospheric flares [18]. The probability of chromospheric flares does not depend on the maximum value of the field strength; more important are the variations of the magnetic flux and the magnetic polarity [19]. There are cases where rather strong x-ray flares of the Sun are not accompanied by chromospheric flares of corresponding strength, or by the presence of a group of sunspots with large area [20,21]. It is not very probable that ordinary chromospheric flares and x-ray flares on the Sun should be caused by different mechanisms; rather these two kinds of flare are related to each other and have the same origin. However, an x-ray flare can easily be detected against the background of a relatively weak x-ray emission of the solar disk.

The production of quite a large amount of energy during solar flares takes place in a very small region of the chromosphere and is not accompanied by a disturbance of the magnetic field in the lower part of the chromosphere. The production of energy must be very efficient since, as shown by the observations, the flare is often accompanied by ejection of high-energy charged particles (cosmic rays) and gaseous matter with a velocity of the order of 1000 km/s.

A comparison of the magnetic field distribution with the observed positions of flares on the Sun does not always show that the positions of the flares coincide with the maximum of the magnetic field, i.e. with the regions where the energy is produced. Normally, the energy production may take place in one point, from which the flare spreads and develops.

Very contradictory is also the role of the so-called neutral points or neutral lines--regions of intersection of the magnetic lines of force--in the excitation of flares. According to some data the flares appear precisely at the neutral points or lines [21], but according to other data they can also appear at other places [22]. There are indications that flares can appear in almost any region of the Sun's surface.

The "magnetic" concept for the production of solar flares is based on the assumption that magnetic lines of force between different sunspots can break and close again

[23]. According to the theory of this phenomenon the field can suddenly disappear
and a flare starts to appear. This mechanism leads to a phenomenon which has
nothing in common with a flare, and the destruction of magnetic lines of force has
never even been observed [22].

Another mechanism for the generation of solar flares is the "pinch-effect" [21].
However, the extreme instability of the pinch-effect, and the difficulty of producing
large fluxes in short times make this mechanism very improbable.

Under these conditions it may be better to discuss the inverse problem, namely to
consider solar flares as one class of stellar flares (!). This approach may appear
strange, but for the study of chromospheric flares we only have one object, the Sun,
i.e. a star of class G2, whereas for the complex study of flares quite extensive
possibilities are available, from the point of view of variety in the types of stars,
as well as in the degree of the phenomena. In fact, the strongest x-ray flares of
the Sun can be explained with the fast-electron hypothesis, and we can say with
certainty that in these cases the flares do not have their origin in magnetic
processes [24,20].

The relation of chromospheric flares of the Sun with magnetic fields seems to be an
observational fact. However, it does not follow that chromospheric flares arise as
a result of the annihilation of magnetic fields.

Stellar rotation and flares. It has been suggested that stars with strong magnetic
fields are objects with rapid axial rotation. However, there is no observational
evidence that flare stars rotate more rapidly than non-flaring stars of the same
spectral class. Up to the value v sin i = 10 km/s it is impossible to see a difference between flare stars of class Me and normal objects of class dM. For instance,
for the flare star BY Dra the value of v sini is found to be ∿5 km/s, but for CC Eri
it is ∿15 km/s [25]. It is true that unusually wide absorption lines are observed
in some T Tau type stars, but this widening need not necessarily be due to rotation.

Stars with strong magnetic fields. Babcock's [26] catalogue of magnetic stars contains about 90 objects with very strong magnetic fields, and about 70 objects with
possible magnetic fields. These stars are brighter than 9^m. However, none of these
stars is a flare star. This circumstance is interesting. The point is that in all
cases the magnetic fields of the stars are variable, not only in magnitude, but also
in polarity. For instance, for the star HD 32633 (class B9p) the strength of the
magnetic field was found to be equal to - 3260 G on November 12, 1956, but on the
following day, November 13, it was + 1530 G! For another star, HD 65339 (A2p) the
strength of the field was on November 3, 1957, equal to - 3000 G, and on the following day it was + 2000 G. In these cases we are referring to the value of the magnetic field strength averaged over the entire surface of the star; we cannot understand (from the point of view of the magnetic flare hypothesis) how such enormous
changes can take place in the magnetic field without excitation of an optical or
radioflare of corresponding strength.

Stellar spots and flares. The existence of stellar spots must be accepted as well
established in at least two flare stars, BY Dra and CC Eri, both of class K7e, and
both double systems. These were found by Bopp and Evans [25] as a result of careful
analysis of the photometric behavior of these stars. In these stars brightness
fluctuations were discovered at a time when there was no flare activity. These
fluctuations were periodic in time and sinusoidal in form, which can be explained
if there exists a cold spot on the surface of the star which rotates around its
axis. The dimensions of the spots are striking: they cover up to 20% of the
surface of one hemisphere of the star. The effective temperature of the spot is
estimated to be of the order of 2000 K, and the effective temperature of the photosphere is 3750 K.

These observations seem correct. But then a chain of assumptions and estimates are made--none of which is confirmed by observations [25,27]. The purpose of these assumptions is to "impose" the idea that these spots are the generators of optical flares. These unrealistic assumptions result in a magnetic field strength of ∿10000 G.

The existence of stellar spots is confirmed by observations, but we do not have any observational confirmation of the presence of strong magnetic fields in flare stars. Accordingly, to propose that these spots are responsible for the phenomenon of stellar flares must at least be premature.

As to apparent brightness the two stars, BY Dra and CC Eri, belong to the brightest flare stars--brighter than 9th magnitude (in V wavelength). It should be possible in these cases to perform high accuracy photometric observations, without which no information can be obtained about the presence or absence of stellar spots. However, for a correct estimate of the role and the importance of stellar spots in the flare problem we should seek ways of making such observations for the most typical flare stars, the intrinsically faint ones (UV Cet, CN Leo, etc.), as well as the intrinsically bright ones (AD Leo, YY Gem, etc.). One should also try to establish the presence or absence of dark spots even in one M-type star.

The problem of the annihilation of magnetic fields. The mechanism by which magnetic energy is transformed into radiant energy is completely unclear. It is thought that first the magnetic energy is transformed into heat. In this connection ambipolar diffusion and various forms of the "Sweet" mechanism are considered as well [16]. In the latter case it is necessary that the magnetic fields are antiparallel, and that the medium is highly compressed--conditions which are in general hard to satisfy. Meanwhile the "Sweet" mechanism is thought to be the most effective, at least theoretically.

But when the magnetic energy is transformed into heat energy, the radiation itself must be of thermal origin, i.e. it must be represented by Planck's law radiating at an even higher temperature in the stellar photosphere (since the annihilation of the magnetic field takes place in the photosphere itself). None of these are confirmed: the energy distribution in the flare spectrum cannot be represented by Plancks' law, and the increase of temperature of the photosphere is in general insignificant.

One of the imperfections of the magnetic hypothesis, according to Parker [16] is the low rate of the assumed annihilation of the magnetic field, which is not compatible with the observed high rates of growth and development of flares.

The above considerations are sufficiently well founded and deserve attention. At present, however, the data do not argue in favor of the "magnetic" concept of the origin of flares in stars (and possibly in the Sun).

Magnetic fields and fast electrons. However, the stars and the Sun do have magnetic fields, local ones and also general ones--nearly dipole fields. Though it seems to us that these fields are not the direct sources of energy for the excitation of flares, they may exert a certain influence on the development of the flares even after these have appeared. In the case where the flare is caused by fast electrons, this influence may be felt specially in the following forms:

(a) The magnetic fields may control the motion of the fast electrons and their displacement (drift) towards the outer regions of the star. As a result the fast electrons produced in the star's atmosphere will practically wrap up the entire star almost instantaneously.

(b) In favorable cases part of the electrons may obtain additional energy by means
of acceleration in the local magnetic fields of the star (Sun). As a result their
energy can increase from a value of 10^6 eV to 10^7 or 10^8 eV and perhaps even more.
The energy spectra of the electrons may be significantly extended, and similarly
the spectra of the radiation of the electromagnetic fluctuations (of Compton or
synchrotron origin, in the optical, x-ray and radio-spectral ranges, etc.). Hence,
it is possible to understand that the hardness of the x-ray radiation generated
during solar flares is not always the same and may vary from one flare to the next.

Still another alternative exists, namely that the fast electrons arise due to
acceleration of ordinary <u>thermal</u> electrons in the magnetic field of the star, or in
the magnetic fields of the spots, and that the optical flare itself is excited by
the interaction of these fast electrons with photons emitted by the star (inverse
Compton effect). In that case two difficulties are solved at once: the origin of
the fast electrons becomes clear, and a "non-thermal" way is indicated for the
annihilation of the magnetic energy. Such a way of producing fast electrons is
tempting since it involves magnetic fields more actively.

However, the "magnetic" production of fast electrons cannot be very efficient,
because the "efficiency coefficient" of the transformation of magnetic energy into
energy of electromagnetic radiation is very low. Only 10^{-5} of the total energy of
the magnetic field of the star (or of the spots) can be transformed into energy of
flare radiation. Computations show that for the excitation of one ordinary flare
it is necessary that the strength of the magnetic field be of the order 10^5 G over
the entire surface of the star (!) and in a layer of thickness 1000 km!

The qualitative character of these discussions suggests unusual strengths of magnetic
fields in astrophysics in general. The essential role of magnetic fields in some
problems cannot be denied. Yet caution should be exercised in all those cases where
the magnetic fields are considered as potential sources of emission of radiation of
enormous power.

8. THE COSMOGONICAL SIGNIFICANCE OF STELLAR FLARES

Thus, the stars show flares. They do this often especially in the early period of
their life, during their formation. In the course of time this flare activity
weakens, and later disappears completely. Some kind of weak flare activity--of the
"solar" type--appears or remains, but it has no cosmogonic significance from the
point of view of energy. The flare activity of the solar type should rather be
considered as a secondary product or ordinary thermal or non-thermal processes,
occurring in the atmospheres of stars which are completely formed and have already
stabilized.

Solar and stellar type flares are different phenomena as can be deduced from Fig.
6.39. This diagram gives the empirical relation between the energy in U wavelengths
$E_f(U)$, emitted on the average during one flare and the absolute magnitude of the
star M_V. It was derived from observational data of flare stars of UV Cet type.
The extreme point in absolute luminosity in this diagram represents the star YY Gem,
for which $M_V \approx 8^m.5$. Extrapolating this empirical relation to larger absolute
luminosities, up to the value of M_V for the Sun, we get a value for $E_f(U)$ which is
at least three orders of magnitude larger than the energy emitted in the most powerful solar flares. One may, however, object that we do not have the right to make
such an extrapolation without sufficient grounds, and certainly not over 4 stellar
magnitudes.

A more convincing proof that flares of stellar and solar types are essentially
different from each other, and that they have completely different energy scales is

given in Fig. 11.11 (similar to Fig. 6.39), constructed for flare stars in Orion and
in the Pleiades. Here the limit of the empirical $E_f(U)$ vs. M_V relation, at large
values of absolute luminosity, extends up to $M_V \sim 5$, i.e. to the luminosities of
stars of class G, or the Sun. For such stars $E_f(U) \approx 4 \times 10^{33}$ ergs, i.e. once more
1000 times larger than the strongest flares of the Sun.

Of course, the difference between stellar and solar flares is not restricted only
to this fact, though these arguments seem convincing.

Hence, two stars--the Sun and some young star of type G in an association--<u>with
the same absolute luminosity</u>, but at different stages of development, and with ages
differing from each other by at least three orders of magnitude, have essentially
different flare activity. Evidently this is only possible when the circumstances
causing the appearance of such a strong flare in the young star, stop working or
existing when the star has covered the long evolutionary path from the moment of its
birth to the state of the Sun, i.e. till it has reached the main sequence. What
should be considered as the main factor leading to such a large difference in flare
activity between the two age categories of stars?

If we assume that the primary source of the flare is the spontaneous appearance of
fast electrons in the outer parts of the star, then the answer to this question is
unambiguous: in young stars the required conditions for the generation of fast
electrons are available; in old stars these conditions are either not present at all,
or they have been nearly exhausted. By "conditions" we mean a supply of intra-
stellar matter from which in some way or another the fast electrons are emitted.

There are those who think that stars are formed from special, unknown, superdense
matter, which could be the carrier of the intra-stellar energy. All forms of instability in young stars are connected with the outward transfer of matter, and the
liberation of energy in the form of "continuous emission". It is further assumed
that in the course of the development of the star the reserves of this matter are
drained, and slowly the general macroscopic instability of the star must gradually
disappear. We remark that no concrete mechanism or process leading to the liberation of energy from the stellar interior in the form of continuous emission is indicated in this theory.

Thus, all instabilities, particularly those of flare stars, according to this idea
[28] are related to the presence of prestellar material of high density in a quite
special state, as yet unknown to us. The fact that instability phenomena are
connected with young stars means that the amount of this special matter in the
stellar nuclei must be small.

Here we consider two questions which in our opinion are the most important in
connection with this problem.

The first question concerns the proposed properties of the matter in the stellar
interior. If this intra-stellar matter really has properties which have nothing
in common with those of the ordinary matter, then this does not imply that this
extraordinary matter has no relation at all with ordinary material. But in these
considerations no attention has been given to the fact that <u>ordinary matter in a
state such as exists in the central parts of the stars becomes degenerate</u>, i.e. it
gets properties completely different from those it had initially. Any transition
from the degenerate state to the state of ordinary matter will be accompanied by
processes widely different from those of ordinary matter. Such a transition takes
place in particular when separate parts or pieces of intra-stellar matter somehow
leave the star.

If fast electrons are necessary for generating the ordinary optical flare, and if

these may appear as a result of β–decay of some nuclear-active matter in a superdense
state, then from the point of view of the generation of the flare, it does not matter
how exactly this nuclear-active material appears in the nuclei of the stars.   In
other words, this matter may appear, or rather may be formed, even as a result of
gravitational contraction of ordinary diffuse matter, followed by strong compression
to the state of superdense degenerate matter.

Certainly, it does not follow from the above that the possibility of forming stars
out of some superdense prestellar bodies is excluded in principle. We are simply
establishing the fact that an analysis of the problem related to the appearance of
a stellar flare does not lead to any internal contradictions with the notion that
stars originate as a result of condensations of diffuse matter.

Thus, from the fact alone that intra-stellar matter brought to the outer regions of
the star produces energy in the form of continuous emission does not result in an
unambiguous choice between two alternatives, namely whether this intra-stellar matter
is ordinary matter brought to a degenerate state by gravitational condensation, or
whether it consists of clots of superdense prestellar bodies with unknown origin.

The occurrence of "flare" matter within a star is necessary, but not a sufficient
condition for a flare to start. An additional important factor is the duration or
"longevity" of such processes, as compared to stellar evolution.

Flares can take place as long as the possibility of transfer of the intra-stellar
energy exists. Therefore we could still interpret the weakening of the flare
activity of a star as a phenomenon of "inability" to bring the "flare" matter to
the outside, and not as a sign that the stocks of this matter are exhausted.

The phenomenon of a stellar flare has nothing in common with the outburst of a
Nova or a Supernova. The appearance of a Nova means a fundamental rearrangement
of the whole internal structure of an old star, and a strong disturbance of the
inner energy equilibrium. Nothing of the kind happens during a stellar flare.
The loss of energy by flare activity is not an important fraction of the energy
balance of the star.

It is not astonishing to think that the flare activity depends on the efficiency
of the processes of transfer of the intra-stellar matter outwards. We know very
little about the processes of transfer of this material to the space surrounding
the star, but there is hardly any doubt that convection plays an important role.

The separate "clots" of intra-stellar matter formed somewhere in the star's
nucleus can be moved by convective currents and transported to the outer regions.
Hence, the higher the convective activity of the star, the larger will be the total
amount of intra-stellar matter transported outwards. When the convective currents
are exhausted, further transport of intra-stellar matter ceases or strongly
diminishes, which in its turn leads to the disappearance of flares.

The fact that part of the young dwarf stars in the Pleiades are certainly not flare
stars (Ch. 11, §14) can be explained in the following way: in these stars the
convective activity has decreased, and the process of transfer of intra-stellar
matter has been strongly diminished. However, the assumption that the flare
activity has ceased in these stars as a consequence of the depletion of intra-stellar
sources of energy appears to be rather improbable, first because the ages of stars
showing flares and of stars not showing flares in this aggregate are about equal,
and second because one cannot understand how the star will radiate further after
it has reached the main sequence if it has exhausted its energy sources so quickly.

With regard to the suppression of convective activity, this is a consequence of the evolution of the star. At a certain stage of stellar evolution the radius of the convective zone becomes smaller than the stellar radius itself. In that case the convective zone does not reach up to the surface of the star, thus making the transfer of nuclear-active matter outwards more difficult. This conclusion agrees with the observations: in the Pleiades the non-flare stars are mainly the bright ones, thus the stars of largest linear dimensions.

The evolutionary significance of stellar flares leads to the phenomenon of convection in stellar nuclei as the main factor determining their macroscopic instability. In proto-stars the convective activity is very high, and the transfer of intra-stellar matter in them must be of a permanent character, and flares must take place exceedingly often, practically without stopping. This is what is observed in stars of NX Mon type, T Tau type, etc. As the star calms down, its convective activity also must get weaker; consequently, the cases of transport of intra-stellar matter and the flare processes become less frequent.

It is well known that convection or convection zones are the main features of stellar configurations in a state of compression, which arise as the result of gravitational condensation of diffuse matter [29]. And if we arrived at the conclusion about the exceptional role of convection in the maintenance of flare activity in stars, then we arrive involuntarily also to the conclusion that these stars must be formed by gravitational condensation.

Thus from an analysis of the intra-stellar matter, and from the means by which it is transferred from the star's nucleus to the outside, we reach the same conclusion, namely that the first stage of star formation is most probably a state of gravitational condensation of diffuse matter. This conclusion agrees with that made when considering the problem of the space distribution of the same flare stars in associations (cf., e.g. Fig. 11.9).

It is not our intention to discuss problems of star formation. We have referred to this problem only to that degree to which it is related to the phenomenon of stellar flares and to their evolution. The conclusion that gravitational condensation plays the predominant role in the process of star formation seems inevitable. The extension of this theory to other types of objects, such as Herbig-Haro objects and T Tau type stars, dust nebulae, infrared objects. etc., would only strengthen this conclusion.

At the same time we consider our conclusions to be the result of further development and broadening of the concept of superdense prestellar bodies, a concept which was, generally speaking, our starting point in the present book. However, our assumption requires confirmation.

The "clots" of matter, initially in a diffuse state, which have been brought into the degenerate state by gravitational pressure, are pushed by convective currents, and after some time they appear in the outer regions of the star. Here the energy accumulated in them is set free, in our opinion in the form of fragments due to β-decay, in particular in the form of fast electrons (β-electrons). Analysis of this problem (cf. §3 of this chapter) led us to the conclusion that the β-decay of a hypothetical nucleus, if not one of the isotopes of helium, namely He^6, must resemble He^6 very much as regards its properties. Evidently the question is whether He^6 nuclei can be formed out of some kind of amorphous mass of nuclei. In this process the He^6 nuclei themselves are probably fragments of more complex nuclear systems.

Together with the transfer of intra-stellar matter outwards, some gaseous matter will also be ejected. Since flare stars do not have a constant gaseous envelope, let alone an extended one, the amount of this latter material must be very small.

The observed emission lines originate mainly in the chromosphere, and the role of
the gaseous envelopes formed during a flare is unimportant. The behavior of the
emission lines can be explained relatively easily, as we have seen in Chapter 9, by
processes taking place in the chromosphere of the star.

The successes in the domain of β-decay, of the theory as well as of the experiments,
are generally known. Physicists know 5 to 6 types of β-decay processes and possibly
new types of β-decay will be discovered in the future. One of the interesting types
of β-decay--the spontaneous β-decay of excited nuclei--was discovered nearly fifty
years after the discovery of natural radioactivity. Therefore there is no pessimism
indicated that actually we cannot indicate the type of β-decay, or the type of spontaneously decaying nucleus, which produces first the β-electrons and finally the
optical, radio, x-ray and other types of flares. As far as we know, the problem of
β-decay has not been considered by anyone from this point of view, even qualitatively
and therefore it would be premature to draw hasty conclusions on this question.

Regardless of these uncertainties, the following statement can be made on very good
grounds: a stellar flare is a grandiose event, truly stellar in its scale, an event
that, each time it appears, encompasses huge volumes of space around the star, and
develops with fabulous large speed. An event the consequences of which repeat with
striking constancy and in many-sided forms - of continuous emission, spectral lines,
radio emission, emission of X-ray photons, possibly the production of elementary
particles, nucleons, etc. An event in comparison with which all forms of appearance
of instabilities in the atmosphere and chromosphere of the Sun--flares, active
regions, radiobursts and so on--seem only weak echoes. An event that the stars
exhibit in the earliest periods of their formation and development, which weakens
as they grow older, and therefore has a clearly outlined evolutionary significance.

Therefore, it is entirely clear that a phenomenon with such extraordinary properties
cannot be explained by means of ordinary thermal processes. Only nuclear processes
taking place in the outer regions of the star's atmosphere, processes of the type
of the decay of unstable nuclear configurations, processes in which sophisticated
forms of exchange of energy and interaction predominante, can bring us closer to
understanding the true nature of this striking and universal phenomenon in the
Universe.

REFERENCES

1. C. S. Wu, S. A. Moszkowski, Beta Decay, New York (1966).
2. J. L. Greenstein, R. S. Richardson, Ap.J. 113, 536, 1951.
3. G. H. Herbig, Ap.J. 140, 702, 1964.
4. K. Hunger, A.J. 62, 294, 1964.
5. W. K. Bonsack, J. L. Greenstein, Ap.J. 131, 83, 1960.
6. G. Wallerstein, G. H. Herbig, P. S. Conti, Ap.J. 141, 610, 1965.
7. G. A. Gurzadyan, Doklady Acad. Nauk SSSR 176, 291, 1967.
8. R. D. Hudson, V. L. Carter, JOSA 57, 651, 1967.
9. R. E. Gershberg, N. I. Shachovskaya, Bamberg IAU Variable Star Colloquium
   No. 15, 1972.
10. V. L. Ginzburg, S. I. Syrovatskii, The Origin of Cosmic Rays, Pergamon Press,
   1964.
11.   N. Durgaprased, Astr. Astrophys. 12, 98, 1971.
12. Ch. Hayashi, Ann. Rev. of Astr. Astrophys. 4, 171, 1966.
13. A. Poveda, Nature 202, 1319, 1964.
14. G. Haro, E. Chavira, Bol. Obs. Tonantz. Tacubaya 5, No. 31, 23, 1969.
15. O. C. Wilson, Ap.J. 138, 832, 1963.

16. E. N. Parker, Ap.J. Suppl. 8, 177, 1963.
17. H. J. Smith, E. P. Smith, Solar Flares, New York, 1962.
18. K. O. Kiepenheuer, The Sun, Ed., G. P. Kuiper, Chicago, 1953.
19. A. B. Severny, A.J. USSR 35, 335, 1958.
20. G. A. Gurzadyan, Doklady Acad. Nauk Armenian SSR 43, 28, 1966.
21. A. B. Severny, Izv. Crimean Obs. 22, 12, 1960; 27, 71, 1962.
22. H. Zirin, The Solar Atmosphere, Blaisdell Co., New York, 1966.
23. T. Gold, F. Hoyle, M.N.R.A.S. 120, 89, 1960.
24. C. de Jager, Space Res. First COSPAR Symp., Amsterdam, p. 628, 1960.
25. B. W. Bopp, D. S. Evans, M.N.R.A.S. 164, 343, 1973.
26. H. W. Babcock, Ap.J. Suppl. 3, 141, 1958.
27. T. J. Moffett, D. S. Evans, G. Ferland, M.N.R.A.S. 178, 149, 1977.
28. V. A. Ambarzumian, Non-Stable Stars, University Press, Cambridge, p. 177, 1957.
29. Ch. Hayashi, Ann. Rev. Astr. and Ap. 4, 171, 1966.
30. M. Garcia-Munoz, G. M. Mason, J. A. Simpson, Ap.J. Lett. 201, L145, 1975.

Index

337

OTHER TITLES IN THE SERIES IN
NATURAL PHILOSOPHY